D0734302

# Organic Struggle

**Food, Health, and the Environment**
Series Editor: Robert Gottlieb, Henry R. Luce Professor of Urban and Environmental
Policy, Occidental College

Keith Douglass Warner, *Agroecology in Action: Extending Alternative Agriculture through Social Networks*

Christopher M. Bacon, V. Ernesto Méndez, Stephen R. Gliessman, David Goodman, and Jonathan A. Fox, eds., *Confronting the Coffee Crisis: Fair Trade, Sustainable Livelihoods, and Ecosystems in Mexico and Central America*

Thomas A. Lyson, G. W. Stevenson, and Rick Welsh, eds., *Food and the Mid-Level Farm: Renewing an Agriculture of the Middle*

Jennifer Clapp and Doris Fuchs, eds., *Corporate Power in Global Agrifood Governance*

Robert Gottlieb and Anupama Joshi, *Food Justice*

Jill Lindsey Harrison, *Pesticide Drift and the Pursuit of Environmental Justice*

Alison Alkon and Julian Agyeman, eds., *Cultivating Food Justice: Race, Class, and Sustainability*

Abby Kinchy, *Seeds, Science, and Struggle: The Global Politics of Transgenic Crops*

Sally K. Fairfax, Louise Nelson Dyble, Greig Tor Guthey, Lauren Gwin, Monica Moore, and Jennifer Sokolove, *California Cuisine and Just Food*

Brian K. Obach,*Organic Struggle: The Movement for Sustainable Agriculture in the United States*

# Organic Struggle

## The Movement for Sustainable Agriculture in the United States

Brian K. Obach

The MIT Press
Cambridge, Massachusetts
London, England

MIT Press books may be purchased at special quantity discounts for business or sales promotional use. For information, please email special_sales@mitpress.mit.edu.

This book was set in Stone Sans and Stone Serif by Toppan Best-set Premedia Limited. Printed and bound in the United States of America.

Library of Congress Cataloging-in-Publication Data.

Obach, Brian K. (Brian Keith), author.
    Organic struggle : the movement for sustainable agriculture in the United States / Brian K. Obach.
        pages cm. – (Food, health, and the environment)
    Includes bibliographical references and index.
    ISBN 978-0-262-02909-4 (hardcover : alk. paper)
1. Organic farming–United States. 2. Sustainable agriculture–United States. I. Title. II. Title: Movement for sustainable agriculture in the United States. III. Series: Food, health, and the environment.
    S605.5.O23   2015
    631.5'84–dc23

                                                        2014042768

10  9  8  7  6  5  4  3  2  1

This book is dedicated to all who work to create a just and sustainable food and agriculture system

# Contents

Series Foreword   ix

Acknowledgments   xi

1   Introduction   1

2   The Birth of the Organic Movement   27

3   Certification and the State: The Dilemmas of Growth   47

4   The Organic Coalition: United and Divided   81

5   Are We Better Off? Movement Achievements and the Threat from Big Organic   127

6   Searching for Social Justice   161

7   Strategic Innovation: The Three Trajectories of the Organic Movement   181

8   The Road Not Taken and the Road Ahead   211

Notes   243

Bibliography   289

Index   319

# Series Foreword

I am pleased to present the tenth book in the Food, Health, and the Environment series. This series explores the global and local dimensions of food systems and examines issues of access, justice, and environmental and community well-being. It includes books that focus on the way food is grown, processed, manufactured, distributed, sold, and consumed. Among the matters addressed are what foods are available to communities and individuals, how those foods are obtained, and what health and environmental factors are embedded in food-system choices and outcomes. The series focuses not only on food security and well-being but also on regional, state, national, and international policy decisions and economic and cultural forces. Food, Health, and the Environment books provide a window into the public debates, theoretical considerations, and multidisciplinary perspectives that have made food systems and their connections to health and environment important subjects of study.

Robert Gottlieb, Occidental College
Series editor

# Acknowledgments

There are many people to thank for their direct or indirect support for this project. Liana Hoodes played a valuable role in putting me in touch with many of the key figures cited in this book. I am indebted to all those who granted interviews and who were so generous with their time. Several students assisted with this project including Carolyn Burgess, Annie Courtens, Jenna Dern, Marigo Farr, and Chris Utzig. I want to thank Clay Morgan and Deborah Cantor-Adams at the MIT Press and the series editor, Bob Gottlieb.

I draw great inspiration from those around me who work to advance a just and sustainable food system including Dan and Ann Guenther and the Climate Action folks; Students for Sustainable Agriculture; Ariana Basco and the Environmental Task Force; and my coauthor, KT Tobin at CRREO.

Many people provide the support I need to carry on, including my great colleagues in the Department of Sociology at SUNY New Paltz; my mom, Linda Horowitz, and the rest of my family in the United States and in Turkey; Questionable Authorities; the Kindy Freemen; Tim Lefebvre and the DSO crew; Stuart Eimer, Kim Akins, and a bunch of other people dear to my life. Most important, I want to thank İlgü Özler, who read many drafts and generally provided the love and encouragement I needed to keep working even when I was grumpy.

This book is based upon work supported by the National Science Foundation under Grant No. 0550550.

# 1 Introduction

Beginning in the 1980s, Liana Hoodes grew organic vegetables, and raised sheep and chickens, providing friends, family, and members of the community with organic eggs and meat. She also started a farmers' market in her small town in upstate New York, thus helping supply others with access to fresh organic produce. Now she spends more time in meetings and on the computer than she does growing food, yet she has more effect on what people eat than she ever did as a farmer.

For many years, Hoodes served as the organic policy coordinator for the National Campaign for Sustainable Agriculture (NCSA), a loose coalition of organizations seeking to promote sound federal organic policy. She spent hundreds of hours in meetings working with this diverse array of farmer advocates, environmentalists, consumer groups, and organic industry representatives. They shared ideas and tried to hammer out policy positions that they could bring to elected leaders and staff members within the US Department of Agriculture (USDA), the federal government entity that oversees the National Organic Program (NOP). Hoodes coordinated lobbying along with letter writing and petitions, all designed to defend organic integrity when big agribusiness and government officials seemed intent on undermining the organic system of which she has been a part for so many years.

Eventually the coalition that Hoodes helped oversee proved to be too loose for the task at hand. When she started, the organic community was "more potluck than policy," as one insider put it.[1] But by the time the federal government program went into effect, more focused and coordinated political action was needed. She moved on to become the executive director of a group formed in 2003, the National Organic Coalition (NOC), a more formal organization complete with its own office in Washington, DC, and a professional lobbyist, who Hoodes coordinates with regularly from her upstate New York home.

Dan Guenther is considered the Johnny Appleseed of community supported agriculture (CSA) in the mid-Hudson region of New York State. After getting his engineering degree from Columbia University and working for a few years on high-rise buildings in New York City, Guenther and his wife, Ann, fled the rat race and retreated upstate to homestead. They lived the simple life, at times so simple that it did not include indoor plumbing or electricity. Homesteading and living off the land suited them fine, except that life in modern civilization sometimes requires purchases that cannot be paid for with canned tomatoes. Guenther began doing odd jobs and eventually built an independent construction company. But Guenther and his wife retained a vision of the kind of life they wanted to lead as well as the kind of world in which they wanted to live, and they were not happy with some of the changes they saw occurring in the upstate region north of New York City. Farms were giving way to housing developments, shopping malls, and big-box stores. Guenther wanted to find a way to preserve farming in the region. "I wanted to have farms close to home, and to live in an area that was surrounded by a working landscape. I'm not very political. I'm not very good at writing letters, contacting senators, and whatnot. I am very hands on. So the idea came to me of starting an educational farm, to raise consciousness and to reconnect people with the earth."[2]

Guenther found his calling in trying to reconnect people to their food and save what was left of the region's agricultural economy. He concluded that CSA programs were the way to go. This innovative business model, which was beginning to catch on in the early 1990s, links consumers directly to a farm by selling seasonal "shares" of the harvest; members typically pick up their portion on a weekly basis. He started with a farm on land made available to him by friends who saw promise in his vision of sustainable local agriculture. Once that was well established, he went on to create another one, this time on land owned by Vassar College, where college officials were willing to take a chance on this unconventional approach to producing and distributing food. Guenther established a third CSA in New Paltz, the small college town in which he had spent his summers as a child. In the meantime, he helped several other small farmers get started.

Guenther isn't farming now, but he is still spreading the gospel. He regularly gives talks on agriculture for schools and community organizations.

He has been working with various groups to address distribution challenges that small farmers face when they try to do anything other than direct-to-consumer sales like CSA programs or farmers' markets. If local agriculture is to thrive, farmers will need to get their goods to a broader market. Guenther is working with schools, hospitals, and colleges along with local distributors and warehouse owners to try to provide institutional customers for local growers.

Guenther holds industrial agriculture and the global food system in disdain. He is convinced that a decentralized system of local agriculture will become a necessity in the face of growing ecological crisis. He is a committed environmentalist, and never used synthetic pesticides or fertilizers on his fields. Most of the farmers who Guenther works with use sustainable farming practices as well. But they're not necessarily certified organic. Guenther himself never had any dealings with the NOP, and the question of whether or not his goods were officially USDA-certified organic rarely came up in his work.

Ron Khosla and his partner, Kate, also run a CSA farm in New Paltz.[3] While his partner can usually be found in the fields planting Swiss chard or driving the tractor, Khosla can more often be seen sitting at his computer writing about certification systems. They grow food organically, but they would never use that word. And legally, they are forbidden from doing so. Ever since the NOP went into effect, farmers cannot use the term organic unless they have been inspected and approved by a USDA-accredited organic certifier. But Khosla has no interest in seeking federal government approval for anything, and he's not really interested in fighting to reform federal organic policy. Although he consults occasionally with the Food and Agriculture Organization of the United Nations, he has a general distaste for large, centralized bureaucratic institutions.

Similar to Guenther, he thinks that farmers should be in charge of farming. Yet Khosla recognizes the value and need for the oversight of farming practices to provide assurances to consumers and weed out fraudulent actors. Instead of federal inspectors, though, he supports a "participatory guarantee" system for small farmers—one in which groups of farmers consult with and inspect one another's farms to share knowledge of the local conditions as well as assure that healthy, sustainable practices are being used.

As an expert on such systems, Khosla consults with the United Nations on this issue. In fact, he created his own participatory guarantee system here in the United States called Certified Naturally Grown (CNG). This system uses essentially the same standards as organic ("better than organic!" he insists), but inspections are carried out by fellow farmers rather than by professional certifiers working under the USDA's authority. Khosla sees this as a way to strengthen local agriculture while providing consumers with the assurances they need, all without the involvement or cost of a distant bureaucratic federal program.

Hoodes, Khosla, and Guenther all want essentially the same thing. They want an agricultural system in which people can get fresh healthy food grown in an environmentally sustainable way. They are all ultimately tied to a common tradition and movement whose roots date back almost a hundred years. For much of the twentieth century and to a large extent today, that movement has taken place under the banner of organic agriculture. Yet Hoodes, Khosla, and Guenther each have a different relationship to the term organic. For Hoodes, it is a crucially important and carefully distinct term that, if legally defined in subtly different ways, could either provide the basis for transforming the industrial food system, or spell disaster for the environment along with farmers and consumers alike. For Khosla, the organic concept is significant, but the term itself is not, at least when it comes to small farmers. Organic, as currently used, represents for him a co-opted label designed more for agribusiness corporations and meddling bureaucrats than for farmers. Guenther does not concern himself much either way. He still buys certified organic for those items that he gets in the grocery store, but his main interest is in knowing who is growing his food. He believes that if people know their farmers, then they can trust in their food regardless of whether it is labeled this or that.

Although they currently have distinctive relationships to the term organic, and despite the fact that they are engaged in different activities, the work that Hoodes, Khosla, and Guenther do ultimately grew from shared roots. The divergences are the consequence of a movement that has passed through many stages. It is one composed of diverse groups that have a range of interests, and frequently differ in terms of ideology and strategy. It is also a movement that has had to cope with "success," the rapid growth in the organic sector, and the broad acceptance it has won among policy makers, conventional food and agriculture corporations, and the public at large. This has created strain and tension within a movement that was never fully united or explicit about the goals it was seeking to achieve.

Moreover, it has led some to celebrate the mainstream embrace of organic, while leaving others feeling that the movement has been co-opted and the promise of organic severely undermined. Conflicting interpretations about the present state of the organic industry and organic movement suggest different strategies for moving forward, as evidenced in the paths that Hoodes, Khosla, and Guenther have taken. In order to understand the current state of the organic movement and where it is headed, we must examine how we arrived at this point. Such is the intent of this book.

Any exploration of a social movement's trajectory requires that we consider the context in which it is occurring. This means looking at the broader political, economic, and social conditions under which the movement developed, and how institutional forces shaped the movement at crucial junctures. As described below, there are sharply contrasting perspectives on the manner in which structural conditions enhance or limit the potential of the organic movement or any movement oriented toward fundamental environmental and social reform. These "macrolevel" factors will be given consideration throughout this analysis and revisited in more detail at the end, but much of the focus throughout the book also will be on the "meso-" and "microlevel" developments within the movement. This means examining the organizations that make up the organic movement and, in some cases, the actions of individuals who played a decisive role at key moments.

At the organizational level, two factors are of particular interest: the way in which organic advocacy groups formed and developed, and the diverse coalition of interests that compose the organic movement. The creation of organizations and the way in which they tend to formalize over time has long been of interest to movement scholars.[4] These developments are vividly apparent within the organic movement, which began as a particularly decentralized and loosely coordinated effort, and took decades to evolve into a movement with any significant organizational base. Coalition formation is another dynamic that has been recognized as a fundamental element of social movements.[5] Movements are, in effect, coalitions of groups with sometimes varying interests and goals. This is especially true of the organic movement—a movement composed of farmers, consumers, environmentalists, small businesses, animal welfare advocates, rural preservationists, and others. Much can be understood about the present state of affairs by looking at the way in which these diverse interests converge and diverge around central questions.

Some of these questions are matters of ideology and movement strategy, which represent another important focus of this analysis. What is it that organic advocates hope to achieve, and how do they believe they can advance their values? The organic movement has gone through many stages. Goals and strategies were never static, nor has there ever been consensus among the various constituencies about what exactly organic advocates hope to achieve or how to accomplish it. The biggest divergence in this regard is directly tied to the question of whether "going mainstream" is desirable. This, in turn, links back to those broad structural and historic forces that shaped the organic movement's trajectory.

An overview of each of these central areas of inquiry—institutional opportunities and limitations on reform, organizational and coalition development, and movement strategy and ideology—will be provided below in addition to an assessment of organic's status as a social movement. The subsequent chapters will examine each of these issues in detail along with other factors relevant to understanding this complex, fascinating movement that is reshaping the way people in the United States eat as well as how they think about food and agriculture.

## Movement Success and the Limits of Reform

Both the organic movement and organic industry have undergone profound changes in recent decades. In the United States of the early 1970s, organic farming was an obscure agricultural practice carried out by a small number of counterculture farmers and a smattering of commercial growers who recognized that there was something wrong with the dominant industrial food model. Few in the general public were familiar with the term organic, and fewer still had ever purchased an organic product. The small handful of scholars and university-based scientists interested in alternatives to the industrial agriculture model were marginalized by their peers. Organic practices were not recognized by the federal government, and food industry scientists dismissed organic methods as a hoax promoted by subversive elements opposed to science and social progress.

Less than forty years later, organic agriculture has grown into a multibillion-dollar industry. Organic food can be found on the shelves of nearly every supermarket in the United States. Seventy-five percent of US consumers report buying organic products. The size of the organic market reached

$1 billion by 1990 and grew to $28.6 billion in 2010.[6] Several major universities have entire programs dedicated to the study of organic agriculture. The federal government not only recognizes organic practices as distinct, but no products can be labeled as organic without meeting federal standards and being approved by federally accredited certifiers. The adoption of organic production methods, once referred to by a USDA secretary as a harbinger of mass starvation, now yields products that bear the official USDA seal. On several counts, the organic movement has been incredibly successful, arguably among the most effective movements of the late twentieth century.

But successful at what? Advocates tout many benefits of organic agriculture, from improved consumer health to environmental preservation, from rural revival to social justice. Some proponents believe the organic movement is bringing about a fundamental transformation of the food and agriculture sector as well as society as a whole. The growth in the organic market is evidence of the profound impact the movement is having. Some do not measure their success in market share or dollars spent, however, and the movement's achievements are more ambiguous by other counts. Skeptics doubt that as practiced today, organic is delivering the health, environmental, and social benefits envisioned by early proponents.

This debate rages in the popular media as well as among food activists and scholars.[7] Does the growth in organic agriculture represent the start of a far-reaching social transformation along with a reawakening of popular sentiment about the place of food and agriculture in society? Or is it simply a consumer trend propagated by a industry cynically targeting a niche market of elite consumers longing for the simplicity of a mythical agrarian past, while yielding few, if any, real health and environmental benefits? Evidence to support both perspectives can certainly be found in any examination of organic issues today.

These arguments fit within a larger debate among scholars about the extent to which social institutions are amenable to fundamental reform, especially as it pertains to curbing environmentally destructive practices. These perspectives can inform how we interpret the effectiveness of the organic movement, and the changes we have observed in the food and agriculture sector. Movement outcomes are, after all, largely matters of power and position as mediated by social institutions, such as the political and economic systems, within which struggles take place.[8] Some analysts view the development of the organic industry as a shining example of how major reforms are achievable in a relatively short time, thus offering hope that daunting ecological problems in other areas can be overcome. Ecological modernization theorists, as these optimists are known, argue that since the 1980s, real institutional changes have taken place in regard to how we understand and relate to the natural environment.[9] In government, in industry, and among the public at large, a grand awakening has changed the way in which we approach resource use and the ecosystems of which, we have come to recognize, we are a part. These theorists point out how concern for sustainability is now a routine consideration in decision making at all levels. Major social institutions are not only amenable to reform but they also help to foster it. This hopeful development is validated by the trajectory of the organic movement along with changes in the food and agriculture industry.

Ecological modernization theorists credit environmental activists in the 1960s and 1970s for having raised awareness about ecological concerns. But they also applaud the manner in which radicals and agitators of that era eventually adopted more "constructive" approaches to these issues. Rather than remaining outside critics of major institutions, environmentalists today often work hand in hand with government and industry leaders to identify practical means of reforming environmentally harmful practices. In looking at the organic cause, these analysts see a movement that matured over the years, changing from a ragtag collection of counterculture radicals living on communes to successful organic entrepreneurs and professional advocates heading respected nonprofit organizations that are willing to work with political leaders as well as the conventional food industry to reform agricultural practices.

Ecological modernization theorists also see a positive change in the way governments address environmental issues. Consideration of environmental impacts is now a regular aspect of almost all policy making. Then, when problems are identified, policies today tend to "steer" industries and individuals in positive directions, rather than imposing heavy-handed "command-and-control" regulatory measures implemented by cumbersome state bureaucracies.[10] Hence we see, for example, the creation of carbon-trading markets for polluting industries and the use of financial incentives to encourage homeowners to weatherize their homes, instead of the imposition of strict emission limits or rigid building codes. Again, organic policy reflects this model of a new and supposedly more effective regulatory approach. The USDA does not mandate that all food is grown organically, but policy does create a framework in which organic production can spread. Proponents of this approach would point to the fact that the organic industry grew rapidly following the implementation of the federal government's NOP, through which the state rationalized the organic market and offered support for its expansion.[11]

According to this theoretical perspective, the free market and private industry are also central actors in advancing ecological sustainability.. Private firms, responding to consumer demand, bring more ecologically sound goods to market. But they are not just reacting to the market; they work to build it. Enlightened entrepreneurs and business leaders, as ecological modernization theorists contend, have internalized the need for more environmentally sound production. They use their vast resources and advanced technologies to develop innovative new ways to produce more sustainably. They actively help to educate consumers by advertising their "green" goods and promoting the virtues of ecologically conscious consumption. From this vantage point, conventional food and agriculture companies should be praised for building the organic market and making organic goods accessible to a broader consuming public. Farmers' markets and CSA programs are quaint supplements to the food industry that serve a small clientele, but the scale of organic production as well as the size of the organic market would not be possible without the major corporations that ventured into this sector, using their financial might, advanced technologies, and expansive distribution networks to bring these goods to market, and sell them at a price affordable to a broader swath of consumers.[12] In other words, those corporate actors once characterized as villains bringing ecological ruin can now be seen as helpful allies in the shift toward a more sustainable social order.

Through the lens of ecological modernization theory, the rapid spread of organic in the last couple of decades is a story of triumph, a demonstration of the success of an innovative movement and the ability of dominant institutions to reform themselves in ways that advance human health and ecological sustainability. It is part of a larger dynamic that can be seen in many other sectors in which environmental reform is advancing, from recycling to renewable energy. Certainly many organic trade representatives subscribe to the ecological modernization narrative, at least when it comes to their own industry. But so do many organic activists. Many who have toiled for years against all odds finally have a sense of vindication as their values and beliefs and the practices they advocated have been embraced by the general public and by those at the highest levels of government and industry. Organic, once only of interest to a tiny fringe, has become mainstream. These activists feel they have won.

Yet not all observers share this rosy view of organic or the essential processes that play out as activists seek to change institutions. Some have concluded that the system is largely immune to fundamental reform, especially as it concerns fostering ecological sustainability. Treadmill of production theorists believe that capitalism, and the unwavering quest for profitability through technological efficiency and increased productivity, will ultimately prevent the reforms necessary to stave off ecological collapse.[13] They argue that ecological sustainability is antithetical to a system geared toward economic expansion, and all major social institutions are, in the final analysis, structured around this growth imperative. The wealth concentrated in the hands of large corporations, an electoral system in which candidates are dependent on private donors, a regulatory bureaucracy in which government agents maintain cozy relationships with regulated industries, a privately controlled media reliant on industry advertising and other features of the dominant political-economic order all play a role in shaping the outcomes of political struggle and limiting the changes that might harm the interests of power holders, or threaten the workings of the economic treadmill. From this theoretical perspective, most grassroots movements seeking fundamental change are, in the end, marginalized or co-opted in the face of these powerful forces.

The transformation of the organic industry can be seen in this light. Political leaders and conventional industry actors may have embraced organic practices of some sort, but what we have is not the organic system that activists envisioned. Under USDA control, organic agriculture, as practiced today, is in most cases a mere shadow of what advocates had hoped to see.

In the public mind, many still associate organic production with small-scale family farming. Organic is thought to represent the return to an agrarian ideal where independent farmers feed the local community by producing food in accordance with the natural order.[14] Indeed, industry marketers seek to exploit that sensibility.[15] It is rare to find an organic product that does not present images of red barns and cows grazing freely on green rolling hills. The growth in organic market share, however, does not necessarily mean that there is a true shift to that type of farming. Much of the increase in organic production has come from large conventional agribusiness entities that have entered the organic market in force. "Big Organic," as the corporate organic sector is known by its critics, has superimposed industrial-style agriculture practices onto the organic framework, for what some refer to as the "conventionalization" of the organic industry.[16] Using this approach, growing organically is simply a matter of "input substitution," identifying materials that are allowed under organic rules to use in place of the synthetic fertilizers and pesticides used in conventional farming. These growers meet organic standards as defined by the USDA, but they utilize industrial practices such as large-scale monoculture crops, the heavy use of off-farm "organic" inputs, and the intensive confinement of livestock.

The agribusiness corporations that own these operations or contract with large growers then manufacture goods that would not have been recognizable as organic (or even as food) to the movement's original visionaries. Today one can purchase organic versions of everything from microwavable burritos to frozen pizza. If prohibited ingredients are needed to make these highly processed products, industry lobbyists work to pressure malleable USDA officials to add another exception to the rules, thus further undermining organic ideals.

Likewise, the distribution of organic food does not reflect the close-knit networks between farmers and consumers that early advocates took for granted. Today, most organic sales do not occur at local food co-ops, farmers' markets, or roadside stands but instead at national chain supermarkets and big-box stores that began introducing organic food during the first decade of the twenty-first century. Nothing in the USDA regulations prohibits the entry of large corporations into the organic market, and many conventional agribusiness players have quietly bought up the most successful independent organic companies and taken them national or even global.[17] Their size and international reach allows them to capture consumer dollars more efficiently than their smaller counterparts. In the meantime, small farmers and independent organic businesses, those who once formed the base of the organic movement, are forced to develop new ways to market themselves in order to survive.

Critics of the current state of affairs add that as organic production has come to mimic the industrial agriculture system, the health and ecological benefits that motivated many advocates have been lost. Economic and political power holders may pay lip service to the ideals espoused by activists, but ultimately, treadmill theorists contend that profitability and the maintenance of the existing power structure trump all efforts to bring real change. Corporate co-optation and the conventionalization of the organic industry are the result of the inevitable all-consuming capitalist treadmill, which is capable of absorbing and redirecting any authentic movement that seeks to advance an alternative.

Significant segments of the organic community hold beliefs that reflect the analyses put forth by treadmill theorists. Their experiences working with the federal government and witnessing the incursion of conventional industry players into the organic market have left them cynical about the hope of reforming the system from within. Many of these activists have given up on direct challenges to dominant institutions and instead have returned to an approach used by early organic proponents: creating an alternative parallel food and agriculture system based on traditional organic ideals.

It is imperative to examine social movements within the institutional context in which they operate. Different features of the political and economic systems provide opportunities, create limitations, and generally shape the way movements develop. The contrasting views expressed by activists on the ground mirror the conflicting theoretical analyses put forth by scholars about the institutional context within which the organic movement unfolded. The sharply contrasting views expressed by activists can help to explain the divergent paths chosen by different segments of the organic movement today. Meanwhile, the theoretical perspectives offered by scholars can inform our understanding of the movement and provide some insight into what we can expect to come of it in the future.

## Organization, Strategy, and Ideology in the Organic Movement

Whether it is possible to fundamentally reform the food and agriculture industry within the given political and economic order is an open question. There is strong evidence to support the claims of both ecological modernization and treadmill of production theorists. Yet while the potential and limitations imposed by social institutions are important considerations, the outcomes of social struggles are never determined solely by the system's structure. At times, organized and mobilized publics have overcome the most powerful actors and most resistant institutions. The actors that compose social movements have agency. The choices that people like Hoodes, Khosla, and Guenther make, along with the actions they undertake matter, as do those of the thousands of people who are in some way part of the organic movement. The strategies they adopt, the manner in which they organize themselves, and the policies they advocate must be examined, even as we recognize that all this takes place under institutional and historical conditions that shape their actions and limit the outcomes of their efforts.

Through this type of analysis we can see beyond the two extreme characterizations of the organic movement implied by macrolevel theories and those articulated by activists and observers. The outcome of the movement to date should be seen neither as a complete victory for the forces of ecological and agricultural reform, nor as a co-opted failure with little to show for decades of struggle. There have been tangible positive changes in the food and agriculture sector as a result of the work of organic activists, although perhaps not all those for which some had hoped. These outcomes are the product of complex dynamics within the organic movement, including the organizational and strategic matters cited above. And the process is still unfolding. But before giving more consideration to the factors that led to the present state or prospects for the future, a more fundamental question must be answered: Is the organic movement a movement at all?

## Is There an Organic Movement?

Today one can find organic products being promoted by some of the largest corporations in the world. Political power holders all the way up to the White House have championed the organic cause. It is difficult to rectify visions of social movements as masses of relatively powerless people marching in the street demanding change with the reality that organic agriculture has the support of the most powerful elites in the United States. With such establishment embrace, one has to ask, Is there now or was there ever an organic movement that actually contested the dominant order?

Scholars debate what exactly constitutes a social movement. One commonly used definition states that a social movement is "a collective, organized, sustained, and noninstitutional challenge to authorities, power holders, or cultural beliefs and practices."[18] Despite its embrace by some authorities and power holders, and the relatively broad acceptance of organic by the general public, in the past and still today, action surrounding the advance of organic agriculture meets that definition.

Organic itself refers to an agricultural practice. In the United States, it is officially defined as "an ecological production management system that promotes and enhances biodiversity, biological cycles and soil biological activity. It is based on minimal use of off-farm inputs and on management practices that restore, maintain and enhance ecological harmony."[19] As a method of agricultural production, it is something that anyone can utilize, from backyard garden hobbyists to peasant coffee growers to multinational agribusiness corporations. Vegetables raised on communal farms and sold at local food co-ops as well as frozen dinners distributed through international big-box chain stores may both bear the organic seal.

Although organic refers to an agricultural process, since its inception it has been imbued with a set of social values and goals that extend beyond specific production techniques. These beliefs have served as the basis for popular mobilization for several decades. Thousands of activists organized into hundreds of organizations did indeed carry out a "sustained … challenge to authorities [and] power holders." They confronted dominant cultural beliefs while seeking to bring about societal transformation along with a host of environmental, health, and social benefits through the promotion of organic practices. Organic activists have used an array of institutional and noninstitutional means to challenge state authorities, the agriculture industry, and entrenched beliefs and practices about how food should be produced. Although mass demonstrations were not among the tactical repertoire of the movement, collective action was not uncommon. Organic activists have protested, lobbied, petitioned, and filed lawsuits in the same manner as most other social movements throughout US history.

There clearly was and still is an organic movement, but that image is complicated by the fact that today, there is also a burgeoning organic industry. While longtime activists speak fondly of the "organic community," if all those presently involved in organic are to be counted, that community includes such members as Cargill, ConAgra, General Mills, and Coca-Cola. In addition to critically evaluating the role that institutional forces played in yielding this outcome, we must take a closer look at the dynamics of the movement itself, and how strategic decisions led to this unusual and in some ways vexing movement-industry hybrid.

## Strategy and Ideology in the Organic Movement

The fact that there is an organic industry is the product of a key facet of the social change strategy adopted by organic proponents—one that establishes this movement as a forerunner of many contemporary movements. While common today, the central approach used by those seeking to advance the organic cause has always been market based. Like many movements, advocates must educate members of the public about a specific issue. Typically this is done to help mobilize citizens to pressure lawmakers for legislation that would institutionalize the desired change. But in the case of organic, for much of its history, proponents were not primarily seeking to mobilize citizens but rather to educate consumers—consumers who are then directed to purchase goods on the market that are produced according to particular standards and in some way embody the values that advocates are trying to promote. Social change is pursued not through government intervention but instead by creating a parallel system of production that, it is hoped, will eventually displace the dominant dysfunctional order. Certification systems and product labels that verify that desired standards are met have been central components of this market-based strategy. Organic proponents have long sought the elimination of the industrial agriculture system via these means.

Certification and labeling is now used to promote numerous causes, from fair trade to animal rights. A multitude of organizations now exist to certify goods according to standards they have developed. "Political consumerism"—that is, expressing political values through the purchasing of particular products—is now commonplace.[20] But the organic movement was in many ways a pioneer of this strategy. Organic certification was being practiced almost twenty years before most major certification systems were in place.

To some extent, the adoption of certification and labeling by organic movement actors can be seen as the product of structural forces. Political conditions and market dynamics led early proponents to create certification mechanisms largely as a defensive move when they felt that the organic practices they were developing could be co-opted or exploited by corrupt actors. Once established, though, this system would shape the course of the movement forever.

While the creation of a certification system was motivated by structural conditions, it was also a conscious strategic decision made by early organic movement leaders, even if those actors were not fully aware of the profound implications of their actions.[21] This was also true of a major shift in strategy that took place after the movement was more fully developed and when the organic industry was just beginning to take hold: the decision to involve the federal government in organic certification. That choice represented another strategic crossroads that would have far-reaching ramifications. It should be noted that this was not a departure from the essential market-based strategy of the movement; the goal to displace the conventional industrial food system by recruiting consumers into the organic system remained. Market conditions nevertheless led some movement leaders to conclude that federal involvement was necessary in order to make organic standards legally enforceable, thereby providing a more powerful shield against corruption or the misuse of the term. Yet others feared that this move would lead to the very outcome it was intended to prevent. They felt that government control would pave the way for conventional agribusiness to enter the organic sector and corrupt the system from within—a system that had been built and controlled by organic activists independently for many years. This was a key point of dissension that sharply divided the movement in complicated ways.

Part of this division hinged on more basic questions of what the movement was seeking to achieve and what real advantages organic methods offered. There was never a clear consensus on these matters. Early organic proponents in the United States often emphasized the health advantages of organically produced foods. Later, environmental benefits would come to the fore. Organic proponents in the 1960s and 1970s who really propelled the movement saw organic agriculture as part of a larger project of basic economic, political, and social transformation.[22] They sought to advance their vision outside conventional institutions or state policy. Utilizing a prefigurative political strategy, activists saw themselves as creating a new order from the ground up. The intent was that new, cooperative, egalitarian, and ecologically sound institutions would come to displace the dominant exploitative, corrupt, ecologically destructive, and violent order. Organic agriculture would be a part of that grand project. It was not just a method of growing; it had significant social purpose.

Some organic activists still hold that the organic movement has great transformative potential, that it is part of a struggle for a just and sustainable social order. Others have more modest goals, seeking only to make agricultural production less ecologically damaging. This distinction is directly tied to the matter of state and corporate involvement in the organic sector, and constitutes the most profound division within the organic movement today. It is a divide that parallels the contrasting perspectives of ecological modernization theorists and those who see a troublesome treadmill of production driven by powerful actors. The divide can be characterized as that between those who see value in the rapid growth in organic production, even if it does not yield all the benefits originally conceived, and those who see organic practices as part of an alternative system that advances broad social and ecological goals. Borrowing agricultural terminology, I refer to these ideological and strategic factions as "spreaders" and "tillers."

Spreaders seek rapid organic growth and value the state's role in addressing flaws in the private certification system. They welcome the entrance of corporate producers that have the resources and capacity to implement organic practices on a mass scale. At times, spreaders are willing to make some adjustments to organic standards in order to accommodate the large players. For them, the environmental and health benefits associated with organic are worth some minor concessions in standards.

Tillers feel that the transformative potential of the movement is lost through the involvement of big organic and the state agents that accommodate their quest to maximize profitability. These activists resist any changes that they consider a weakening of organic ideals, even if that means slower progress toward a full-scale agricultural conversion.

Spreaders, despite their willingness to work with organic agribusiness and prioritize more narrow gains, do not see themselves as devoid of the broader social concerns that occupy the tillers. Both of these camps make appeals about food justice—a burgeoning movement in its own right.[23] Seeking to overcome the image of elitism associated with higher-priced organic goods, those who promote the rapid spread of organic through conventional channels see themselves as making products more affordable and available to lower-income consumers. In contrast, those favoring small-scale local production believe that they are the true proponents of social justice through their advocacy on behalf of independent farmers and for a decentralized social order in which people are not exploited by massive economic powers.

The social justice issue has presented a challenge for all organic advocates. Many have pondered long and hard about how to incorporate principles of fairness and equality into a farming system that does not explicitly include social justice criteria. While always a central concern of those striving to promote reform in the agriculture sector, social issues were rarely included in organic standards, and the NOP does little to address these matters.

### Coalition and Division in the Organic Movement

Both the development of a market-based certification strategy and the decision to involve the federal government in that system had profound effects on how the organic movement was organized. That transformation is evidenced by the contrasting images of organic proponents in action today and in the past. A 1977 photo of the inaugural regional planning meeting for Oregon Tilth, one of the first organic farming associations in the United States, depicts a couple dozen young people in blue jeans and flannel shirts sitting in a circle on benches in a dilapidated barn. At that time, farmers would regularly meet to share new ideas about farming methods and collectively decide what practices should appropriately be defined as organic— the early stages of the development of a certification system. But as that system was formalized, and especially following the move to involve the state in organic certification, organic proponents had to create organizations that were more professional in form and style.

Today, discussions about what defines organic take place at the meetings of the National Organic Standards Board (NOSB), a federally appointed advisory body for the NOP. There are typically more people wearing ties than blue jeans at NOSB meetings. There are fewer farmers than there are marketing consultants, food scientists, and consumer products specialists. Yet even those representing the organic movement are of a different stripe. Some of them also have offices in Washington, DC, and rarely are these people seen in overalls.

Scholars have identified the tendency for social movements to develop more formal organizations over time.[24] The pros and cons of this organizational transition have been debated among activists and scholars alike. By some counts, organization is key to advancing social change as it allows relatively powerless groups to pool resources and coordinate collective effort. For others it portends bureaucratization, self-interested leadership,

and the loss of true commitment to movement ideals.[25] There is some valid-
ity to both perspectives, and that debate will not be resolved here. The link
between organizational form and movement strategy nevertheless is important for understanding the organic movement's trajectory.

The organic movement was decentralized and slow to formalize. From
the 1940s through the 1970s, organic practitioners in the United States
were primarily individuals connected only through publications distributed by Jerome Irving Rodale and later his son, Robert. Eventually, organic
adherents in the United States began to form associations for the purposes
of sharing ideas about growing and promoting the benefits of organic
methods. Most participants at that point were gardening hobbyists, but
there were more and more farmers taking part. Advocates believed in the
importance of not simply growing but also eating organic, and thus took
an interest in bringing organic goods to the consuming public. Given that
advancing the cause involved the distribution of products on the market,
there was eventually a need to clarify just what defined those goods as distinct from other commodities. As those standards became more formalized,
so too did the organizations that were developing them and taking responsibility for certification. Although this process began in the early 1970s,
even into the 1980s these groups remained fairly informal, and little in the
way of national coordination existed.

The prospects for federal government involvement in the organic system
changed all that quite rapidly. It was in the early 1990s, right around the
passage of the Organic Foods Production Act (OFPA), when we witnessed
what scholars refer to as the professionalization of the movement. Organic
activists recognized that in order to engage effectively in policy making at
the federal level, they had better be more organized and professional in
their approach.

While organic advocates themselves began to organize more formally,
another crucial development was also occurring. The cause was being
joined by well-established groups that already had formal structure. These
other constituencies, recognizing the link between their goals and those
of organic proponents, became more involved with the issue. Consumer
groups, environmentalists, and others brought energy and influence, and
strengthened the movement in a number of ways. But they also complicated matters for the loose collection of associations that had made up
the whole of the movement for almost two decades. The movement

became a coalition of a number of organizations representing a wide range of interests. Organic farmers and others who had been the core of the movement were, by this point, just one of a number of constituencies, and a relatively weak one compared to these more experienced and established DC players.

The fact that the organic movement grew into a broad coalition of groups is not unique. In some respects, all social movements are coalitions. Though we may talk of a movement as a whole or hold idealized images of a movement as a mass of individuals spontaneously assembled in the street to express their shared dissent, on closer examination it is clear that movements are largely operated by organizations, and rarely is it just one. At any given moment there are usually several organizations working together on various campaigns within any movement. In many ways, the organic community includes an even more diverse array of interests than most.

The NOSB, the official advisory body to the NOP, has designated seats for representation by farmers, consumers, retailers, processors, environmentalists, and scientists, all formally recognized as fundamental stakeholders in organic issues, and each also has a host of organizations through which advocacy is carried out. Added to these voices are those of health advocates, certifiers, animal rights groups, rural preservation organizations, public interest groups, and others. They come together in various configurations to advocate for particular issues, but overall can be considered a broad coalition that makes up the organizational component of the organic movement.

Having a diverse set of voices rallying behind a cause adds weight to a movement's demands, yet it also heightens the possibility for internal conflict and division. Each group within this broad coalition supports the advancement of organic agriculture, although these constituent subsets often prioritize different goals. At times these priorities come into conflict with one another. What goals should take precedent when policy nuances may favor environmental sustainability or the interests of small farmers? Consumers may demand assurances that they are getting what they pay for, but when does that oversight become too expensive or burdensome for organic food producers? How are concerns about farmworker wages ranked relative to the need to contain food prices and make healthy organic goods available to low-income consumers? These are just a few of the policy issues on which elements of the movement can differ.

How these differences are resolved and the policies that emerge out of this process are of profound importance. It was not just federal involvement and the creation of the NOP that propelled the tiny organic niche into a multibillion-dollar industry. The specific policies and rules that defined the new organic system would determine whether the system would be inviting to conventional agribusiness, or remain largely the province of small farmers catering to a local consumer base. Those details would be worked out among the complex array of players that came to make up the organic movement. Thus, understanding where we are today and where the movement is heading requires that special focus be placed on how the interests of these different sectors align as well as diverge, and the coalitions that emerge out of this field of actors.

## Where Is the Movement Moving?

The organic story is still unfolding. While there is now a sizable organic industry, there is still a significant organic movement too. This includes everything from professional DC-based lobbying groups to mobilized CSA members to an expansive network of activists linked through several Web-based organizations. Organic advocacy organizations regularly hold meetings throughout the United States, bringing together movement participants to not only share the latest organic farming techniques but also to strategize to defend and promote the values that bind them. "Action alerts" regularly mobilize thousands of Internet activists to bombard the USDA or government officials with letters making demands about organic policy. CSAs and farmers' markets are sites at which community groups do education and outreach about the organic issues of the day.

The movement is clearly still alive and well, and perhaps stronger than ever. Yet at this point, and in some ways throughout its development, the organic movement can be considered a movement for sustainable agriculture. An agricultural system that is sustainable, ecologically, economically, and socially, was always the goal for many in the organic community. For historical and strategic reasons, organic simply became the primary vehicle for advancing those values. The term and the concept united the movement for several decades. But as indicated in the opening vignettes, the movement is no longer united behind the word organic, and proponents have adopted varying strategies for advancing their goals in light of the changes that have taken place within the organic system. Despite some core unifying values, this movement is now moving in several different directions.

The efforts of Liana Hoodes, Ron Khosla, and Dan Guenther are indicative of the diverse approaches that have been adopted by those who carry on the struggle for a just, sustainable food and agriculture system. Many, like Hoodes, remain committed to defending the integrity of the organic system as manifest in the NOP. Activists fought hard for many years to develop a set of federal guidelines and procedures that they felt embodied the central principles of organic philosophy, and they are deeply invested in the term organic and the NOP, despite its flaws and shortcomings. They faithfully turn out at NOSB meetings and fight on every proposed rule amendment to ensure that the original spirit of organic is retained in the ever-growing guidelines that delineate what legally qualifies for the label. These activists are deeply engaged in every element of federal policy as it pertains to food and agriculture.

While some remain committed to organic as it has been formally established in federal law, others, more troubled by the conventionalization of organic, have shifted strategies and de-emphasize the NOP. Such people have sought other banners under which to advance their values and goals. This has resulted in a broadened and arguably strengthened movement, even as the term organic itself has lost its universal appeal.

As described earlier, some sustainable agriculture advocates like Khosla favor new forms of certification outside the USDA structure. Khosla's CNG label seeks to create a system similar to that in place before the OFPA set in motion the development of a fully government-supervised model. Under the CNG rules, farmers themselves are once again active in the creation and enforcement of standards, and the rules are structured to favor a locally oriented food system based on a network of small farmers, as originally conceived by many early organic proponents.

Others have also sought to capitalize on the private certification strategy pioneered by early organic proponents. While avoiding state entanglements, these activists have utilized private certification systems in order to address issues that were not incorporated into the federal standards, such as those involving fair trade, social justice, and the humane treatment of animals. Although many of those committed to the federal system are wary of certification schemes that might compete with the NOP, even many NOP defenders believe that such private labels can, in some instances, serve as useful supplements to the federal system.

Yet others steer clear of formal organization and certification systems of any kind. Dan Guenther represents another significant and growing subset of activists within this movement. Many sustainable agriculture advocates place increasing emphasis on simply sourcing foods locally, with no label necessary. Organic practices are still valued by these local food advocates, but many assume that farmers serving their local communities will act as good environmental stewards, whether or not government-accredited organic inspectors or independent certifiers review their practices.

Local food advocacy still represents a movement of sorts, even though the character of much "locavore" or "slow food" activism is quite different from that of its organic counterparts. There is less need for national mobilization or the targeting of government officials. There are fewer membership-based organizations designed to advance the cause. "Leaders" of the movement are more likely to be independent authors and spokespeople who espouse the benefits of a locally based diet.[26] Advocacy efforts are primarily educational, and for individual consumers, participation in the movement simply means buying local. For farmers taking part in this effort, advancing the cause differs little from marketing. They merely target local consumers through food co-ops, farmers' markets, local grocers, or CSA programs, stressing the benefits of their locally produced food. Restaurateurs partake by sourcing ingredients locally and advertising their "farm-to-table" approach. There are no meetings to deliberate about standards or inspectors to monitor their practices.

The local food segment of the sustainable agriculture movement, though, resurrects the question of what constitutes a social movement. In a way, the lack of concerted collective action suggests that this development should not qualify as a social movement at all but instead a loosely coordinated consumer trend. Yet it is a trend still imbued with certain social and ecological values. And it is more coordinated than most other socially informed consumer behavior, such as the purchasing of hybrid vehicles or recycled paper. We must also bear in mind that local food advocacy is still relatively new. There are already some more formal local food advocacy organizations, and they and other segments of the sustainable agriculture movement have identified policy approaches designed to bolster local food systems.[27] Therefore, over time, the local food wing may undertake greater coordinated action and pursue policy reform in ways more characteristic of a traditional social movement.

The movement to advance organic ideas is alive and well, but the changes that have taken place in the policy realm and within the organic industry have transformed the movement and created several fissures. Some consider the movement's greatest achievement to be the official recognition of organic agriculture by the federal government and the NOP's creation. The phenomenal growth in organic sales and widespread public acceptance that organics have achieved since that time confirm the wisdom of their approach. For others, the usurpation and control of the word organic by the USDA was the final turning point in which hope for real progress under the organic banner died. The years since have been characterized by ongoing struggle over the "soul of organic."

The creation of national standards represents neither the final triumph nor ultimate failure of the organic movement. The movement lives on in many ways, but these conflicting perspectives, even about the very meaning of the word organic, reflect the wide diversity within this community. Consensus may not be what is needed, however. Movement participants always strive for unity. Efforts to share ideas and resolve differences are important for maintaining the ties that allow for cooperation along with collective mobilization around common concerns. But as goals are achieved, segments within any movement will always seek to push further.

Although expansion is slowing, the organic industry is likely to grow for many years to come, advanced by the private actors who profit from the system and activists who tout its many advantages. Organic advocates hope that organic practices will one day be the standard agricultural production method. Still, as can already be seen, those practices, while greatly improved over conventional chemical intensive production, will not fully reflect the fundamental goals of the original organic proponents, even if the new ways do become dominant. This is where the other manifestations of the sustainable agriculture movement push to advance "beyond organic." Whether these new efforts will enable us to develop a truly sustainable, healthy, and socially just food system is yet to be seen. There are certainly significant structural barriers that would have to be overcome in order to achieve this ideal. But a careful examination of the course of the organic movement may provide us with the insight needed to arrive at that point more rapidly. It is hoped that this book will help to provide some of that insight.

## Overview

This work is the result of several years of observation and analysis of the organic movement. The primary data come from interviews with thirty leading figures in the organic movement past and present along with observations of dozens of NOSB meetings and those of several organic advocacy organizations. In addition, movement documents including newsletters, Web sites, magazines, and campaign materials were used, and some archival materials such as internal correspondence among movement actors were accessed. Much was also drawn from the work of many outstanding scholars who have examined organic issues, most of whom are cited here.

The chapters presented here are organized thematically and historically. Each chapter focuses on issues that emerged at various stages in the development of the organic movement, while also utilizing themes that run throughout its history. Chapter 2 summarizes the early years of the organic movement, from its roots in the early twentieth century through the late 1960s and early 1970s, when the counterculture breathed new life into the cause. This chapter also identifies themes that have reemerged throughout the history of the organic movement, such as the role of science and spirituality, and how precisely to define and measure organic practices. The third chapter concentrates on developments in the 1970s and 1980s, especially the issues that compelled movement actors to become more organized and establish standards for organic production. This chapter examines the use of market-based certification as a movement strategy and how organic advocates considered the role of the state given problems that were developing within the industry. Chapter 4 takes a closer look at the diverse constituencies that make up the organic movement and how they dealt with the many dilemmas they faced during the NOP's development. Chapter 5 considers the organic system that arose out of the NOP, how we can measure the progress that has been made in terms of advancing organic values, and how some of those values may have been lost or undermined in the course of creating a national system. Chapter 6 considers the question of social justice. In many ways, environmental and health issues have come to the fore in debates about the benefits of organic. But the movement can trace its roots to struggles for social justice, and despite waves of interest in these other aspects of organic, the question of social justice continues to provide a vexing challenge for organic proponents. Chapter 7 discusses the current state of the organic movement; what are organic activists doing now and how do they hope to advance their agenda in the face of a system now largely under the influence of big agribusiness players? In some ways the movement is more powerful than ever, yet in other respects it is fractured and lacks a unified strategy. The three individuals introduced at the start of the book are revisited as the three distinct strategies now evident are reviewed. Finally, chapter 8 explores the future of the organic movement. In addition to assessing the progress that the movement has made and the momentum it has built, this chapter weighs different paths that organic advocates could have taken. In particular, the strategy of spreading organic agriculture via a market-based certification scheme is critically evaluated.

## 2 The Birth of the Organic Movement

The foundations of all good cultivation lie not so much in the plant as in the soil itself.

—Sir Albert Howard, *The Soil and Health: A Study of Organic Agriculture*

For decades following its inception and introduction in the United States, organic agriculture remained below the federal policy radar. It would not be until the 1970s that organic practices and products would garner the attention that would eventually compel a policy response. Yet the early years of development laid the groundwork for the policy battles to come, and actions taken by some early organic proponents set the movement on a course from which later organic activists could not diverge. The primary focus of the analysis presented in this book is the later years, when the movement was maturing into a political force with policy goals that would solidify the status of organic agriculture and give rise to the industry configuration that we see today. Nonetheless, a review of these early years is essential to understanding that trajectory.[1] An examination of the history of the organic movement reveals a number of themes and controversies that arose with the concept of organic itself in the early twentieth century—discussions and debates that continue to this day. This chapter provides an overview of the early development of the organic cause in Europe and the United States, and introduces some of the themes that will reemerge throughout the analysis of the more recent period.

Perhaps the most fundamental and enduring question facing the organic community is that of defining what constitutes organic agriculture. The meaning of organic and specific practices that defined organic agriculture have long been the subject of deliberation as well as contention.. In 1942, in the first volume of *Organic Farming and Gardening*, Rodale, the United

States' chief organic proponent, could only say that the various organic practices in use at the time "have one thing in common, which is that they frown on the use of so-called chemical fertilizers."[2] Rodale was referencing the use of synthetic materials in agriculture—a practice that was already widespread at the time he began to champion the organic cause.

The aversion to the use of synthetic fertilizers is tied to the notion that an organic approach is one that works in conjunction with natural processes, and constitutes a more environmentally sound means of growing food and other agricultural products. But agricultural production, by definition, involves a reengineering of naturally occurring ecological processes. In constructing a more "natural" agricultural method, such as organic production, where one draws the line between working *within* an ecosystem and *intruding* on it is obviously riddled with ambiguities. Organic methods are designed to mirror natural processes, but the concept of nature itself is socially constructed and not simply a matter of empirical inquiry.[3]

While social theorists contemplate the larger question of humans' place in nature, organic farmers and their associates would be forced to wrestle with these matters in the context of how best to grow food in ecologically sound ways. For decades, these issues were debated in various publications, within organic gardening clubs and associations, and eventually, in the halls of government. These arguments came to the fore much later when specific regulatory standards had to be constructed, but dissension existed from early on in the development of the concept of organic growing.

A second recurring theme is that of the role of science versus spirituality in organic meaning and practice. Spirituality was at the heart of some of organic's progenitors, and a spiritual dimension to organic still persists for many of its contemporary advocates. Today many may associate organic proponents with pagan, Eastern, or new age forms of spirituality. For a time, critics of organic exploited this link, charging that organic was little more than a hoax perpetrated by quasi-religious hucksters preying on misguided cultists. Biodynamic agriculture, another form of alternative agriculture closely associated with organic, does include some explicit mystical elements. In addition to these spiritual influences, many early organic advocates were adherents of Christian teachings who saw a natural farming approach as tied to their religious values.[4] Yet even as natural farming techniques were informed by Christian beliefs, science also played a significant role early in the development of organic philosophy.

Some practitioners have always sought to utilize scientific methodologies to test and improve organic techniques. During the period when organic proponents were considered a marginal fringe and ridiculed by the scientific community, adherents established their own sites for experimentation and empirical study. Decades later, with the acceptance of organic's legitimacy by governments and universities, conventional scientific research has taken a more prominent role. In some sense, this fosters tension with the spiritual connection to the earth as well as the countercultural identity that some organic practitioners and consumers nurture through their embrace of organic ideals. Some have suggested that when challenges to conventional practice are defined in scientific terms, their potential for social transformation is weakened or lost.[5] Thus, one might ask, Is organic agriculture truly an empirical practice that can be scientifically scrutinized? Or is it a value system that serves as an abstract guide for believers? Can organic practitioners be defined by their practices alone, or is organic an identity embraced by those who subscribe to some larger belief system of which agricultural methods are just a part? These questions stand at the heart of the battle for what some movement actors refer to as the soul of organic—a fight that still rages today.

A final theme found throughout the history of organic is that of the connection between organic agricultural practices and larger social issues and political values, such as nationalism, social justice, independence, and freedom. To many, organic is more than just a set of beliefs about how to treat the land or grow food; it is a guiding philosophy for social and economic relations as well. A libertarian-leaning agrarian vision is common among many organic farmers, but organic methods have also been incorporated into the agendas of those ranging from socialist-inspired proponents of social justice fighting against imperialism or economic globalization, to fascists who sought to strengthen national ties of blood and land. Paralleling the questions raised about organic as an empirically observable set of agricultural practices versus a spiritually guided set of principles, organic proponents have wrestled with the question of whether organic can be defined by farming methods in isolation, a simple matter of land use and material inputs, or whether the place of land in the economic order and relations between farmers and others in a broader social system must be included in a more encompassing organic whole. These issues of how social and economic factors are tied to organic philosophy have served as points of contention from the earliest years of the movement to the present day.

From its founding, the organic movement would be divided by some of these concerns. But the disputed issues were also key features that served to motivate some of the movement's most ardent proponents, be it a commitment to nature, God, country, or other philosophical principles. In addition, disagreements on most issues did not pose a threat to a movement that was only loosely coordinated to begin with. Competing claims made in the editorial pages of organic publications or conflicting ideas espoused by different spokespeople on a lecture circuit did not risk fracturing a movement that existed only at the margins and had no coherent political organization. Only later, when the movement matured and started to take on a more formal organizational structure, would these differences present real challenges. And it was not until later still, when organic practices achieved official recognition and proponents entered the policy arena, that some of these fissures had the potential to truly tear apart the movement. These themes will be analyzed more thoroughly when examining the contemporary organic movement in later chapters, but the seeds of dissension can be seen as they arise in the history of the movement reviewed here.

### Organic Roots in Europe and Asia

The origins of the organic movement can be found in the more general opposition to industrialization and modernization that emerged in the late nineteenth century. Nascent movements began to develop at that time, offering a broader social critique of which food and agriculture issues were but a part. In both the United States and Europe, reform movements opposed modern technology along with the loss of connection between people and the land.[6] Groups espoused various principles of "natural living" in opposition to the encroaching modernity. The philosophies embraced by some of these movements included dietary practices, sometimes involving vegetarianism, and support for various farming methods, including the use of compost and a rejection of synthetic inputs.

Support for alternative agricultural methods was bolstered by the serious problems with farming practices that developed during the first half of the twentieth century. Yields were found to be declining in parts of Europe and the United States. The dust bowl in the US Great Plains was a vivid

indication of problems in the conventional agricultural system.[7] The depletion of topsoil and the dust storms that swept away tens of thousands of
acres of what had once been fertile productive land were warnings that
our ability to produce enough food to feed the nation was in danger. Some
viewed this as a sign pointing to the need to work more harmoniously with
natural processes and reverse course away from intensive modern farming
methods. These people were the ideological founders of what would eventually become the organic movement. Yet most who controlled resources
or held positions of power had a decidedly different perspective. The loss of
topsoil, crop failure, and other problems plaguing the farming sector were
indicative of the need for *more* scientific study and engineering. Those with
their hands on the controls of agriculture policy sought to double down on
the technology gamble in order to coax more productivity from the land
and produce the food needed for a growing population.

Those advocating technological solutions have always had the upper
hand. Increases in productivity evidenced in the short term with the aid
of technology tend to neutralize the warnings of naysayers, who appear
primitive and fanatic in their unverifiable warnings about long-term repercussions. Empirical gains today always trump the theoretical problems of
tomorrow. This dynamic, combined with the resources available to powers seeking to benefit from technological fixes, led to decades of industrial
domination in agricultural development.[8]

Ideas about how to use more technology to facilitate agricultural production have been under investigation for decades, even before the problems
that emerged in the twentieth century. In 1840, Justus Von Liebig published
*Chemistry in Its Application to Agriculture and Physiology.*[9] He introduced the
idea that plant growth could be facilitated by the application of inorganic
fertilizers instead of the manure that had traditionally served as the basis of
plant nutrition. Von Liebig argued that manure simply contained the base
elements that fed plants, and that the crucial ingredients could be isolated
and applied directly to the soil without relying on animal waste. Critics of
this reductionism would later dub this the "NPK mentality"—a reference to
the elemental symbols for nitrogen, phosphorous, and potassium, the key
chemical ingredients considered necessary for plant nutrition.[10]

The use of chemical fertilizers went hand in hand with the development
of other agricultural technologies such as mechanization and irrigation.
Although manure from livestock was traditionally the primary source of

fertilizer on diversified farms, as machinery was introduced, farms grew in size and farmers increasingly specialized in particular crops. As a part of this transition, the livestock portions of farm operations were eliminated, and so too was the source of natural fertilizer, and hence, synthetic off-farm inputs were introduced.

At the same time, these synthetic substances had become widely available. Advances in the production of these chemicals for military purposes had expanded greatly during World War I. Manufacturers of these materials sought alternative markets for surpluses following the war and found a ready outlet in commercial agriculture.[11] These forces greatly advanced the trend toward industrial farm production during the first decades of the twentieth century.[12]

These trends were disconcerting to some. Rural preservation movements and others opposed to various aspects of modernity had been active since the nineteenth century. They commonly included food and agriculture issues within their oppositional ideologies, but specific production techniques were rarely at the center of such movements. It was not until after World War II that organic agriculture advocacy would emerge as a distinct reform effort. Following the Second World War, the industrialization of agriculture as a policy came into force in the United States and Great Britain. Broader interest began to coalesce at that time around organic agriculture, specifically among communities of technology skeptics.

Albert Howard, a British agricultural scientist, is credited with laying the foundations of organic philosophy. After earning degrees from the Royal College of Science and Cambridge University, Howard was appointed as the imperial economic botanist to the government of India at the Pusa Agricultural Research Institute. Howard was critical of reductionist approaches to research conducted in isolated experimental stations by disconnected "laboratory hermits."[13] He described the "analytical methods of Science" as "both piecemeal and incomplete."[14] As he observed, "We must emancipate ourselves from the conventional approach to agricultural problems by means of the separate sciences and above all from the statistical consideration of the evidence afforded by the ordinary field experiment."[15]

Howard saw more value in observing operating farms and soliciting knowledge from active farmers who had developed their practices over time.[16] He was given great latitude in his position in India and utilized this opportunity to focus his attention on the farming techniques of the indigenous population. Howard wrote, "I decided that I could not do better than watch the operations of these peasants, and acquire their traditional knowledge as rapidly as possible. For the time being therefore, I regarded them as my professors of agriculture."[17]

What Howard was learning during his time abroad was that the local traditional farmers had maintained productive, fertile soil for centuries despite regular cultivation. They honored what would be referred to as "the rule of return"—that the nutrients drawn from the soil by plants must eventually be returned in order to maintain soil health. Howard determined that "the foundations of all good cultivation lie not so much in the plant as in the soil itself."[18] He considered humus, the rich soil with all its interacting biological elements, as the key to healthy plants and effective agricultural production. Through the 1930s, Howard conducted research in India, eventually directing the Institute of Plant Industry in the state of Indore. It is there that he developed what would be called the Indore process of composting for making humus.

Howard was not alone in developing ideas about soil health and sustainable agriculture. One of his contemporaries, Rudolf Steiner, offered a series of lectures in 1924 in a German-speaking area of what is now part of Poland. Drawing on his knowledge of science and observations of the intuitive wisdom of peasant farmers, Steiner presented a series of eight lectures in which he espoused a philosophy of agriculture that incorporated both contemporary scientific research and some esoteric spiritual beliefs about the relationship between earth and the cosmos.[19] Steiner's followers utilized his ideas as a basis for detailing a "biodynamic" approach to farming.[20] In 1928 Demeter, the organization formed around Steiner's teachings, codified these methods, thereby creating the first explicit set of standards designed to define sustainable agriculture practices.[21]

Biodynamic agriculture emphasizes the farm as a self-contained holistic living system and shuns the use of off-farm inputs. This was a more spiritually infused approach to farming—one influenced by the notion of a "cosmic correspondence" that allowed agriculture to be carried out in harmony with natural processes and more metaphysical patterns thought to order the universe. Biodynamic farmers utilize a number of "preparations"—precise mixtures of various plant and animal materials designed to aid in soil health. Phases of the moon and the positioning of celestial bodies dictate the timing of these applications to maximize their effect.

In today's parlance, Steiner's "new age" approach was considered unscientific and irrational by Howard and other advocates of natural farming methods. Yet biodynamic agriculture is still practiced by many farmers following the principles of Steiner's philosophy, and the standards and certification system used by biodynamic practitioners preceded those of organic agriculture by many years.[22]

Both Steiner's biodynamic approach and Howard's Indore Method came to be referred to as organic agricultural practices in the United States. This was based on the terminology used by another contemporary, Walter Northbourne. In Northbourne's 1940 book, *Look to the Land*, he conceived of the farm as an "organic whole" in which elements were cycled through the natural system.[23]

While biodynamic agricultural practices were based almost exclusively on the specific teachings of Steiner alone, Howard was but one important figure laying the groundwork for the more dominant stream of organic development. Not all organic proponents embraced Howard's disdain for controlled experimentation, and others sought to advance the organic cause through the more traditional application of scientific methods. Eve Balfour was among the most prominent of these organic advocates. While Howard is often credited with the development of organic agriculture as a philosophy and practice, Balfour was among its most enthusiastic proponents. She played a key role in spreading organic philosophy globally through her numerous talks and writings. Her influential 1943 book, *The Living Soil*, inspired the founding of the Soil Association in Britain in 1946, and she served as its first president.[24]

The Soil Association, still the most important organic organization in Britain today, was an incubator for organic philosophy and experimentation. The association conducted and supported research as well as distributed information in an effort to spread organic practices. According to organic historian Philip Conford, by the mid-1950s the Soil Association had over three hundred members from fifty countries.[25]

Balfour had previously begun the Haughley experiment in 1939—the first scientific test directly comparing the results of organic versus conventional practices.[26] The study, which would be carried on for decades, was taken over by the Soil Association in 1947. As evidence of the scientific disputes even in those early years, Howard never joined the Soil Association and was openly critical of the research methodology deployed by association investigators. These early differences highlight disagreements that still resonate among organic proponents today. Is organic a philosophy or spiritual belief system, or simply a farming method to be understood and perfected through rigorous scientific investigation? Can organic be reduced to a set of specific practices whose benefits can be isolated and methodically assessed, or can it only be understood and practiced in context by farmers with close ties to the land and soil? These questions about science and experimentation directly tie into issues about philosophy and spirituality associated with organic agriculture, and debate continues to this day.

While most organic proponents never embraced the unconventional spiritual beliefs and practices connected with Steiner's biodynamic approach, almost all early organic philosophy was informed by religious, primarily Christian, beliefs. For the most part, organic adherents supported some form of scientific inquiry and strived to discern exactly how organic methods yielded what they believed to be superior products produced in a sustainable manner. But for many, scientific inquiry was considered a means to discover the God-given natural order that gave rise to this bounty. A Christian component can be seen in some of Howard's writings. He links his discovery of the rule of return to the God-given order of the universe: "The depletion and recuperation of a fertile soil were not devised by man. They are a part of Creation."[27] Thus, from the start, organic was always more than merely an approach to farming. It was a philosophy tied to broader understandings of how the world works and the rightful place of humans within it.

Conford describes the underlying spiritual foundation of early European organic proponents:

Chiefly ... their opposition to industrial agriculture was rooted in a belief in a natural order whose limits could not be exceeded with impunity. [Proponents of industrial agriculture] ... appeared to believe that the soil was an inert material to be goaded by chemical whips and spurs into ever-increasing productivity.... Such an attitude broke a God-given law of life, the Rule of Return. The organic criticism of orthodoxy was essentially prophetic: it predicted problems and ultimate disaster for any agricultural system which flew in the face of the natural order, it read the signs of the times and called for a return to God's ways.[28]

Yet Howard did not focus on Christianity exclusively when discussing organic principles. His beliefs were, after all, based on what he had learned while studying the practices of indigenous people in India. He made linkages between ancient Eastern religious beliefs and organic practices. Howard referenced the Buddhist "Wheel of Life" in some of his writings about organic practices—a symbol that some organic associations adopted.[29]

Spiritual beliefs were hardly the only basis of attraction to Howard's methods. Disruptive social changes enhanced the appeal of ideologies that challenged dominant trends. Many people were troubled by the transformation of agriculture they were witnessing during the first half of the twentieth century. Larger farms, crop specialization, and mechanization were all becoming more common, as were the use of synthetic pesticides and fertilizers. But concerns were not limited to the changes in agriculture alone. Social and economic issues were also tied to resistance to industrial agriculture.[30] By the end of World War I, global industrial capitalism had taken root. The international trade of agricultural goods and other products coupled with the growing power of finance and industrial capital threatened the overall social order. Not unlike sustainable agriculture advocates today, early proponents incorporated these social concerns into their work, and supported efforts to protect small independent farmers and rural communities generally.

Organized resistance came in many forms and reflected diverse ideologies. As Conford points out, these social concerns were at times linked to nationalist movements that saw ties between the native soil and the strength of the race. Some fascist-leaning proponents of organic saw it as a means to return to the agrarian social foundation as well as a way to resist the threats posed by international trade along with the growing strength of manufacturing and finance capital. Although some extreme nationalists were associated with the organic cause, the broader, usually Christian-based ideologies connected with the organic movement more commonly included progressive notions of social justice for farmers. But like their nationalist counterparts, they were equally skeptical of industrialization and the power of urban-based financial institutions that were gaining dominance at the expense of the countryside.

In addition to worries about the ever-encroaching industrial capitalist order, most organic proponents were opposed to the centralized government controls associated with state socialism. Foreshadowing the political orientation of environmentally focused and organic-friendly Green parties that would emerge decades later, one early British organic association identified its political views as "Not Right nor Left, but Straight."[31] The broader ideology linked to organic during this era is one of a devolution of political power in which largely autonomous farmers with close ties to the land could engage in farming practices consistent with a spiritually guided natural order. Decades later, these sentiments would clash with the drive to create uniform enforceable standards and a centralized regulatory structure to guide organic practice.

## Organic Takes Root across the Atlantic

The United States had its own agrarian movements prior to the development of organic philosophy. In the late nineteenth century, US farmers drawn into the world agricultural economy often found themselves squeezed by intermediary agents, banks, railroads, and wildly fluctuating market prices. The Populist movement was the manifestation of this anger against the dominant institutions of the early industrial period.[32] Unlike some of its European counterparts, this movement called for more centralized control over large institutions, including the public ownership of railroads and banks, but continued to value the independence of small farmers. Despite some commonalities between the Populists and the nascent organic movement, there are no clear ties linking them.

More in keeping with the lineage of the organic movement was agrarian theorist Ralph Borsodi, an early proponent of simple living and self-sufficiency who inspired many urban dwellers to go "back to the land" during the Great Depression.[33] Rodale was among those taken with Borsodi's writings. While Howard, Balfour, and others were leading the charge for organic agriculture in Europe, across the Atlantic in the United States, Rodale became the champion of the organic cause.

Rodale was an entrepreneur as well as a publisher of health- and science-related magazines when he read some of Howard's work on the connection between compost and health. He was impressed by the implications. Rodale had personally suffered from poor health attributed to his childhood in the highly polluted city of Pittsburgh, and was interested in the association between health and the environment.[34] After corresponding with Howard, Rodale sold his other magazine interests, moved to a farm in Emmaus, Pennsylvania, and in 1942 launched a new publication, *Organic Farming and Gardening*.[35] He created the Soil and Health Foundation (later the Rodale Institute) a few years later in order to support organic research.

For over two decades, Rodale's magazine would serve as the central vehicle for spreading organic ideology and practice in the United States. Rodale coordinated closely with Howard, who frequently wrote for Rodale's publications and served as an editor for *Organic Gardening*. While Rodale was partial to Howard's Indore composting method, he also regularly printed articles by biodynamic proponents, although devoid of some of the philosophy's more esoteric spiritual elements.

The magazine served as a forum for discussions about organic methods. Its pages were filled with assessments of particular organic techniques, and readers regularly wrote in with questions about how to grow organically, or suggestions and advice about what they were doing on their own farms or in their gardens. While there was much dialogue and sharing of ideas, conflicts about what specifically defined organic agriculture were still far off. Advocates and practitioners shared fundamental views about synthetic chemicals, and other than the more formally delineated guidelines for biodynamic agriculture, there were no fully developed criteria used to certify organic at this juncture.

Rodale's magazine was slow to turn a profit and for a long time was supported by his other business enterprises, but by 1949 its circulation had reached a hundred thousand.[36] Along with it, community-based organic gardening clubs began to form around Rodale's message. The first was founded in Maine in 1945. Rodale facilitated the spread of such organizations by providing technical advice and networking services, first through the magazine and later with the publication of a directory of organic organizations. By 1955, there were over a hundred such clubs throughout the United States and Canada.[37] These groups helped to further promote organic ideals through events and exhibitions as well as via their own publications.[38]

Organic philosophy was spreading among home gardeners while at the same time growing commercially. The first commercial organic farm, Walnut Acres, was started in 1946, but many other enterprises would come on line in the decade to follow. Rodale included organic business advertising in the "Trading Post" segment of his magazines, and mail-order services for organic products began to appear. The first organic market was organized in 1951 in New York City, and grocery stores specializing in organic goods made their way on to the scene.[39]

Foreshadowing conflicts regarding the definition of organic practices, fraudulent profiteering, and the loose application of organic methods that would eventually lead to the development of a legal standard, there were worries expressed about these risks as early as the 1950s.[40] But at this stage, neither the movement nor the market were developed enough to necessitate technical definitions, or for fraud to rise to the level of widespread concern.

Although internal divisions about the finer details of organic methods were still well below the surface, the developing organic community did come into conflict with agribusiness interests and other establishment powers fairly early in its history in the United States. Some agricultural scientists associated with established research stations challenged claims being made by organic proponents about the effects of synthetic pesticides and fertilizers.[41] In 1951, congressional hearings were held on the subject of chemicals in food at which Rodale testified, but where experts extolling the benefits of modern farming techniques far outnumbered critics. Despite the fact that organic proponents were not politically mobilized in any significant way, Rodale encouraged organic gardening club members to be engaged in relevant issues. He called on his readers, for example, to speak out against the aerial pesticide spraying for gypsy moths and fire ants that was common during this period.[42]

It is crucial to acknowledge this advocacy so as not to misconstrue organic as a purely commercial venture. Even at this stage, although without mass demonstrations or any highly coordinated political action, a budding movement nonetheless was behind the organic cause. It was more than simply a fad or marketing trend in that these phenomena lack the elements of grassroots public advocacy that have characterized the organic movement from early in its development. Organic proponents had a mission; they sought to transform the way agriculture was done. They had an ideology, organizations, and goals that called not just for personal transformation regarding certain individual behaviors or purchasing habits but also broad social changes that ultimately involve public policy. True, there was a commercial element of organic oriented toward private profit—one that would become a significant force in its own right in later years. But to this day, private business interests have squelched neither citizen mobilization nor the efforts of organic movement activists to fundamentally reform the food and agriculture system.

Although one can identify the political nature of the organic cause, the movement was slow to develop from its founding through the 1960s. In the prosperous postwar years, progress through science and technology was the dominant cultural sentiment, and its promotion was central to public policy and industry pursuits. Government, private corporations, and knowledge producers in universities and research institutions were united in their support for the full application of mass production as well as chemical technology to all areas of life, including food production.[43] The organic movement would remain on the margins largely unnoticed by mainstream America. It was not until the late 1960s and 1970s that signs of its political potency began to emerge.

### Enter the Counterculture

A convergence of a number of factors help to explain the transition of organic from a loose collection of gardening hobbyists held together primarily by a publication and a small number of advocates to a genuine social movement. Environmental issues were of growing concern by the 1960s. In prior decades, environmental thinking was primarily characterized by the desire to preserve wilderness or conserve natural resources, but by the 1960s, the public was increasingly concerned about the threats posed by the toxic by-products of industrial society.[44] Rachel Carson is often credited with giving rise to the modern environmental movement through her book *Silent Spring*, a work that focused on the threat posed by synthetic chemicals in the environment, particularly those used in agriculture.[45]

The link between synthetic chemicals and ecological degradation was not the only growing public concern that provided impetus for the organic movement. During the same period the United Farm Workers (UFW), led by Cesar Chavez, was pressing for the rights of agricultural laborers. While compensation was frequently the primary grievance of these highly exploited workers, in the 1960s and especially in the 1970s, many of these labor struggles also involved issues of worker health and safety, including risks posed by exposure to toxic pesticides.[46] UFW activists did much to publicize the harmful effects of these materials on worker health. Consumer boycotts were a common tactic used by the union, and thus efforts were made to educate the public about the toxins being used on their food. While the ties between these labor struggles and the organic movement were indirect, like Carson, the UFW did much to raise public consciousness about problems posed by synthetic pesticides and fertilizers. Yet in the midst of other political turmoil during the 1960s, sustainable agriculture issues would not come to the fore for a number of years.

The counterculture movement of the 1960s, including a revived back-to-the-land element, would ultimately bolster the organic cause.[47] The civil rights movement and anti–Vietnam War mobilization raised the awareness of many about the broader social problems afflicting modern industrial society. The conventional food and agriculture system would eventually come under attack along with every other dominant social institution.

Members of the Diggers, a radical group based in the Bay Area, were the first to embrace the counterculture's use of food as a medium to defy convention by advocating for the development of alternative institutions. The Diggers regularly distributed food for free in a direct challenge to the for-profit industrial food system. Food also played a role in the famed struggle over People's Park in Berkeley. It was there on the University of California campus that activists seized a vacant lot in order to establish a public garden and grow food.[48]

General discontent with modern social institutions considered oppressive and violent led some to seek a simpler agricultural existence along with a social order reminiscent of the agrarian Jeffersonian ideal. Others envisioned a radical social transformation in which communal enterprises would displace industrial capitalist institutions. While the activists of this period cited diverse ideologies, many spurned industrial technologies and gravitated toward organic practices as a natural alternative to a food system increasingly dominated by corporations and dependent on synthetic pesticides and fertilizers.[49]

Ronnie Cummins, the national director of the Organic Consumers Association (OCA), explains how the organic cause became integrated into the radical politics of the era:

In the late 1960s, those of us coming out of the civil rights movement and the anti-war movement saw that building a counterculture or an alternative economy was a political strategy for radical social change. We realized that protesting in the streets and even lobbying for civil rights legislation wasn't enough, and that we needed to have fundamental changes throughout the institutions. One of those things that started as a political tactic was to build food cooperatives all over the nation and have these communes, these back-to-the-land communes. That was really the impetus behind the modern organic, buy local movement. It was started with a political holistic vision.[50]

Counterculture radicals were introduced to organic through alternative publications such as the *Whole Earth Catalog*, which gave the movement new credibility and importance by referring to *Organic Gardening* as "the most subversive publication in the country."[51] The discovery of organic by this highly mobilized population gave a significant boost to the organic movement. Circulation for *Organic Gardening and Farming*, now under the editorship of Rodale's son, Robert, shot up 40 percent in a single year, from 1970 to 1971.

The established organic community of the 1950s and 1960s was not necessarily ready for the influx of counterculture radicals that came to embrace organics. According to historian Warren Belasco, "In 1969, with the hip turn to ecology … *Organic Gardening and Farming*'s readership suddenly broadened…. [T]he younger Rodale cautioned his 'middle aged frustrated farmers' against reacting too negatively to the weed eating 'social rebels.'"[52] Yet it was these social rebels who would come to dominate the organic movement in the coming years.

In his analysis of the radical food movement's history, or what he calls "counter-cuisine," Belasco describes three organizational manifestations that became integrated into the organic movement. The first were the communes and farms that adopted natural farming practices. Belasco cites the fivefold increase in country communes between 1965 and 1970, with the total number reaching roughly thirty-five hundred. Most of these proved unsuccessful within a few years as the harsh reality of farm labor and complex collective social relationships intruded into idealistic visions of the new world. While most commune dwellers returned to more conventional lifestyles, others stayed with farming, developing skills and establishing enterprises that would further the vision of a decentralized, sustainable agrofood system.

Food co-ops distributed some of this farm bounty, and served as another means by which key social ties and institutional networks necessary for effective political action would form. As Belasco points out, "In addition to linking producers and consumers, co-ops provided advice, moral support, and living examples unavailable through regular media and markets."[53] These types of gathering and networking venues were critical to the formation of politically mobilized publics, and co-ops played a vital role during this phase. Similar to farm communes, not all food co-ops proved successful in the long term. But they were essential in establishing network ties and advancing ideas about alternatives to the industrial food system. There are some highly successful co-ops still in existence from this period, such as the Rainbow Grocery in San Francisco, the Honest Weight Food Co-op in Albany, New York, and the Brattleboro Food Co-op in Vermont. These continue to serve as important centers of organic movement activity given the bridge that they gap between producers and consumers.

In addition to these cooperative production and distribution enterprises, private alternative restaurants and food stores supplemented the communes and co-ops as centers of organic movement activity. Alice Waters's famed Berkeley restaurant, Chez Panisse, was founded in 1971. While not all such establishments (including Chez Panisse) were started to explicitly advance a political cause, independent restaurants represented a reaction to a food system increasingly characterized by standardized mass production and industrial processes.[54] Although McDonald's had yet to serve its first billion customers, the first generation to be raised in a world with national fast-food chains was already reacting against this "plastic" fare.[55]

Together, these alternative enterprises not only bolstered the number of organic adherents; they provided a space in which like like-minded people could share ideas, form networks, and further organize themselves into a more coherent political force. Informal centers for organizing such as these also serve important culture- and identity-building functions for any movement.[56] In addition, unlike the bookstores, bars, barbershops, cafés, and other social centers that have facilitated informal networking among activists in other movements, the economic ties among organic producers, distributors, and retailers, by necessity, created a convergence of people with shared interests in advancing the organic cause. Whether that economic interest came to supersede social and environmental concerns is a matter of some debate today.[57] But certainly, at this stage of the movement and for many organic activists to this day, the underlying motivation has always been to create a just and sustainable system of agriculture. The organizational forms that developed in order to more effectively advance this vision in the face of growing opposition from dominant institutions will be examined in the next chapter.

## Conclusion

The organic movement developed slowly over the course of decades, yet the issues described here that were present at its founding would persist in various forms to the present day. From its start organic was something more than an agricultural method for many of its adherents. While some, especially the biodynamic wing, explicitly incorporated a spiritual element into their philosophy even many proponents of less esoteric organic methods believe that their practices cannot be properly understood through traditional reductionist science. This latent spiritual component would remain in tension with the role of science within the organic community even in later years when universities and government agencies embraced the scientific study of organic practices.[58] Technical analyses examining the biochemical processes that occur while using organic methods have supplied some with the assurance that there is something empirically beneficial about the practice and its products, yet others believe that through this scientific scrutiny, the soul of organic has been lost.

These competing sentiments about organic are directly related to a second issue—one that would become central in the following decades: how to define and verify organic practices. Ironically, the more spiritually infused biodynamic wing of the organic movement was the first to have clear definitions and prescribed practices. Others sought to embrace an elusive idea of natural. How best to work with nature while clearly manipulating it for instrumental purposes poses vexing questions under any circumstances. In the absence of a pressing need for certification, defining organic was an informal dialogic process primarily carried out in the pages of *Organic Farming and Gardening*. Yet the development of the organic market would soon necessitate a more explicit definition. This was especially true once organic proponents sought to codify and regulate their practices. At that point, organic practitioners would be forced to articulate exactly what defined organic growing. These debates have since moved from the pages of *Organic Farming and Gardening* into government-sanctioned committee meetings, yet the specific meaning of organic remains highly contested.

A third issue, still tied to the first two, is how organic agriculture relates to broader social concerns. Resistance to industrialization, urbanization, modernization, and rationalization can all be found in various elements of the organic movement. The embrace of organic agriculture by the counterculture in the 1960s and 1970s infused an anticapitalist component within the movement. In many ways organic was viewed, first and foremost, as a social issue during this era. While social justice issues would never disappear completely from the organic agenda, they would later recede in significance as environmental concerns grew in the public consciousness of the

1970s.[59] Contemporary organic proponents still struggle with how this agricultural practice can be more closely allied with movements more explicitly focused on issues such as poverty, workers' rights, small farm preservation, and racial justice.[60]

While all these issues were present from the start, the tensions would take different forms at each stage of the movement's development. Coming out of the 1960s, the movement would develop a more structured organizational form. With this formalization, the movement was forced to confront issues in a way that it had not when it was a loose collection of adherents with a handful of ideological leaders. Contradictory or ill-defined practices and beliefs can coexist among informal networks. But organizations tend to establish more formal positions and policies.[61] This shift would compel organic adherents to address the most challenging aspect of organic agriculture: defining what it is. In addition to tracking organizational development and strategic evolution within the organic movement, the next chapter examines how economic and political dynamics compelled organic proponents to more formally define the practices they advocate.

# 3   Certification and the State: The Dilemmas of Growth

I have read with much interest your various articles. Would you please tell me which are chemical fertilizers?

—Roy Roeschley, letter to editor, *Organic Gardening and Farming*, December 1942

The 1970s would be a period of growing momentum for the organic movement. While marginal from the time of its introduction in the United States through the 1960s, the 1970s was a period of maturation and organization. Still demonized by establishment proponents of conventional agriculture in government and industry, the ragtag collection of counterculture farmers began to organize themselves in associations of mutual support. These farmers were isolated from the mainstream of agricultural production, but they experienced growing acceptance among consumers, who were increasingly suspicious of chemical pesticides, fertilizers, and food additives.

This marginalization from dominant institutions coupled with the embrace of consumers created dilemmas for organic activists and policy makers alike. These challenges were heightened during the 1980s as the organic market grew in size and complexity. Organic consumers, and even some farmers long wary of centralized institutions, began to recognize that there could be a role for government in rationalizing the market and protecting against fraud. Established consumer and public interest groups, already oriented toward policy approaches, started to call for federal action on organic, catching many farmers and other organic activists off guard. In the face of these developments, government officials, most of who had been hostile or at best indifferent to the organic sector, had to come to terms with the fact that organic agriculture was becoming a socially accepted practice. Organic was no longer something they could suppress or dismiss as the irrational obsession of a fringe element. Confronted with occasional chemical food scares that drove millions to seek out organic alternatives, government was ultimately brought in to play an active role in overseeing the organic industry.

The transformation that took place within the organic movement and organic industry over these decades was dramatic. How that transition occurred is revealed through an examination of the movement's development through the 1970s and 1980s.

## The Food Establishment Responds

The social turmoil of the 1960s created skepticism about dominant social institutions that extended well beyond the activist core. Many people in the United States lacked faith in what authorities were telling them, including in the areas of food, nutrition, and health. Growing environmental awareness and concern about the effects of synthetic chemicals led many to reconsider their diets. The processed, packaged foods popularized following World War II were looked on with suspicion by a growing segment of the population. As a result, the demand for natural and organic foods was expanding rapidly during the early 1970s. *Barron's* called the health food industry the "nation's new 'glamour' business" and reported annual growth rates of 30 percent.[1] Some analysts predicted that health foods would compose 40 percent of the national food market within a few years' time.[2] While these forecasts did not fully bear out, they indicate the enthusiasm for alternatives to conventional food and the speed with which the market was growing at the time. As it did, the conventional food industry, government officials, and the food and nutrition science establishment stepped up its criticism of what they considered to be a dangerous fad—a fad that also happened to pose a grave threat to their economic and political interests.

President Richard Nixon's colorful and controversial secretary of agriculture, Earl Butz, summed up the administration's position on organic agriculture in a 1972 feature story on him in the *New York Times*. He propagated the idea that modern technology, including the use of synthetic chemicals in agriculture, was essential for meeting the nutritional needs of the world's people—a claim commonly echoed today in regard to biotechnology.

Too many people, too many of the environmentalists, many of the food faddists, think we ought to go back to organic foods and organic agriculture. We could go back to an organic agriculture in this country if we had to. We know how to do it.

We did it when I was a kid. We didn't use any chemicals then. But before we go back to organic agriculture somebody is going to have to decide what 50 million people we are going to let starve.... You simply could not feed 206 million Americans even at subsistence levels with the kind of agriculture we had fifty years ago. It would be impossible.

Butz went on to blame naive and disillusioned youths for undermining faith in conventional agriculture.

I think our basic problem with respect to all this is that two-thirds of us in this country are so young that we never had the experience of biting into a wormy apple, looking at the worm hole and wondering, "Is he in there or did I get him?" Two-thirds of us are so young that we think that God and nature automatically made nice, red, juicy, plump, healthy, tasty apples. As a matter of fact, He didn't. Mother Nature put the worm in the apple and man took him out.[3]

Butz was the most visible critic of organics within the Nixon administration. The USDA was firmly allied with big agribusiness and committed to a path of intensive industrial agriculture development. Butz himself served on the board of Ralston Purina, a giant food corporation. He dismantled programs from the New Deal era designed to manage supply and stabilize prices, and instead admonished farmers to "get big or get out."[4] While federal officials were openly hostile, attacks on the organic and health food industries could be found throughout government institutions.

Criticism did not just come from government officials. Food industry leaders also routinely denounced organic and natural foods as a hoax perpetrated on a gullible public by profiteering swindlers. Given the relatively low cost of conventional food achieved through government support and economies of scale, organic goods tended to cost more. Although organic premiums were low or nonexistent in most places during the early 1970s, there was a growing market through which organic producers could demand more for their products. Critics took advantage of this vulnerability to charge that organic was more about making money than advancing health or environmental protection. Henry Heinz, chair of the H. J. Heinz Company, stated, "We are a nation of nutritional illiterates. Food faddism advocates are persuading thousands to adopt foolish and costly eating habits."[5]

While organics were vulnerable to critique on the basis of price, debate centered primarily on whether there was anything to truly distinguish organic products that would justify paying more. Food and agriculture business leaders clearly had an interest in denying any benefits of organic methods. Even though natural and organic foods made up only a tiny fraction of the overall food market, the speed with which that market was growing posed a threat to conventional producers, and this elicited the expected counterattack.

Along with political leaders as well as conventional food and agriculture industry representatives, many established nutrition and health experts and those in agricultural sciences denounced what they considered to be a food fad.[6] These critics cited the lack of evidence that organic foods were of any greater nutritional quality than their conventionally grown counterparts. They argued that synthetic pesticides and fertilizers were benign, and posed no threat to human health or the natural environment. In the face of growing public skepticism, they sought to reaffirm the cultural narrative that had developed in the postwar era: that the deployment of technologies in all aspects of life represented social progress. Health and nutrition were no exception, and experts were only increasing abundance and improving the human condition through the application of food science.

A drumbeat of denunciation from these experts was maintained throughout the early 1970s. Kenneth Beeson, a professor of soil science at Cornell University, concurred with Butz's view that if "extremists" were able to limit the use of inorganic chemicals in agriculture, it would result in "widespread malnutrition and even starvation."[7] Jean Mayer, professor at the Harvard School of Public Health and chair of the White House Conference on Food, Nutrition, and Health, claimed that while it was reasonable to assess the effects of certain synthetic substances, there was no need to follow organic advocates "back to the Stone Age."[8] Another critic, George Christakis of Mount Sinai Medical Center, went so far as to call organics a "public health nutrition threat" given that the health food movement was steering people away from recommendations based on established nutrition science.[9]

Among the most vocal critics of the organic movement was the head of Harvard University's nutrition department, Frederick Stare. Like many who shared his views, Stare had extensive ties to the food industry. Conventional food corporations contributed a great deal to support his research and public campaign against the alternative food movement. Stare regularly spoke out against "health quacks" and other alternative food advocates. He coauthored a full-length book, *Panic in the Pantry*, in which he denounced health food proponents and argued that the conventional food supply in the United States was the safest and most nutritious in the world.[10]

The mainstream media gave voice to and often amplified the criticism coming from government officials, food industry leaders, and conventional health and agriculture experts.[11] Organic received some support in the alternative press and from the underground newspapers that had emerged during the 1960s, and some influential public figures such as Wendell Berry, Jim Hightower, and Barry Commoner also touted the benefits of organic methods. But in mainstream newspapers and magazines, organic and health food advocates were commonly derided as fanatics. In a feature article on his life's work in the *New York Times*, Rodale himself was labeled the "guru of the organic food cult." Organic proponents were variously characterized as "reactionaries yearning to turn back all clocks," "Dr. Strangelove paranoids who read poison plots on the ingredient labels of pancake mixes," and "a cult of organic food devotees, who proselytize with all the fervor of converts to astrology or the mystic rites of the East."[12]

In addition to official pronouncements, in some instances government officials took action targeting the organic industry. Hearings were held by the attorney general in New York City in December 1972 to investigate claims of fraud associated with natural and organic foods. A study conducted by the Department of Agriculture and Markets found pesticide residues in 30 percent of the organic products it sampled. The investigation concluded that there was "significant misbranding" in the organic market.[13]

Given the level of hostility toward organic among government agents tied to conventional agriculture, it was expected that the state would soon act to reign in this budding industry. A 1971 article in *Advertising Age* magazine predicted that USDA regulation of the organic industry was imminent.[14] Federal officials did consider taking measures against organic practices during this era. Attention was focused on the idea that gullible consumers were being misled by the implied superiority of organic products—an assertion that, in the view of federal officials, had no scientific basis. The Federal Trade Commission (FTC) undertook a study to assess the issue. Its report cited research that found that "only 23% of the general population ... claimed to 'generally understand' the term 'organic.'"[15] Officials concluded that "there is significant consumer confusion regarding the use of the terms 'organic food' and ... 'organically grown food'"[16] An

FTC staff report in 1974 recommended that the use of the term organic be banned in advertising given the risk of consumer deception.[17]

Yet organic was never intended to mean that foods were completely free of pesticide residues. Given the pervasiveness of synthetic chemicals in the environment, this is a guarantee that no one could make. Federal officials struggled with how to regulate a practice for which the actual meaning was elusive. Officials ultimately balked at implementing the proposed FTC ban on the term, and in 1978 the commission reversed its stance. Trade officials reluctantly accepted that the meaning of organic was sufficiently established within the industry to allow for the term's use in advertising.[18] The definition that the FTC concentrated on was drawn from Rodale's *Organic Gardening and Farming* magazine: "Organically grown food is produced on humus-rich soil whose fertility has been maintained with organic materials and natural mineral fertilizers. No pesticides, artificial fertilizers or synthetic additives are used in the production of organic foods." This definition sidesteps the issue of whether or not organic foods are free of synthetic chemicals—a commonly held notion, but a claim that organic practitioners do not make. This significant divergence in understanding about the fundamental meaning of organic continues to present challenges to the organic system. The belief that organic foods are free of synthetic chemicals is widespread among consumers, placing growers and even more so regulators in an awkward position as they seek to address the needs of all constituents.

As a method of agricultural production, the definition offered by Rodale was clear enough to pass muster with trade officials, who proposed that organic practices be regulated to ensure that farmers adhere to their claims. This placed political leaders in a peculiar position. Given a commitment to conventional agribusiness, there was pressure to stifle the perceived threat posed by the organic movement. But despite some confusion on the part of the general public, there did not seem to be any basis for preventing organic consumers access to the products they wanted, and no grounds for barring organic farmers from selling their goods and labeling them as such. While officials expressed concern over alleged fraud, Congress demonstrated no interest in establishing a regulatory framework to ensure that products met standards that they considered substantively meaningless to begin with. As a result, no federal action was taken, and organic practitioners would be left largely on their own to define what they were doing and to establish a system to ensure they were doing it right.

## Organic Gets (Sort of) Organized

An obvious problem emerging out of all this was that beyond the general concept described in Rodale's publications, there was no universally accepted definition of organic even among organic practitioners, let alone any mechanism for enforcement for products bearing an organic label. Organic growers had a general sense of what they were striving for, and specific practices had long been discussed in *Organic Gardening* and elsewhere. Certain composting methods were clearly favored and other practices, such as the use of most synthetic pesticides, were explicitly off-limits. Yet the production of an organic product was still not fully defined. According to Lynn Coody, who ran an organic farm in the 1970s and went on to play a central role in standards development, "There were no standards, so there was no guidance about what was organic or not. It was just what *felt* organic."[19]

Given the criticism coming from the conventional food establishment and government officials, organic proponents were going to have to do more than "feel" the organic character of their methods. Even within the organic community, many recognized that some tightening of definitions was in order. Concerns about fraud had been expressed as early as the 1950s in the pages of *Organic Gardening*, but the rise in organics' popularity coupled with the criticism and scrutiny under which the practice was coming demanded a more formal response.

Organic practitioners sought to stave off criticism and charges of fraud by stepping up their efforts to self-regulate. In 1971, Rodale formally established a set of criteria for organic production along with a procedure for monitoring and enforcement. Agri-Science Laboratory was hired to conduct soil testing for farmers who sought certification. The soil tests would reveal whether synthetic chemicals were being used on the fields. There were also record-keeping requirements, and Rodale representatives would interview farmers to determine whether appropriate composting methods were being employed. Farmers who qualified would receive a seal to be displayed at their farms and to attach to produce boxes. By mid-1972, Rodale had officially certified eighty-five farms.[20] Aside from the standards created by Demeter for biodynamic producers, this was the first concerted attempt to rationalize the organic market.

Legitimate players in the organic industry recognized the need for monitoring, but the form that should take was debated from the start. Most organic farmers were still rooted in the 1960s' counterculture movement. Many were wary of any government involvement in their affairs.[21] Their strategy for creating social change was *prefigurative*.[22] That is, their goal was to create the kind of world they wanted by living it, not by legislating it. They wanted to develop their own institutions—ones that were reflective of the social order they aspired to create. For the most part their vision did not include involvement by government, still viewed as corrupt and unlikely to aid in nurturing an alternative to the industrial food system. On the contrary, government agencies appeared intent on undermining their efforts.

The alternative to government oversight and regulation was for organic farmers themselves to band together to advance their cause and create systems designed to protect the integrity of their practices. While Rodale's organization was still the biggest and most visible entity attempting to bring coherence to the organic movement, by this time networks and relationships had been built through the food co-ops, communal farms, and alternative stores and restaurants that developed out of the 1960s' movements.

During the early 1970s, organic growers in some regions went beyond involvement in these loose networks and cooperative business partnerships, and began to formalize their ties as organic farmers. Those who had participated in informal organic gatherings and gardening clubs in Maine formed one of the first statewide organizations, the Maine Organic Farmers and Gardeners Association, in 1971.[23] This was around the same time that some other northeastern states were creating formal statewide organizations, which would eventually be networked through the Northeast Organic Farming Association (NOFA). California Certified Organic Farmers (CCOF) followed in 1973 with fifty-four grower members.[24] European organic farmers were also coming together at this time. In 1972, the Europe-based International Federation of Organic Agriculture Movements (IFOAM) was formed during a meeting at which some US organic representatives were present.[25]

Despite the international coming together implied by IFOAM, the organic movement would remain a decentralized entity for many years.[26] The establishment of organic farmers' associations in other parts of the United States would soon follow, but even within the country, contact and coordination among these groups was relatively rare. In spite of these grassroots efforts to formalize and regulate practices, the absence of national coordination would ultimately create the conditions whereby movement centralization and government involvement was virtually inevitable.

Notwithstanding the lack of coherence as a national or international force, this organizational development in the form of creating organic farmers' associations represented a maturing of the organic movement. Social movement scholars have long recognized the tendency for movements to transition from a mobilized but loose collection of adherents, toward more structured and organized forms.[27] The 1970s saw the beginnings of that formalization among organic proponents. Yet their organizational structure remained highly decentralized and other characteristics commonly associated with formalization, such as direct political engagement and a turn toward policy goals, was still a way off for organic advocates.

The lack of policy focus among organic associations at this time does not suggest that these groups were apolitical. Many had never lost sight of the notion that agriculture could be transformed one farm at a time as consumers and farmers themselves recognized the underlying costs of industrial agriculture along with the need for a better way. Few believed that their ends could be achieved through policy, especially given the hostile attitude toward organic held by many in positions of political power. In the meantime, these newly formed organic farmers' associations primarily served as venues through which farmers could share ideas and best practices while strengthening their identity as change agents.

Through the 1970s and even most of the 1980s, there was no institutional support for organic agriculture from the state or universities.[28] According to Elizabeth Henderson, a leader of the New York NOFA chapter,

In the northeast ... these farming associations were the groups that spread the know-how about organic farming before the universities or the government got into it. Back in 1979, '80, when I was getting interested in organic gardening first and then doing it on a bigger scale, there weren't any courses that you could take in the United States. And if you called up the extension with a question like, "how do you do organic raspberries?" they would laugh at you. So then you didn't call again.[29]

Coody, one of the founding members of the organic farmers' association Oregon Tilth, had the same experience on the West Coast. "At that time, it was very hard to get information about how to solve problems and what to expect.... The extension service, and anybody that you would get information from as a farmer, did not have information about organic farming."

Organic farmers' associations would come to serve as clearinghouses for information on organic methods. Coody added that the associations played an important role not just in a practical sense; they supplied coherence and an institutional presence around which a sense of identity could form. "The goal, in my view … was to provide a support network for people who were interested in organic farming.… [I]t was a farmer-to-farmer network to problem solve, and really, just to have friends who were doing the same thing you were. That was a big part of it, creating a community around organic farming."[30]

This marginalization from mainstream institutions and the creation of organic organizations reinforced an alternative identity among organic farmers.[31] As will be discussed later, this in some ways helped to strengthen and unify the movement, while in other ways limiting its ability to effectively engage with conventional institutions at times when that could be beneficial. In any case, farmers were left to their own devices in terms of organizing an organic system during this period. The formation of organic associations, while still only loosely coordinated, was a crucial step in the development of the movement and formalization of the whole concept of organic.

Clarifying what exactly was meant by organic agriculture and what specifically defined an end product of that process was central to the agenda of many of these groups. In addition to sharing ideas about what worked well, organic farmers' associations set out to codify what they were doing and what should qualify as organic, and to develop procedures for certifying authenticity. There was already an established body of information about what constituted organic agriculture in Howard's writings and the publications put out by Rodale. Early on, some organic associations directly adopted Rodale's guidelines. The Maine Organic Farmers and Gardeners Association originally used Rodale's organic standards when it began to offer its own organic certification in 1972. Others used guidelines developed by IFOAM as a starting point and then added their own rules in accordance with local conditions.

A key factor at this stage of organics' development is that organic farmers themselves were setting the standards through their associations and doing so primarily for their own purposes. Although there was some participation by individual consumers who were involved in the organic associations, this was in sharp contrast to later efforts in which consumer organizations and other constituents would be heavily involved in shaping organic policy. Yet at this time it was primarily organic farmers who were coming together and collectively defining what qualified their practices as distinct. According to Enid Wonnacott, executive director of the Vermont NOFA branch, "The process … was always very much farmer driven.… We'd meet in the fall, we'd have … committee meetings and [discuss] 'what does it really mean to farm organically?'"[32] Out of these conversations, standards were established and modified over time.

It was clear that throughout the early certification period, philosophy was privileged over technical detail; the focus was placed on the value of seeking to grow food in ecologically harmonious ways. Coody describes the evolution of organic standards during the early years of Oregon Tilth:

By the end of the '70s we started trying to write down standards. We still did not have a certification agency. We wrote the standards just to try to answer questions that were coming to us in droves about what is organic and what is not. So this was more of a philosophical and standards document. And then from there, there were a number of major rewrites over time.… [W]e went to meetings and just argued all the time about "should this be in or should that be in?" It was the farmers who were de-bating philosophy and turning it from philosophy into practice.… That continued until the mid-'80s.… [B]y that point, we started creating a lot more professional set of standards that didn't have the philosophy in it anymore, and included a policy manual that told how the certification program actually worked.[33]

Organic farmers across the country were engaged in a similar process. They were organizing themselves, creating associations, and setting about defining what it was that united them and identified their practices as unique. Throughout the 1970s and 1980s, more state and regional organic farmers organizations were formed, each creating their own standards and certification procedures.[34] By 1975, there were eleven organizations in the United States, each offering organic certification based on their own distinct set of standards. Programs differed from state to state in many respects, but most involved inspectors affiliated with the organic associa-tion visiting farms to assure compliance in order to receive certification. In 1978, IFOAM created standards that served as the baseline for organic certification throughout much of Europe.[35] In the United States, indepen-dent certifying regimes would continue to proliferate. Eventually by 1989, when the federal government was getting involved, there were over forty certifying organizations.[36] While these systems in some ways helped to provide organic with a greater degree of legitimacy and limit fraud, the decentralized nature of the organic movement at the time set the stage for problems as the industry grew.

## From Prefigurative Politics to Policy: Organic Strategy Takes an Accidental Turn

While the organic movement was entering a phase of organizational growth and formalization at the state and regional levels, there was still almost no national organizing throughout the 1980s. By the end of that decade, however, developments largely outside the control of the movement drove activists to organize nationally and take defensive action to protect their achievements. This ultimately took the form of a federally supervised national organic system—something that few organic advocates sought or even conceived of just a few years prior. Three factors in particular converged to compel this action in the late 1980s: the growing size and complexity of the organic market, cases of fraud that threatened the integrity of the organic name, and some high-profile food scares that resulted in droves of consumers heading to the organic market. After almost a decade of grudging tolerance and neglect, some federal officials began to take more notice of organic agriculture, and act on some of the problems and opportunities it presented. In the meantime, organic farmers were forced to redirect their energies from internal organizational development and self-monitoring toward engagement with the federal government as well as the public policy world.

### Growth and Complexity in the Organic Market

During the 1980s, growing concern about personal health and environmental protection bolstered the popularity of organic foods. Some chic restaurants were featuring organic items, drawing in upscale consumers.[37] As new demographic groups were brought in, the market for organics grew, and distribution extended beyond the direct-to-consumer and co-op sales of fresh produce that characterized earlier years. The supply chains between organic farmers and the growing number of retailers that carried organic products became longer. Larger outlets for organic goods began to appear on the scene. Whole Foods Market opened its first natural food supermarket in 1980 and rapidly grew into a national chain in the following years. There was also increasing international trade in organic products, further diminishing the personal relations that originally existed among organic farmers and close networks of retailers and consumers. In addition, a greater variety of organic products started to appear. More processed organic foods involving multiple ingredients were being marketed, not simply the whole fruits and vegetables that characterized most of the organic market since its birth.

At first, industry growth was tied to small players in the organic whole-sale and retail trade. Several entrepreneurs had begun as grassroots activists, but their businesses were growing at a rapid pace. Gene Kahn founded Cascadian Farm in 1971 and built it into a highly profitable, vertically integrated national distributor of packaged organic foods (the multimillion-dollar company was sold to General Mills in 2000). Mark Retzloff, a cofounder of San Francisco's Rainbow Grocery, also started Eden Foods and eventually Horizon Dairy, among other highly successful organic enterprises.

The patchwork of organic standards and certification systems created by state and regional organic farmers' associations were ill suited to handle some of the problems that grew out of this increasingly complex market.[38] Grace Gershuny worked for NOFA Vermont starting in the 1970s and went on to work for the federal government, aiding in the development of organic policy. She describes the difficulties that were cropping up during the 1980s:

Part of the problem was that there were beginning to be processors and manufacturers and people making multi-ingredient products, or even single ingredient products from different producers. They were getting certified for their products, and then they'd find that they had to have their growers, their producers, certified by the same certifier, otherwise the certifier wouldn't accept them.... Then they were selling to people who had different certifiers ... and it was a mess. It was a total mess.... And the businesspeople, the traders, and the manufacturers knew that there was no way that they were going to be able to expand their companies or their production or their markets unless they had some kind of consistent standard.[39]

Most of the differences in standards used by the array of certifying organizations were not that significant, but as Gershuny describes it, some were important enough to inhibit certifying groups from reciprocal recognition of one another's products. For example, most organic standards in the Northeast required a three-year transition period for conventionally farmed land to be converted to organic. Farmers would have to adhere to organic rules during this time, but their products could not be certified until after the transition was complete. This period was designed to both build the soil based on organic practices and reduce synthetic chemical residues. California standards only required a one-year transition period, though. Thus, California-certified organic goods were considered suspect and not counted as certified by other organic associations using the longer transition period. Processors using some ingredients certified by CCOF could not hope to get the finished product certified by groups on the East Coast.

The lack of a single standard led to tensions between organic certifiers around the country. Organic certifier Jake Lewin said that organic associations began "playing a very complicated game of, 'who's better than who?'" Gershuny recalls that "nobody trusted each other. They'd say, 'No, ours are better than theirs. We have higher standards.'"[40]

The multitude of certifiers was also problematic for consumers. A columnist for *Mother Earth News* wrote of the organic market, "The proliferation of groups, the differences between respective standards, and the potential for confusing the food-buying public are all making this an intolerable situation."[41] All this led some to conclude that the organic market had reached a point where uniform standards along with some kind of centralized control and oversight were necessary—a role that would eventually be played by the federal government.

There had been some attempts to bring national coherence to the organic sector prior to federal government involvement. As indicated earlier, the Rodale organization had originally attempted to set the national standard. Although Rodale's standards served as a basis for some organic organizations, at least initially, variation proliferated when independent organizations built on Rodale's foundation. Aside from Rodale's early efforts to provide a single standard for organic growers, the first significant step toward national coordination was set in motion in 1984. That was the year that representatives of a Maryland produce distributor came to Vermont seeking organic lettuce and other crops that could not be grown in the South at that time of year. They insisted that the crops be certified organic, and entered into an arrangement with NOFA and a private entity called the Organic Crop Improvement Association, which they hoped would eventually serve as a unitary national certifier. Gershuny later wrote of the arrangement that she helped set up with the Maryland business owners, "The idea they were pitching was a new, producer-controlled national certification programme. That first year, all certification costs were covered by the produce distributor, and 23 growers signed in—about as many as the total number that had been certified by the Northeast Organic Farmers Association, the region's leading certifier, since it began in 1977."[12]

Those involved viewed the Organic Crop Improvement Association arrangement as a pilot effort for building a truly national system. The idea took root with a number of others active in the organic community who were brought together at a meeting organized by IFOAM in 1984. Participants included eighteen representatives of certifying groups, processors, and academics associated with organic agriculture in the United States. The need for greater national coordination was discussed, and participants agreed to create an organic trade association, the Organic Food Production Association of North America (OFPANA), later the Organic Trade Association, to serve as a vehicle for rationalizing the national organic system. Processors and distributors clearly had the greatest interest in bringing more coherence to the organic marketplace, and they were the key figures in OFPANA, although certifiers and farmers organizations were also represented.[43]

Gershuny was hired on by the new trade association to establish a process for evaluating the myriad of certifiers operating in different regions throughout the United States and develop what would serve as national "Guidelines for the Organic Industry." According to Gershuny, "The idea was to have an accreditation program for certifiers ... not assuming that all the standards would be the same but that they'd be consistent with the guidelines ... [T]he idea was that we were looking toward self-regulation, but understood the need for consistency at the national level."[44]

The hope was that by, in a sense, certifying the certifiers, each regional group would be more willing to recognize one another's products as legitimately organic. This would ease the burden for those who sought to make processed organic goods using ingredients from growers certified under different systems. OFPANA's endorsement was also designed to work the other way: to allow farmers to avoid the need for multiple certifications if they sought to sell their goods to buyers in different parts of the country. This was part of a larger effort to heighten recognition of organic as a whole. Katherine DiMatteo, the first executive director of OFPANA, explained that the idea was to "have an identity that everyone who was in the program could use nationwide. And then an [organic] brand identity could be recognized on labels as opposed to individual certifier seals. It would be something that you could promote generically as opposed to helping one certification organization over another certification organization."[45]

While rationalizing the organic system nationally was the goal, OFPANA was still careful to respect regional variation and avoid alienating members of organic associations who had spent years developing their own standards. Suspicion of centralized authority was common among many growers used to setting their own organic standards. OFPANA intentionally issued general guidelines rather than specific standards in order to allow local groups some flexibility and autonomy, while trying to make the system workable at the national level.

OFPANA's attempts to draw regional certifiers into a nationally coordinated system failed. As DiMatteo portrayed it,

There was a lot of resistance.... People weren't really excited about the idea of a self-declared group saying, "We want you all to buy into this concept that we have, and let us be the regulator." ... People were like, "Why should we do this for you?" A lot of the certifiers liked the fact that they had their own seals. They wanted to keep their own seals. They wanted to promote their own seals. They wanted to have their own standards, and they wanted to compete on their standards being better than somebody else's standards or them being a better certifier than somebody else.[46]

Although OFPANA's attempts to create a national private certification system under its supervision failed to gain traction, the association would grow in importance within the organic movement. The Organic Trade Association, as it was later renamed, would come to play a critical and at times controversial role in the development of the federally run NOP. The processor-dominated organization would be a driving force for standardization in its effort to foster expanded national and international markets for organic goods.

## The Threat of Fraud

Conflicting standards and the lack of a clear definition of organic were only two of the challenges facing the organic community. Processors and distributors required a more coherent system for their business model to succeed, and thus they were motivated to press for a single national system. But other groups were also negatively impacted by the chaotic and inconsistent rules. Organic farmers recognized the need for action based on the growing threat of fraud. This was a concern for consumers, who could easily fall prey to acts of misrepresentation given the confusing array of organic certification labels and lack of clear standards enforcement. Yet it was farmers and certifiers who recognized that fraudulent operators could not only undercut legitimate organic growers; fraud could undermine the integrity of the entire enterprise in the minds of consumers and destroy the whole of the organic industry.

Although fraud had long been a worry, for a while there was not a big incentive to take advantage of the vulnerable organic system. One report found that only 20 percent of organic growers were receiving a premium for their organic products in 1979.[47] During the 1980s, however, growing demand changed the equation, and the level of profit available in the expanding organic market was attracting some unscrupulous actors. Imported organic goods with questionable certifications were also appearing on store shelves. Jim Riddle, an organic inspector in the 1980s who would later play an important part in professionalizing the inspector's role, saw the need for state involvement as the industry grew. "As both a farmer and inspector, I saw the level of commitment that real organic farmers were making, and felt that they needed to be protected. There needed to be enforcement against fraudulent claims, and without law, without it being codified … it's small claims court and "he said, she said.'"[48]

Organic oversight was still by and large a private affair where credibility was based on consumer trust as well as the established reputations of certifying organizations. But in some instances, state governments began to get involved to try to deal with concerns about organic fraud. Government engagement at the state level was largely welcomed by organic farmers, at least to the extent that it would help address fraud and protect those who were truly practicing organic methods. According to Coody, farmers in Oregon were pleased by the state's involvement: "Everybody was really pretty excited about it, because we had started to feel the pinch of fraud. … [W]e had significant organic premiums, and so we started to feel people were cheating by selling conventional products as organic. At that time, the farmers felt like, 'Gosh, if we can have our state government watching out for us!' That's how it was perceived."[49]

Oregon and several other states passed some kind of organic legislation during the 1980s, but clear standards and enforcement mechanisms were rare.[50] California's Organic Food Act of 1979 was the first that sought to define organic.[51] This act was just two pages long, and contained no means of enforcement or any funding for oversight.[52] By 1989, almost thirty states would have laws or regulatory provisions for organic.[53] Only four (Texas, Washington, Minnesota, and New Hampshire) provided direct support for certification, though, and almost none had any strong enforcement provisions to guard against fraud.[54]

The threat of fraud presented a dilemma for those within the organic community. Exposing suspected cases would draw attention to the flawed system of oversight, and posed the risk of damaging the reputation of organic while reinforcing the charges of detractors critical of all organic claims. But failure to act heightened the possibility of a major scandal and total loss of faith in organic integrity. Concern about fraudulent actors was growing among authentic organic practitioners, yet challenging fraud presented its own perils.

In California, organic proponents took the chance of confronting a fraudulent operator. CCOF leaders decided in 1988 to expose Pacific Organics, a distributor that was flagrantly repackaging conventionally grown carrots from Mexico and selling them as organic. Through their investigation they were able to secure photographs of the actual repackaging taking place. CCOF staff members sought to have legal action taken against the distributor on the basis of the 1979 California Organic Food Act. Yet the law authorized the California Department of Heath Services to oversee the law's provisions. The department lacked viable enforcement mechanisms, leaving state officials with little ability to act. Given the egregiousness of the violation, CCOF staff members were able to persuade officials to pursue the case as a consumer complaint, but the incident revealed that there was need for clearer and more enforceable legislation.[55] The fraudulent carrot case received attention throughout California and elsewhere. In a sense it demonstrated that the system was working; established organic certifiers were capable, with some help from the state, of policing their own. But it also exposed the weaknesses of the private organic system.

Not only was there the growing risk of fraudulent sales of bogus organic goods, but aside from OFPANA's attempts to serve as a national accreditor, there was no oversight for organic certifiers themselves. There was no way to prevent any private entity from declaring itself an organic certifier and issuing organic seals on the basis of its own standards. Lewin of the CCOF claimed that some companies were starting to do just that: inventing their own standards and utilizing in-house certifiers. Many established certifying organizations already felt that some of the newer certifying groups were not upholding the stringent standards that had been established by their predecessors. Wonnacott of NOFA Vermont noted that by the late 1980s, "there were some poor certification offices out there that were just starting up. Fly-by-night, cheap operations. So I could see that the writing was on the wall."[56] What Wonnacott saw written on the wall was that some kind of enforceable national system was needed. All this added to the pressure for federal government involvement.

## Food Scares

The growing threat of fraudulent actors seeking to cash in on organics and the problems associated with market complexity all stemmed from the fact that organic food was growing in popularity. This can be attributed at least in part to the general "fitness craze" that sprouted up in the 1980s, but environmental awareness and health concerns specifically related to toxic chemicals also played a significant role.[57]

These fears were propelled to new heights by high-profile food scares that grabbed national headlines. One significant incident occurred when watermelons contaminated with a pesticide known as aldicarb sickened over fourteen hundred people in several western states over the Fourth of July weekend in 1985. This was the largest outbreak of food-borne pesticide illness ever to take place in the United States.[58] This incident drew attention to organic agriculture and the farmers who were growing food without the use of synthetic inputs like aldicarb. In California, where watermelons statewide had to be destroyed because of the contamination, the CCOF was just establishing a formal office. Founding staff member Mark Lipson reported that the incident set off a flurry of calls from farmers and a dramatic jump in the number of certified organic operations in California.

The greatest national food scare came a few years later on February 26, 1989—a day that organic activists would come to call Alar Sunday. That was when the television news program *60 Minutes* aired a report on the health threats posed by Alar, a plant growth regulator commonly used on apples. The fact that this important cultural symbol could be treated with a carcinogenic chemical elicited a reaction not unlike the extinction threat faced by the bald eagle as a result of the use of the pesticide DDT. The alar exposé set off what *Newsweek* magazine called the "panic for organic."[59] Conventional distributors and retailers scurried to find organic suppliers in order to meet the surging public demand.[60] While few conventional supermarkets were carrying organic products at the time, the dramatic increase in demand led many to begin stocking their shelves with organic offerings.

Natural foods supermarkets, such as Whole Foods and Wild Oats, which had appeared on the scene earlier in the decade, saw a huge growth in popularity. As Coody observed,

When Alar happened, the organic industry started going ballistic. We suddenly had to ramp up supply. The market was just insane. It was boom and bust. Some people went out of business. Some people became millionaires. It was really hard to figure out what was going on. And we realized then that the writing was on the wall. There was going to be a federal law, and we better be involved in the process of making it workable.[61]

### Enter the Feds

The convergence of these factors would bring about significant federal involvement in the organic industry by the decade's end, but some modest steps preceded the 1990 passage of the landmark legislation that would serve as the basis for the NOP. Following the FTC's failed attempts to establish a regulatory framework for organic in the 1970s, in 1980 there was a glimmer of hope that the federal government might actually offer some support for organics.

The energy crisis of the 1970s had spurred new interest in exploring opportunities to reduce dependence on fossil fuels and petrochemicals. This was also a time during which soil and water contamination from agriculture was recognized as a serious threat. Moreover, the agriculture sector in general was in crisis. Falling commodity prices led to a dramatic loss of family farms and the continued restructuring of the industry.[62] It was in this context that the USDA, under the Carter administration, sponsored a study on organic farming.

The extensively researched "Report and Recommendations on Organic Farming," released in 1980, concluded that there were "serious misconceptions" about organic farming among its critics and significant promise to the organic approach.[63] Among the report's suggestions were that the USDA launch an aggressive research agenda on organic agriculture and the hazards posed by pesticide residues on conventionally grown food. It also called for support for farmers seeking to convert to organic production or otherwise reduce their dependence on synthetic chemicals. The report gave a big boost to organic proponents, and one of its authors, Garth Youngberg, was appointed to serve in a new position as the USDA's organics resource coordinator.[64]

The hope for new respect and legitimacy was nevertheless short lived. The Fertilizer Institute and pesticide companies reacted swiftly and forcefully, condemning the report. With the election of Ronald Reagan that same year, the report and its findings were suppressed. Its recommendations were ignored as the new administration took a decidedly hostile stance toward organic.[65] By 1982, Youngberg was fired and the organic resources coordinator position at the USDA was eliminated.[66] Some organic supporters in Congress sought to include provisions recommended in the organic report in the 1981 Farm Bill, the primary federal agriculture funding and policy legislation enacted every five years, but opposition from the USDA squelched this and all other legislative attempts to advance such policy.[67]

It was not until 1985 that organic proponents saw any successful action at the federal level. The Rodale Institute had long called for federal funding for organic research. Lacking any such support, for decades the institute carried out its own research with private funds and donations. But given growing public concern about synthetic chemicals being used in farming, Congress made a token gesture toward this interest in the creation of the Low-Input Sustainable Agriculture program (LISA, renamed the Sustainable Agriculture Research and Education program or SARE) in 1988—a grant program included as part of the 1985 Farm Bill.[68]

By this time, the term sustainable was commonly used among both policy wonks and many organic activists. It was considered more politically palatable among those seeking to avoid the countercultural trappings associated with the word organic and the political firestorm that erupted following the USDA's report on organic agriculture in 1980. It was also more accurate given the recognition that moves toward agricultural sustainability could include measures short of a full-scale organic conversion, such as integrated pest management—a pest control approach that avoids, but does not eliminate, synthetic pesticide usage.[69]

Vermont senator Patrick Leahy, who would eventually play an instrumental role in the NOP's creation, championed the research program. Vermont is a progressive state with many small farms, a sizable dairy industry, and a long history of organic agriculture, and as such, Leahy had reason to support the organic cause. Powerful conventional industry players opposed the SARE program from the start, however, and successfully prevented the allocation of any funding until 1988, when SARE received a paltry

$3.9 million out of the roughly $1 billion allocated for education and research by the USDA.[70] Although SARE would remain a relatively small and marginal program, increased action on organics at the federal government level served as further impetus for activists to pay closer attention to these developments.

While federal funding for organic research was not allocated until 1988, that was also the time that real federal action on organic picked up. The farm crisis of the 1980s exposed the problems with USDA policy and the industrial agriculture system as a whole. Overproduction and falling prices left conventional farmers deeply in debt, and continued to drive many small and midsize farms out of business. In the meantime, the US Environmental Protection Agency had identified agriculture as the primary cause of nonpoint source water pollution, soil erosion continued to advance, and there was growing insect and fungal resistance to pesticides.[71]

While serious problems within the industrial agriculture system were increasingly evident, a report on alternative agriculture issued by the National Academy of Sciences was also getting widespread attention in the news media.[72] The authors of the extensive study reaffirmed the 1980 USDA findings that alternative agriculture methods, such as organic, offered promise for addressing many of the problems resulting from conventional approaches.[73] Pressure for federal action was again building, but the state of California would lay the groundwork for expanded government involvement in the organic market.

California was home to the largest organic sector in the country, and developments there had a critical role in shaping events at the federal level. Cases of fraud made it apparent to CCOF staff members that stronger legislation was needed to combat corrupt players. According to CCOF director Bob Scowcroft, "Fraud would have brought us down. Fraud was happening in the farmers' markets. It was emerging in the national distribution network. We needed a law; we needed a penalty."[74]

Lipson, CCOF assistant director at at the time, remarked that organic association leaders were especially motivated to act because conventional agriculture players in California were becoming more engaged with these issues. Conventional agribusiness actors were worried that food scares and the growing interest in alternative agriculture might set in motion laws that would restrict the use of the synthetic inputs on which they were dependent. Agribusiness lobbyists tried to preempt any strict limitations, and there were indications that their efforts would impact organic law. Responding to threats of industry action and growing worries about fraud, the CCOF mobilized organic proponents throughout California to defend and strengthen the state's otherwise-toothless organic law. These advocates were successful in securing enforcement provisions and the formal adoption of organic standards based on CCOF rules in the California Organic Foods Act of 1990.[75]

Although strengthening the California law would provide organic with some official oversight and state enforcement capacity, this and other state measures elsewhere would do nothing to address the need for a coherent national system with clear, credible standards. CCOF leaders feared that even as they successfully strengthened and defended the state law, they could easily face a much greater challenge if federal officials got involved at the behest of big agribusiness. Lipson articulates their reasoning in taking the struggle to the next level: "Looking at what happened in California, where there was this initial preemptive move by agribusiness interests to assert control over the term, we decided, let's get our foot in the door in Congress.... There was this recognition that if we didn't control it, it would be given to us. That it's going to happen one way or another, so we might as well do our best to be in control of it."[76]

## Organic Goes National: The OFPA

In 1988, Lipson was in Washington, DC, attending a conference of the National Coalition against the Misuse of Pesticides (later renamed Beyond Pesticides). Scowcroft sent him to meet with Kathleen Merrigan, the senior staff member of the US Senate Committee on Agriculture, Nutrition, and Forestry, which was at this time headed by organic movement ally Senator Leahy. Lipson described for Merrigan what was happening in California, and they discussed the prospects of creating a federal organic program. As Merrigan put it,

Mark Lipson was persuasive. He made a compelling case. It was something I was interested in and it was something that worked well for the politics of Vermont.... One of problems was that there were unscrupulous certifiers out there, and we need to rein them in.... [T]he sudden profit margins were appealing to a whole lot of people who had no background in organic. They're coming into the market to fill the demand. [There was concern that] the whole thing would be lost.[77]

Merrigan began gathering information on the myriad of standards and certification systems that existed throughout the country in hopes of identifying commonalities along with best practices. On learning about the problems that the industry was facing and interests of those involved, Merrigan had originally considered pursuing a path similar to that used for kosher foods. Federal food safety officials do not set standards or inspect food to ensure that it is produced in accordance with kosher rules, but they do have role in addressing fraudulent claims.[78] Merrigan thought that this might be viable for a national organic system that would rationalize the industry and provide credible oversight, while not placing fundamental control over the meaning of organic in the federal government's hands.

This may or may not have worked in the organic case. Kosher rules are stable, unlike the organic ideal of a continuously evolving and improving standard. And presumably a private unitary organic authority would be necessary to establish a uniform standard and avoid the problems that plagued the industry under its more decentralized configuration. This may have given rise to internal struggles greater than those to come under the government-administered system, with no guarantee against capture by big industry or some faction of the movement.

The potential problems associated with that approach proved to be moot. Merrigan ran into a brick wall with the USDA, which was still thoroughly dominated by big agriculture interests overtly hostile to organic. She had little opportunity to even explore a model similar to the kosher standard. Merrigan later learned that the department was so averse to any movement on the organic front that the USDA's point person on organic was instructed by her superiors not to return Merrigan's phone calls and, if she came to the USDA with questions, to provide her with misinformation.[79] Lacking any regulatory alternative, Merrigan undertook the process of crafting legislation.

It turned out that Merrigan's proposal would not be the first federal organic legislation to be put forth. Senator Wyche Fowler of Georgia proposed his own organic legislation in early 1989.[80] His bill, less than one page in length, was offered with little knowledge or input from the broader organic farming community, although it did receive OFPANA's backing.[81] Fowler's legislation would have allowed the secretary of agriculture to provide a uniform definition of organic—a notion that was anathema to the grass roots of the movement. Although the bill was never passed, concerns about the issue getting away from the grass roots were being borne out. There was going to be federal legislation with or without the involvement of organic activists. It was clear to them, then, that they would need to become more engaged or lose the ability to define the practices they had so painstakingly developed for decades.

Merrigan worked closely with consumer groups in developing the legislation. Roger Blobaum, director of the Americans for Safe Food Project at the Center for Science in the Public Interest, a nonprofit consumer organization focused on health and nutrition policies, was heavily involved, as were representatives of the now-defunct group Public Voice for Food and Health Policy. Blobaum's group was already independently collecting signatures on a petition calling for federal organic legislation. The confusion caused by multiple conflicting organic standards and threat of fraud had propelled consumer action. While CCOF staff members had supplied the initial impetus for Merrigan to act, it was an increasingly mobilized consumer base that fueled the push. Consumer groups were much bigger and more organized than the loose patchwork of organic organizations, plus they had a Washington presence, unlike many organic organizations, which had only recently opened state offices.

Although environmental organizations had not made organic agriculture a priority issue, Merrigan and some consumer advocates recruited key DC-based environmental groups to join in the effort to pass organic legislation. Humane organizations also provided input early in the process, although some did so with reluctance due to their general concerns about raising animals for slaughter. Yet there was recognition that organic practices could yield great improvements in livestock treatment and that they could potentially advance the humane cause through participation in the process. These established Washington interest groups would prove decisive to the final bill's passage, but as discussed in the next chapter, their entry into the organic community would also prove challenging for many long-term organic activists.

Ironically, the voices most lacking as the details of the legislation were developed were those of organic farmers themselves—the originators of the movement. Organic farmers were handicapped by a lack of resources and the absence of any national coordination. At the time, there were only a handful of paid staff members in organic organizations across the entire country. The major certifying groups like the CCOF, NOFA, and Oregon Tilth each employed a few staff members, but there was no Washington presence, let alone any professional lobbyists with the know-how to advance their interests in the world of beltway politics. According to Henderson of NOFA New York, during this period state and regional organic organizations "were all hand-to-mouth.... [W]e couldn't get the funding that we needed to be a national group."[82]

Some organic groups were networked regionally. In the Northeast, NOFA served as the umbrella for state-based organizations. On the West Coast, organic groups had established an informal network, comically calling it WACO, the Western Alliance of Certifying Organizations. But Scowcroft of the CCOF remarked, "We had no budget, no leaders. It was really just a collaboration of about twenty people."[83] Sometimes organic advocates would receive financial support from some of the emerging organic business leaders. Scowcroft credits Kahn of Cascadian Farms with funding much of the early effort to build national ties among organic advocates and to coordinate standards and political strategy.

In general, though, organic farmers lacked the ties necessary to formulate any kind of coherent political strategy at the federal level. Coody describes the status of national organic farmer political organization at the time that federal legislation was being considered: "We didn't know anything. We [organic leaders across the United States] had literally never met each other.... You've got to remember, I was milking goats. I couldn't go anywhere for many years. I was stuck. I didn't know anybody. And everybody was like that. We were managing farms, so we couldn't go places."[84] This lack of organization and political sophistication among the core constituency of the organic movement would take its toll. In Lipson's view, "Our movement was not sufficiently prepared to be players on the federal level, and some of the outcomes certainly reflect that. We did not know what we were getting into."[85]

Some recognized that a national voice for farmers was sorely lacking and that this could be detrimental as legislative action picked up momentum. Some OFPANA participants had been seeking to create a "farmers' caucus" within the trade association to ensure that this essential constituency was adequately represented. They secured grant funding to sponsor a meeting in Leavenworth, Kansas, in 1989, bringing organic farmers together to share information and to strategize about how to protect organic integrity and the interests of organic farmers as the federal legislation was crafted. Merrigan was there to present the bill that she had developed by drawing on the various state laws and certification programs she reviewed along with the input she received from consumer groups in Washington.

Many organic activists were still skeptical about involving the federal government in their world.[86] The prefigurative political perspective did not include the state, which was viewed by many not as a force of democratic governance and social coordination but instead as an external corporate-dominated entity hostile to organic activists' efforts. This was especially true among some NOFA affiliates. Henderson, who was present at the Leavenworth meeting, considered a constructive role for government "laughable."[87] Agricultural officials in New York had been particularly hostile to organic, reinforcing the general skepticism of large institutions still found among many small organic farmers. Strengthening the organic market through a federal program was not on their agenda. Henderson's primary goal for even being involved in the federal process was to try to shape the legislation so as to protect small organic farms.

NOFA New York's counterparts in Vermont shared the concern felt by others in the Northeast. In regard to the idea of a federal program, Wonnacott said, "There was right away real skepticism and not support for it, mostly because nobody wanted the federal government to make decisions about what organic is or isn't…. The farmers in Vermont were very fearful that they would lose that voice."[88]

The New York and Vermont chapters of NOFA, along with Demeter, the biodynamic certification organization, were the leading critics of the move toward a federal organic program, although small farmers in other parts of the country also expressed skepticism. In Oregon, Coody was already sensing the divide that would ultimately grow into a major fissure within the organic movement. "Everyone's businesses were growing at that time," she explained, "so we did begin to have a divide between the smaller market gardeners and people who mainly focused on farmers' markets and the people who were then beginning to wholesale their products."[89] Many of those small growers working only with local markets had not directly experienced any of the threats identified by larger players or some staff members at the organic farmers' associations. Amy Little, who would later head the NCSA, a coalition designed to unify voices within the broader sustainable agriculture movement, heard resistance from small organic growers.

"The farmers were saying, 'We already have our standards! Don't tell us what to do! The system is working! Don't [mess] with it!'"[90]

The breadth of opposition to federal action on the part of small farmers is difficult to gauge. But it is apparent that relative to the more organized segments of the broadening organic community, small farmers during that time were weak and not positioned to stop the momentum even if they had wanted to. Wonnacott informally solicited others to join in opposition to the whole federal effort. "I had a sense that there was much more dissent. But when it came down to it, when people would really ... put their name out on the paper, there was not enough dissent to go anywhere with it. So we kind of shifted at that point to say, 'OK, this is going forward, there's no turning back. Let's make it as strong as we can make it.'"[91] According to Henderson, at that point it was time to stop resisting federal involvement and "start building the capacity of the organic groups around the country, so that we could stand up to what was obviously coming."[92]

Although many farmers were fearful of losing their ability to control the meaning of organic if the federal government got involved, they also recognized that they could easily lose control of organic if no action was taken. By this time organic advocates had already seen what could happen without well-defined rules and governance. The fate of the term natural loomed large in the minds of many who advocated for federal action. Fred Kirschenmann, an organic leader from the Midwest and participant at the Leavenworth meeting, observed that

we had just, at that point in history, in the middle of the 1980s and going into the 1990s, we had just come through watching what happened to the natural label. Because there was no regulation around natural, ... it became popular as a market-ing term. Companies started to use the term natural for anything. It didn't really have any meaning anymore.... We were just scared to death that the same thing was going to happen to organic, unless there was some kind of policing of the use of the label. Many of us, myself included, could not see any other avenue for having that kind of policing if we didn't get the federal government involved, despite the fact that we had great reservations about it.[93]

Yet even those supportive of federal action recognized that the path would be fraught with pitfalls. The initial draft of the organic legislation that Merrigan had circulated contained many provisions to which farmers objected. Coody recalls that over three days of intensive discussion, the participants at the Leavenworth meeting "created a list of principles that we felt, as farm groups, had to be incorporated into the law or we would not be able to support it."[94] Although the farmers at Leavenworth were of marginal political significance, Merrigan was responsive to their concerns and vowed to rewrite the legislation with their input in mind.[95] She considered the original bill, introduced in 1989 at the end of the legislative session when it had little hope of passage, as a trial balloon designed to solicit feedback for a revised version to be presented the following year.

For their part, farmers made efforts to organize themselves nationally to ensure that their voices would be heard during that process. Coming out of the Leavenworth meeting, they formed the Organic Farmers Association Council (OFAC). The creation of an organic association specifically for farmers was indicative of the growing complexity of the movement. As will be discussed in greater depth in the next chapter, the notion of a singular organic community masks the diversity of interests that fall under that umbrella. Federal legislation would have a tremendous impact on everyone involved. As such legislation was being developed, segments of the organic community recognized that they had distinct interests to protect and thus organized to do so. OFAC was to serve as the farmers' vehicle. Central to their mission was to "promote the interests of organic farmers" as well as "provide a unified growers' position on the definition and regulation of organically produced foods."[96]

While some were skeptical about involving the federal government, most organic practitioners and advocates were pleased at least with the process that the bill laid out for the creation of the national program.[97] The law provided basic guidelines along with a procedure for how a national organic system should be developed; it did not dictate details about what those standards would be. More important, an inclusive body, the NOSB, composed of representatives from all interested segments of the organic community, would advise the directors of the NOP, housed in the USDA, on the standards and certification system to be created. The fifteen-member NOSB, appointed by the secretary of agriculture, would include four seats for organic farmers, two seats for those who own or operate an organic handling or processing operation, one organic retailer, three environmentalists, three consumer representatives, one scientist, and one certifying agent. Within the board, measures needed a two-thirds majority in order to pass, thereby ensuring that consensus or at least compromise would guide the proceedings, and that the process that would define organic could not be commandeered by a single segment of the organic community. In addition to these measures, the legislation, revised from the earlier bill introduced in 1989, put certification back in the hands of the private organic certifying organizations, rather than having that be a function of the USDA itself. In the new version, the USDA would serve as an accreditor of the private certifying groups.

Feeling at least somewhat comfortable with the proposal, the organic community mobilized around getting the bill passed into law. Established Washington consumer groups, environmental organizations, and others played decisive roles in ensuring the bill's passage. Their lobbyists rallied behind the legislation, which Leahy formally introduced in 1990. The Washington groups supplied funds and support for the organic farmers, who were brought in to reinforce the concerns of local constituents from back in the legislator's home districts, sometimes allowing the cash-strapped farmers to sleep on their apartment floors while in town lobbying. In the end, they emerged victorious.

The passage of the legislation was considered a coup by many, because the House version passed in a floor vote brought by Oregon Democrat Peter DeFazio after it had been voted down by opponents on the Agriculture Committee. Overcoming committee opposition is a legislative feat seldom seen. Jay Feldman, the founder and executive director of Beyond Pesticides, is still awed by the whole experience, especially given the opposition to organic within the USDA. "It was phenomenal how quickly that bill went from draft to law. It's astounding, given that it was so antithetical to the USDA's philosophy."[98]

Thus, against all odds, the OFPA was adopted as part of the 1990 Farm Bill. Given ongoing resistance in Congress and at the USDA, the program was starved of funds for the first two years and no progress was made at establishing a national organic system during that time. Some funds were allocated and the NOSB was assembled once the Clinton administration took office in 1993, but it would still be another ten tumultuous years before the NOP was up and running.

The passage of the OFPA was nonetheless seen as an incredible achievement for a movement that had for years been denigrated by government officials, the food and agriculture industry, and conventional food scientists. Some heralded the passage of OFPA as a historic victory for the organic movement—one that would establish organic agriculture as a credible alternative to the industrial system. Others feared that it was the beginning of the end for a movement that once embodied truly transformative potential. In a sense, there is truth to both perspectives. But regardless of how one views the current state of the organic industry, the OFPA's passage was really just the start of what would be years of struggle and debate.

## Conclusion: The Challenge of Success

The 1970s and 1980s were a time of significant development for the organic movement. The convergence of a number of factors, some actions undertaken by activists, and some external forces drove the movement in new directions. Organizational development, the emergence of allies, and support from some powerful sectors were among the factors that defined this era of transition.

The development of formal organizations is a core focus of movement scholars, who also cite the importance of resources to movement mobilization and success.[99] Despite the long gestation period, by the 1970s organic farmers started to create their own associations as networks of mutual support and vehicles for formalizing and defining that which united them. In the prior period, grassroots volunteers who lacked funds and a coordinated structure were limited in what they could accomplish. Organizations that collected membership dues and fees for certification services provided the organic movement with tangible resources, as did financial support from a growing segment of organic entrepreneurs. Previously, such basic needs as an office, phone, and mailing address were lacking. But by this later period a handful of paid professional staff leading organic associations in several states supplied personnel with the time and resources to make connections with others in the movement, and to more systematically consider their political goals and needs.

Although organic farmers created their own associations to define and advance their practices, organizational maturation came slowly. The creation of national networks and the turn to direct political engagement, common to social movements in later stages of development, did not occur in the organic movement for almost two decades after the post-1960s' surge in interest. In a sense, the process that the movement underwent could be considered "defensive professionalization." This is due to the unique strategy embraced by the movement coming out of the 1960s' counterculture:

a prefigurative approach to social transformation. The organic movement did not develop national networks of professional policy-oriented lobbying organizations as occurs in most movements, because the goal was not to legislate a new agricultural order; it was to create one that would eventually grow and displace the dysfunctional industrial system. The primary reason for the development of formal organizations and state-oriented political capacity was not to advance goals through legislation but instead *to prevent the undermining of goals* that were already being achieved by elevating consumer consciousness and building the organic market.

But with this market success came problems that the movement could not address internally. Organic entrepreneurs sought to rationalize the organic market for the sake of more rapid expansion while farmers and grassroots activists were more concerned with the threat of fraud and cooptation. Together, this is what drove many to accept the need for state involvement. It was defensive action, not an offensive strategy. The move nevertheless fundamentally changed the relationship of the organic movement to the state. This was especially true once the OFPA was passed. The purely prefigurative social change strategy implicit in grassroots private certification systems was altered as organic activists would now have to directly engage with the federal government.

There were other central elements of the movement's transformation during this period, including new support from some key political leaders and the broadening of the constituent base for organic. Movement scholars often cite the benefits of elite support for social movements. Senator Leahy's backing of the organic cause and the active role played by his staffer, Merrigan, were essential in the passage of the OFPA. More than simply responding to constituent pressure, Merrigan helped to coordinate the broader organic community and facilitated allied support for the legislation that she played a central role in crafting.

The fact that the organic movement attracted allies from other more established interest groups such as consumers, environmentalists, and others would also prove to be decisive. Some critics of the NOP contend that the OFPA was the product of the conventional agriculture industry seeking a means to co-opt or destroy the organic movement. Regardless of how one interprets the current state of affairs, the OFPA itself was clearly a movement effort. At the time, most in the organic community saw great promise in this legislation and it passed over the objections of many powerful actors in the conventional food industry and their supporters in government. Eventually more organic proponents would come to believe that the NOP allowed conventional corporate agribusiness to exert undue influence over the organic system, but the evidence does not indicate that the OFPA itself was a conventional food industry conspiracy. Rather, the specific provisions of the law and the subsequent policy particulars created openings that could be exploited by corporate interests not truly devoted to the organic cause.

To understand how these vulnerabilities were created we need to examine the process by which the specifics of organic policy were developed and the role that different groups played in that process. The next chapter will examine the broadening organic coalition, to see how differences and divisions within the movement may have opened the door for the incursion of conventional agribusiness into the organic market and the subsequent transformation of the industry.

# 4    The Organic Coalition: United and Divided

Nowhere has the appeal of decentralism been greater than among agriculturalists.
—Kathleen Merrigan, "Negotiating Identity within the Sustainable Agriculture Advocacy Coalition"

The rules of any game play a large role in determining the winners. Thus, any critical analysis of how the organic market came to be what it is today requires that we examine the nature and power of those who are making the rules. As will be discussed in the next chapter, large conventional agribusiness players now exercise a great deal of control over the organic market. Based on this, one might surmise, as some do, that the whole federal organic project was engineered by these powerful actors—a ploy to strip organic of its transformative potential and convert it into a profitable niche market based on a slightly modified conventional industrial agricultural production model with marginal environmental or health benefits.[1]

The combatants in this epic struggle are often presented in stark terms.[2] On one side is Big Organic, the multinational conventional food conglomerates, opportunistic latecomers who entered the organic market bent on extracting as much profit as they could while watering down standards to the point where the term would eventually lose all meaning. On the other side are selfless, ideologically committed small farmers struggling to advance the ecological and social ideals at the heart of the organic concept. While capturing some of the important dynamics of organic politics today, the reality of how we arrived at the present situation is far more complex.

As explored in the previous chapters, the drive for a national program was not fueled by corporate interests or other established economic or political powers in the food and agriculture world. To the extent that these big players did have an interest in organic, it was to contain and marginalize it in order to prevent threats to their core conventional enterprises. By most accounts, conventional industry actors were not extensively involved in the development of the actual organic rules.[3] We therefore must examine the constellation of interests *within* the organic community to understand present outcomes. The organic case provides insight into the way movement coalitions function—at times in a self-defeating manner.

Through the 1970s and most of the 1980s, the organic community consisted of a relatively small number of organic farmers and business owners who slowly organized themselves into regional farmer associations and a small national organic trade organization. By the late 1980s, when federal legislation was being considered, that community broadened to include consumer groups, environmentalists, and other allied interests. While the early actors had some disagreements regarding standards, they were similar in their outlook. The growing commercial market in California was beginning to set the CCOF apart in some ways, but even through the 1980s, it was still a fairly small, tight-knit group of ideologically committed grassroots farmer-activists. But alliance with other groups during the OFPA effort introduced new types of concerns and interests that did not always align. Part of the reason that the organic industry has developed as it has is not because of foul play by large corporate actors but instead because consumer watchdogs, environmentalists, organic farmers themselves, and others advocated for provisions that, together, created the conditions for the entry of corporate players. Once these larger, primarily conventional food and agriculture firms entered the organic community, divisions became more vivid and internal struggle was heightened.

Yet the movement is not united in its disdain for Big Organic.[4] The battle lines are not drawn simply on the basis of big versus small. The strategy for transforming the agriculture industry is still primarily market based. In this context, private businesses of all sizes can play a role. Of course, no one would argue that Dean Foods or Kraft is embarking on a campaign of social transformation through its organic divisions. In the minds of some organic proponents, though, they are essential players in spreading the organic gospel and bringing the benefits of organic food to the masses. As such, the dominating presence of corporate actors in the organic market is not universally viewed as the unwelcome intrusion of profit-seeking interlopers.

This is a reflection of differences defined not simply by size or position but also by ideology and strategy. Organic proponents, while united in support of certain principles, also hold divergent perspectives on what they are seeking and how they should achieve it.[5] In order to understand the present condition, we must consider all these differences: those based on interests and position, and those premised on ideology and movement strategy.[6] During the development of the organic rules and to this day, in what ways was the organic community allied and in agreement, and where did opinions and interests diverge? How did this confluence of interests within the organic community yield a system that is quite different from what so many early organic proponents envisioned? These are the questions considered in this chapter.

## The Role of Coalitions and Risk of Division in Social Movements

The organic movement achieved unprecedented cooperation and coordination during the struggle to pass the OFPA. Organic proponents who had been operating in relative isolation for years came together in an effort to defend organic integrity, and foster a system that would protect their shared interests and ideals. Organic farmers across the country, after having formed their own state and regional associations in the previous two decades, were finally brought together nationally for the first time. But it was not just farmers. Organic advocates found allies in several other sectors— groups that would become part of the extended organic family. Environmental organizations discovered that organic agriculture offered promise for advancing environmental protection. Groups interested in consumer health and safety realized that the farmers who grow their food have a role to play in advancing their goals. Humane organizations, rural preservation groups, organic businesses, and others all joined in this effort. Back-to-the-land hippie farmers were testifying at congressional hearings alongside professional inside-the-beltway lobbyists employed by established consumer and environmental groups, and even sleeping on their couches at night.

Despite this important victory, the battle was neither won nor lost with the OFPA's passage. The legislation itself was just the beginning. Specific rules and regulations about what would qualify for the official organic seal still needed to be created. Unbeknownst to those in the organic community who thought that the entire process could be carried out in a couple years, the NOP's launch was still over a decade off. It was up to organic activists to come together and hammer out the specifics of the program, and push their plan on a reluctant and at times obstructionist USDA. The interim period would be fraught with painstaking internal debate about technical specifics of organic policy and intermittent mass mobilizations during which grassroots supporters were rallied to defend core organic ideals. These controversies would not end with the implementation of the NOP. If anything, conflict intensified over time. Pitched battles would continue to be fought over the interpretation and modification of organic rules, and to this day, the very meaning of organic remains contested terrain.

The rapid, unprecedented mobilization and coming together of organic activists yielded a historic victory in the OFPA's passage. But the following process would require the various constituents that make up the organic movement to come together in a new way. Having a decentralized collection of self-regulating entities was fine for a grassroots movement utilizing a prefigurative social change strategy centered on a vague concept of agricultural transformation. Ongoing engagement in national politics and policy, however, would require new organizational forms and formal alliances with like-minded political actors. Just as individual organic adherents had started to organize themselves into organic farmers' associations two decades prior, now organizations within the newly expanded organic community began to organize themselves into new coalition organizations such as the NCSA and NOC.

Little, who had extensive experience with coalition work as a veteran activist with a number of public interest organizations, was hired to head the NCSA. She notes the historic significance of this development. "This was the first time in US history when different farming interests, different sectors of farmers … small, large, medium, all different kinds of farmers came together with a broader community." This sort of coalition building is considered essential in national policy politics, and was the explicit emphasis of the NCSA and other coalitions that were coming together at the time. It was crucial in light of the new types of political activity in which the organic movement would now have to engage. According to Little, the NCSA was "focused on federal policy and supporting regional, local, and state policy as well, but [our role was] leading the way with a national united front on agriculture … [with the mission of] using policy as a vehicle to build power, to engage people, and to make structural changes, because policy has a huge impact on the structure of anything, particularly agriculture."[7] Formal coalitions and even interaction with other constituencies was not necessary when organic farmers were solely utilizing a prefigura-tive approach. This turn to policy and coalition building represented a new stage in the evolution of the organic movement, but it came with many challenges.

Coalition formation is a strategy for building political capacity used by virtually all social movements.[8] The need for cooperation and coordi-nated action among several organizations is critical. Movements are, after all, efforts by those who do not hold power and feel that their interests are not adequately represented within governing institutions. Relatively powerless individuals joining in coordinated action to increase their politi-cal capacity and press their demands are what define movements. But just as like-minded individuals must come together to create a movement, in today's political context, amassing power requires the coming together of like-minded organizations. In a political field densely populated with inter-est groups and movement organizations of all sorts, rarely would any single group be capable, on its own, of mustering the political influence needed to advance its agenda.[9] Thus, movement organizations tend to seek allies, and virtually all movements are actually composed of subgroups differentiated on the basis of a wide range of social positions, identities, and ideologies.[10]

Since bringing together diverse organizations is an essential strategy, a major challenge for many movements is preventing differences from becoming divisions. Groups may be united on certain "deep core beliefs," but policy-level specifics may expose fissures.[11] Ensuring effective unified action can be especially difficult when the coalition partners themselves are of unequal size and influence, and when interests are not perfectly aligned.[12]

Prior to the OFPA's passage, the organic movement was primarily divided along regional lines through the various farmers' associations and certifying organizations. Although some rivalries existed among these regional enti-ties, the organic community was still relatively small and its marginaliza-tion within the larger agricultural context enabled it to come together with relative ease for the purposes of passing the OFPA. Disagreements were not uncommon in the hasty discussions leading up to the passage of the legisla-tion, but neither were compromises, and all the actors remained committed to the process, although some did so with misgivings. In the decade fol-lowing the law's passage, overt conflict within the organic community was avoided for the most part, and in some instances the movement achieved spectacular unity of purpose to defend core principles.[13] Yet by this time there were many actors giving input into the policy development process and their interests were not always aligned. The amalgamation of interests shaping the organic system would ultimately provide openings into which private interests, with little true commitment to the original organic principles, could creep.

### The NOSB: Managing Difference?

The OFPA included a mechanism for ensuring that all voices would be incorporated in the policy development process by way of the NOSB.[14] It was because of the myriad divisions within the organic community that the NOSB was given such an important role in the NOP. This body, composed of fifteen representatives from every major organic constituency, was officially charged with developing recommendations for the national program and accorded actual statutory authority under the law—a highly unusual feature in the federal regulatory system. Specifically, the NOSB was given authority over the national list, those materials permitted or prohibited under organic rules. The law states that the secretary of agriculture may remove ingredients deemed allowable by the NOSB, but materials prohibited by the board cannot be added to the list, thereby granting real power to the group.[15]

The NOSB itself was designed in a way to ensure that all voices, including relatively weak sectors of the organic community, were able to influence decisions. Measures required the support of two-thirds of the full board in order to pass, thus forcing representatives of different groups to seek compromised agreement. Given the distribution of representatives on the board, this would require agreement by almost all sectors. No single group, like consumers, processors, or farmers, was positioned to push anything through the board on its own, and no individual constituency had the numbers to effectively veto measures supported by others. Six subcommittees were assigned to study specific policy areas, including livestock, handling, materials, compliance, accreditation, and certification, and bring proposals to the full board for consideration.[16]

In some cases, critics would charge that the careful structuring of the NOSB was undermined by the appointment of individuals who did not truly stand for the interests of those they were to represent.[17] As will be discussed in the next chapter, at times these issues would rise to the level of open confrontation. Despite these concerns, though, most organic advocates were pleased with the structure along with the process laid out for the development of the organic rules. Many did not foresee the challenges that they would face along the way.

The process of crafting these rules got off to a slow start. There were essentially two years of inaction following the OFPA's passage. Although the act was passed in 1990 during the George H. W. Bush presidency, officials were not quick to act. The NOSB would not have members until 1992, when Agriculture Secretary Edward Madigan of the Bush administration appointed the first board.

While the NOSB was officially in charge of developing the rules of the new system, one must consider the makeup of the broader organic movement to understand the interests influencing this process. Board members were in regular contact with others within the broadly defined organic community, and the system allowed for input from anyone during public comment periods scheduled into every meeting of the full board. On examination, differences and divisions within the organic community are myriad, and these manifest themselves in the rules that were developed and the system that ultimately emerged out of the process. We must consider each of these differences in both interests and ideology in order to understand this outcome.

### Spreaders and Tillers: A Strategic and Ideological Divide

In her analysis of the organic movement in California, Julie Guthman identifies a fundamental difference "between those who see organic agriculture as simply a more ecologically benign approach to farming and those who seek a radical alternative to a hegemonic food system."[18] As mentioned in chapter 2, some of these differences have been evident since the early days of the movement. To some degree, one can identify waves of predominance of one type of thinking or the other throughout the movement's history. Many of the nineteenth- and early twentieth-century predecessors of the organic movement were groups seeking broad social change, or acting to prevent changes that they thought were profoundly disruptive to traditional ways of life. While some of these movements helped to inspire the turn to organic, many early organic advocates had much more narrow goals. In Europe, Howard and Balfour were primarily interested in the practice of agriculture itself, as was Rodale in the United States. Rodale was particularly focused on the personal health benefits associated with organically grown food—a perspective that in many ways individualizes the benefits of organic and works against broader social transformation. The 1960s' counterculture once again linked organic agriculture to a deeper set of social change goals—aspirations that would later be abandoned by many contemporary organic proponents, whose hopes are again primarily centered around health or environmental reforms.[19]

At any given time period, both strains of thought could be found among different individuals and groups within the organic movement, and this is certainly true of organic activists today. This fundamental difference in one's understanding of the goals of the movement, and by extension one's whole conception of social change, underlies many of the debates that take place in regard to specific policy measures. As introduced in chapter 1, the two basic camps to which Guthman refers can be thought of as spreaders and tillers. While this distinction relates to the big versus small contrast commonly presented in the popular media, the difference analyzed here is one of ideology and strategy, not size or position alone.

Spreaders, focused on the immediate benefits of organic production and consumption, tend to believe that virtually any measure that will grow the organic market is worthy of support. They emphasize the real immediate and tangible benefits of organic production for both human health and the environment. Even if one considers the contemporary state of the industry to be less than ideal, few would disagree that organic production offers important ecological and health benefits. The reduction in the use of synthetic pesticides and fertilizers and genetically modified organisms (GMOs) alone represents significant progress toward a more ecologically sustainable and healthy society. Every acre converted to organic production is a victory, and means that over the years, tons of toxic material will not be applied to the land. Consumers who ingest far fewer toxins will enjoy the health benefits of that change with every bite, as will the farmworkers who no longer must handle the toxic materials of conventional production.

Spreaders also inject a social justice element into their call for unbridled organic growth. As the organic sector expands, goods become more readily available and more affordable. According to one spreader, "We [the organic community] were accused, and somewhat justifiably, of being an elitist segment.... I want to see *all* agriculture be organic and I don't want to wait two thousand years.... More people, including lower-income people, will gain access to these benefits. Shouldn't the person who shops at Walmart have the choice of buying organic?"[20] From the perspective of the spreaders, to take action that inhibits this popularization of organic is to restrict these benefits to the wealthy and enlightened few who are positioned to access the otherwise-exclusive organic market. From this vantage point, private businesses and even large agribusiness corporations and big-box stores have a role to play in advancing the cause. While some may wince at the prospect, for spreaders, organic goods appearing on Walmart's shelves is not something to lament; it is something to celebrate.

Urvashi Rangan, an organic advocate with Consumers Union, is wary of big corporations when they seek to undermine organic standards, but when it comes to providing consumers with access to organic goods, she is sympathetic to the spreader viewpoint. She believes large operators have a constructive part to play. "Reporters focus on Walmart as the bad guy, like, 'Walmart is selling it, so it must be bad.' I think there are a lot of people in the organic industry who've been at this a long time who resent that Walmart can source this stuff cheaply, get it in cheaper from other countries, and sell it." But Rangan sees the benefits of broadening the market and reaching mainstream consumers, and that involves big businesses, where most consumers buy their food. "I think it's great if Walmart sells organic products. That should be great for the whole industry."[21]

Merrigan, one of those most responsible for the OFPA's passage and subsequent mainstreaming of organic, struggles with the issue. "If it's just people who earn seventy-thousand dollars a year and above going to Whole Foods and spending hundreds of dollars every week on organic, I won't feel very satisfied with what has become a big chunk of my life's work," she explains. "I want organic to be available everywhere for all kinds of shoppers at all different prices. That's not going to happen with a two-acre farm here and there. It's going to involve some of the big players."[22]

Of course, the spreader narrative is most commonly found among large organic industry players who stand to benefit from this thinking. Katherine DiMatteo got her start in the co-op movement in the 1970s, but she eventually went on to serve as the first director of the Organic Trade Association. She came to believe in the vital role to be carried out by big private industry actors, and seeks to drive those entities further down the organic path. DiMatteo unabashedly embraces the spreader mind-set. "The biggest division was between the people who see organic as a niche opportunity for a premium price, so that farmers would have a place to survive, versus the other side of us, including myself. We say *all* farming should be organic. We should be the mainstream. We should be conventional agriculture. And in order to do that, we have to have as much of it as we possibly can."[23]

Spreaders like DiMatteo are more willing to make compromises, both in terms of broader social transformation and the purity of the organic label, in order to hasten the day when all farming will be organic. She believes that organic advocates shouldn't be overly restrictive in terms of materials, especially those used in processing, which is what allows organic to reach beyond the narrow fresh fruit and vegetable market. DiMatteo is also wary of those who would restrict organic production to small producers, which while fitting the idealized image of the organic farm, was never officially part of any organic rule. Attaching additional terms that inhibit big producers may serve the interests of particular farmers, but according to the spreader perspective, it ultimately stifles organic growth and limits its benefits to a small consumer/farmer niche.

Although DiMatteo's views are common among large industry players who stand to benefit from scaling up organic production, it would be unfair to claim that these sentiments cannot also be found among other segments of the organic community. Scowcroft, executive director of the Organic Farming Research Foundation and former director of the CCOF, is critical of the "provocateurs" who offer "utopian visions" of a purely local food system and "make press with dire claims of social disaster if organic goes global via chain stores."[24] Like many with roots that go far back in the organic movement, he sees a place for Big Organic and the part it fills in expanding access to organic goods.

The spreader perspective is also commonly found within the environmental community. Feldman, as noted earlier, is the executive director of Beyond Pesticides. His organization has deep ties to the organic community, and he considers his group to be somewhat of an outlier because it supports the small farm vision. But he commonly encounters the spreader argument from his environmental movement colleagues, who see ecological value in the rapid spread of organic practices. "A lot of our friends … believe that the only thing we … should be concerned about is organic acreage, the extent that we have converted, or transitioned from chemical intensive to organic. That that's really all we ought to be concerned about. Size, that's it.... They say, 'We're reducing pesticide use. We're eliminating pesticide use. Isn't that what you're about? Isn't that the mission?'" Feldman considers his group to be in the minority in the environmental community in that it includes issues of social justice and support for small farmers among its goals, and not simply the reduction of synthetic chemicals in the environment. Unlike Beyond Pesticides, Feldman says, "a lot of ... environmental or antitoxics groups don't take a position on family farming."[25]

In contrast to the spreaders, tillers are seeking a more radical transformation, even if it takes longer to achieve. These activists are more akin to what Guthman considers adherents of the "agrarian dream."[26] They draw inspiration from the likes of agrarian essayist Wendell Berry. They envision a highly decentralized agricultural system characterized by small independent farmers growing a diversity of crops that supply their local communities with all the food they need. Tillers oppose any measure that they believe will lower organic standards and refuse compromises designed to allow for expanded organic production on terms other than those consistent with the small farm ideal. From the tillers' viewpoint, it is better to preserve a smaller but purer organic sector in the short term than to make compromises that would allow it to spread through the involvement of conventional agriculture enterprises.

Michael Sligh, director of the Rural Advancement Foundation International, has been a long-term organic activist and was the founding chair of the NOSB. He describes the disagreement that characterizes the two positions. "The big debate ... that has been there all along, is this question of how do you expand the market. Do you expand the market by lowering the standards to allow a faster growth curve, or do you expand the market by creating a longer runway and providing the necessary infrastructure to bring farmers and processors to a high standard? ... Industry has seen that ... one way to keep the growth curve going ... is to chip at the standards."[27]

Hoodes describes the spreader assertion that she often encounters, especially among organic business proponents. "Our argument with our friends in the trade association is that they say, 'What if we have the opportunity to increase the acreage of organic exponentially, shouldn't we be doing it? If we have to adjust a standard here and there, we are still a significantly better agricultural system than conventional.' That's their argument."[28]

While tillers vary in the extent of their allegiance to the small farm ideal, all reject measures that they see as weakening organic standards simply as a means to grow the market—actions that they feel would endanger the entire organic enterprise. Sligh argues that "the keys to the growth of organic, is that it was this authentic claim to customers that they could trust and believe in, whereas they had not felt that way in a long time about their food supply. This was something they could become passionate about."[29] Tillers are skeptical about whether consumers could ever feel that sense of trust and remain enthusiastic about food that is officially organic, but was produced by large corporations using industrial production methods under a standards regime that diverges from traditional understandings of organic. There is evidence to support this contention. Some studies show that consumers are deprioritizing organic for that reason and instead seeking access to food from small farmers who better reflect their vision of what organic food should be.[30]

Just as tillers are wary of the intentions of spreaders allied with profit-seeking corporations, some spreaders contend that tillers are just instrumentally seeking to protect the economic interests of small farmers. Critics suggest that tillers are damning organic to a small niche market where price premiums are used to prop up small farmers who cannot compete against those operating on a larger scale. Roadblocks to organic growth are thrown up to protect these constituents with little thought as to how the benefits of organic methods can ever get beyond this fringe group of producers, let along the millions of consumers who would benefit from more convenient and affordable access to organic goods.

Although the organic community largely pulled together around the OFPA, signs of the division between spreaders and tillers were visible even in the talks leading up to the historic congressional vote. As discussed in the last chapter, in New York and Vermont, where small farms predominate, organic farmers' associations opposed federal involvement up until the point when the imposition of some type of national system appeared inevitable. They remained reluctantly involved, primarily with the hope of shaping the policy in a way that would protect small organic growers. On the other hand, organic leaders in California, where larger farms are common, were vocal proponents of the federal system, as were most members of the Organic Trade Association (OFPANA at the time), including processors and others who were stymied by the uncoordinated private system.

Although patterns clearly exist regarding which groups are likely to reflect a spreader versus a tiller ideology, both perspectives could be found to some degree within each of the formal subgroups that make up the overall movement. And although this ideological divide is central to many debates on organic and has implications for a number of policy questions, it is just one fissure within the organic community. An examination of the interests and goals of each of the subgroups along with their associated organizations reveals a number of other ideological, strategic, and structural divisions. These divides can be seen in some of the policies that were built into the NOP and those reforms crafted after the NOP went into effect.

## Organic Farmers: The Weakest Voice

Organic farmers were at the heart of the organic movement for much of its history, both in terms of the actual actors who made up the movement and the ideal envisioned by organic proponents. Consumers provided support and resources by purchasing organic products through food co-ops and other movement-related venues. Yet it was organic farmers themselves and the associations they created that formed the movement's foundation, at least through the 1990s, when organized consumer groups, environmentalists, and others joined them. But organic farmers have always been a relatively weak political force. They were completely marginal in agricultural politics through the 1980s. As noted previously, aside from their organic associations, farmers had no organized voice and no real intent to engage politically except through their prefigurative practices. Thus, when an organized political presence was necessary, farmers were the weakest constituency.

Consumers and environmental groups, not farmers, provided most of the political muscle to pass the OFPA. Knowing that organic farmers were politically weak and unorganized, Blobaum, with the Center for Science in the Public Interest, strategically mobilized groups that he thought would be allies in the organic struggle when he was pressing for a federal program. "My basic feeling today, as it was then, was organic farmers alone are never going to be [an effective force]. They were never going to have political clout, the time and the staff, to do what needed to be done.… I have always felt that there was a whole range of allies that wanted organic to happen, they wanted it to be done right and that they would be willing … to make an effort."[31]

From early on in the drive for federal legislation, farmers recognized their own political weakness relative to other groups mobilizing around organic issues. The creation of OFAC, following the Leavenworth meeting at which federal legislation was first discussed, was an attempt to create a stronger and more organized platform for farmers as the process moved forward. Farmers recognized that their voices could easily be drowned out among the more powerful and established constituencies now taking an interest in organic. But that effort was short lived. Kirschenmann, a farmer who helped organize OFAC in 1990 with some onetime grant assistance, explained the association's demise just four years later. "Unfortunately I couldn't get any more funding, and farmers, as you know, don't have enough money to pay their own travel for meetings and stuff, so it just kind of faded."[32]

In contrast to organic farmers, consumers and environmental organizations were well established in national politics by the 1980s. Many had been in existence since the early 1970s, and some of the more established environmental groups dated back almost a century.[33] Farm advocacy organizations had also been around for decades, and some of them represented small farmers, but organic farmers were marginal even among these relatively peripheral groups. OFPANA was essentially the only national voice for the organic community. Yet even this group was relatively new, and was already oriented toward the interests of wholesalers, processors, and other segments of the organic industry. The trade group did not appeal to small farmers in the organic movement, even if more of them had displayed the wherewithal to actively participate. According to Gershuny, "At that time, and even more so now, a lot of farmers didn't trust the traders and the manufacturers and the businesspeople.... A lot of farmers didn't want to be involved with the Organic Trade Association."[34] Those farmers who did participate felt that their voices were drowned out by the growing organic industry actors who were seeking to shape policy to suit their own needs.

This disparity in political power whereby the original founders of the movement were the weakest camp would not necessarily constitute a problem as long as the interests of the other constituencies were perfectly aligned with those of the organic farmers. This was not the case, however. On a number of central issues, the interests of farmers and others within the broader organic community diverged—and still do. Large conventional industry players would eventually enter the organic market, putting price pressure on small farmers and maneuvering to adjust standards to further solidify their position. But they did not orchestrate this outcome from the start. The coming marginalization of small farmers could be read into the structure of the NOP and the policies primarily developed by actors other than large conventional players. The system that emerged from the interest and power configuration of consumers, farmers, environmentalists, organic processors, and others paved the way for the entry of large conventional actors in a system viewed as highly problematic by many small organic farmers and their allies today.

## Farmers and Consumers

Apart from the ideological and strategic differences between spreaders and tillers regarding the nature as well as speed of the transformation that organic activists were seeking to achieve, one can identify fissures when examining the interests of different segments of the organic com-munity. A central division, not often acknowledged by observers of the organic movement, is between two key groups: organic farmers and con-sumers. While sharing many common beliefs and interests, organizations representing these constituencies differ on some central issues such as the importance of farming processes relative to the materials used in produc-tion, the oversight of organic production, and the consistency of organic standards. These differences have significant implications for the structure of the organic system.

### Health and the Value of Organic

In part, the values that motivate organic farmers and consumers underlie the different preferences regarding how the organic system should be struc-tured. For farmers working the land, organic is about the connection with the natural world. They strive to utilize methods that reflect how nature works in order to produce food in ways that protect the integrity of eco-systems and integrate human food production seamlessly into that order. In contrast, surveys repeatedly show that organic consumers are driven by health concerns more than anything else.[35] It is not surprising that organic milk is among the biggest-selling organic consumer items. Parents seeking to protect their children from potentially harmful substances make up a significant segment of the organic milk buying population. Organic con-sumers also commonly cite environmental values and a desire to support small farmers as motivating factors, yet their interest in health and safety is a priority. These policy preferences are not necessarily always in line with those of organic farmers.

Blobaum recognizes that consumers often do not fully appreciate the role of farmers along with the environmental and social benefits associated with organic agriculture. Although he felt that leaders from different segments of the organic community were able to maintain good relations and achieve a high level of cooperation at the policy level during the development of organic standards, he sees a problem in the way ordinary consumers conceive of organic. According to Blobaum,

From a marketing standpoint, I think a lot of consumers come to organic because it's safer and more healthful. I think that companies are using that more.... I think that the environmental benefits and benefits in terms of what farms we'll have in the future are lost on a lot of consumers. They don't know anything about farming. They don't fully understand. I think one of the reasons is, that in the press, organic is almost always described as what we do without, and not what we do.[36]

Health-oriented consumers are primarily concerned with the end product—the food that they will actually ingest. Many worry about the consequences of consuming products containing synthetic materials, be they from chemical pesticides used on crops, antibiotics fed to livestock, or artificial preservatives, color, and flavor added to their food. Yet traditionally, organic food was not defined by its freedom from synthetic substances; it was characterized by the methods used to grow it. Organic had always been an approach to agriculture and the soil. In Blobaum's words, it is not "what we do without" but rather "what we do" that defines organic agriculture and the products of these methods. While the organic approach shunned the use of most, though not all, synthetic materials, the end product was never guaranteed to be free of contaminants that are pervasive throughout the environment. Rodale had stressed this point decades earlier in defense of organic food found to contain pesticide residues. Health claims were always one part of the organic message, especially in Rodale's writings in the early years of the organic movement, but it was only one.

During the organic surge propelled by countercultural radicals of the 1960s and 1970s, social and environmental issues received greater emphasis.[37] But as Blobaum suggests, in recent decades, resulting partly from industry messaging and partly as a consequence of broader cultural shifts, individual health concerns have shaped consumer attitudes, redirecting this important segment of the organic movement away from collective social and environmental goals. This focus on health and the nature of the final product of organic production does not necessarily foster support for policies that would protect farmers, and in some instances, consumer interest in the purity of the end product and regulation of production process conflicts with the interests of small organic farmers.

## Farming Method or Material Outcome?

The fact that health-oriented consumers had come to perceive organic foods as those that did not contain synthetic materials shaped the position of many consumer advocates engaged in organic policy making. There was much less emphasis on the traditional meaning of organic as a philosophical approach to agriculture and much more stress placed on the nature of the materials that went into organic food.

Coody, involved in organic standards writing since she started with Oregon Tilth in the late 1970s, describes her experience with established consumer advocacy representatives in Washington, DC:

Consumers really don't know a whole lot about the specific practices used by farmers.... Consumers thought organic products had no synthetic pesticides and chemicals on them.... There was concern when farmers would say, "Well, actually, we *do* use pesticides. We use things like BT and horticulture oil." ... So we had to have a lot of negotiation about that with them. That was the main sticking point: the materials list. We did not really argue about organic production practices. No one argued about that. That was just a given. That was written, basically, based on the OFAC principles. So it was a farmer-generated system. It was the materials we argued about.[38]

This health-motivated spotlight on materials served as a basis for division when developing the federal organic system.

DiMatteo and Gershuny, who both had been intimately involved in standards development, portray the debate as it played out within the Organic Trade Association during its early attempts to create national organic guidelines.[39]

A fundamental tension emerged between an orientation emphasizing production systems versus the desire of consumers for "food you can trust." ... The first round of guidelines was production oriented.... This gives primary importance to the effects of a given practice on the health of the soil and the farm organism.... By this reasoning, the origin of a given material does not matter. Synthetic compounds might be more environmentally benign than natural ones, and in any case it is often difficult to define clearly what constitutes a "natural" material. Although many consumers clearly believed that organic meant "chemical free" or "no synthetics," all existing standards allowed some synthetic materials to be used, and prohibited or restricted the use of some natural materials, such as raw manure.... An opposing view argued for "origin of materials" as the basis for organic standards, based on the idea that organic certification is primarily a consumer guarantee system. The rationale was that consumers had come to believe that organic producers used no synthetic inputs, and consequently believed in organic products as cleaner, purer and safer than conventional ones.... [F]lexibility and need for judgment ... was considered dangerous because it opened the system up to abuse, as opposed to providing clear, bright lines and allowing for greater consistency in decision making.[40]

Lipson is the senior policy analyst at the Organic Farming Research Foundation, but he previously served as CCOF's first paid staff member. He saw standards gravitate toward a materials focus not only due to consumer concerns but simply because it was difficult to define organic agriculture otherwise.

Because there is such a scientific deficit about understanding biological agriculture, the proscriptive nature of that standard was just kind of an inevitable fallback. We didn't know how to write the prescriptive standard of "this is how you actually do biological agriculture." We could only do it against a backdrop of "this is what you don't do, this is what's prohibited." So that is how all that got routed that way, and the certification system built up around that.[41]

Kirschenmann, one of the original farmer representatives on the NOSB, echoes Lipson's assessment about the challenges of converting the philosophy of organic farming into a legal standard. While serving on the NOSB, he saw the standard become increasingly concerned with materials instead of the farming process.

One of the things that several of us didn't like about the standard was this cut and dry "you can use these materials, but you can't use these materials, and then if you abide by that material use, then you can qualify as organic." Many of us were aware that true organic production was a much more complicated process in that it included ecological principles, and how you manage soil and take care of soil, etcetera.[42]

Some analysts have argued that the rationalization of organic meaning through legislation and rule making inevitably undermined a more holistic organic approach, but it was consumers who insisted that explicit rules be established for materials given their concerns about chemical pesticides and fertilizers.[43] Kirschenmann saw how this rigidity would constrain farmers and their ability to adjust their methods based on particular conditions and circumstances. He and others sought to include provisions that would allow farmers to be guided by organic philosophy, much as the system had operated when it was largely under farmer control. Organic principles would aid farmers in selecting the most appropriate methods to use in a given situation, not standardized, rigid dictates. What Kirschmann came to discover during his time on the NOSB was that "the lawyers at the USDA ... would not allow those kinds of nuanced things.... [T]he lawyers say ... legislation all has to be answered with a 'yes' or 'no.' You can do it or you can't do it. And so, we simply said, 'OK, if that's what it has to be, that's what it has to be.'"[44]

One can discern echoes of past debates about the meaning of organic from decades before. Those focused on materials were looking for a system with empirically measurable standards—a science, not a philosophy. From the perspective of one camp, if organic has real meaning, then one should be able to define it, document its use, and measure its outcomes. Yet for many organic farmers, what they were doing was more an art or philosophy than a science. They were more devoted to a set of principles to guide their practices. Consumers, on the other hand, were interested in guarantees about the end product that they will put in their bodies.

Consumers were not only a loud voice in favor of clearly established rules about materials but regularly advocated for the testing of chemical residues on organic products. For many, organic resided in the product itself, and therefore strict controls should be placed on what can go into that product and scientific testing should be used to verify its purity. Again, this reductionist scientific approach did not square well with organic growers motivated by philosophical or even religiously rooted values. According to Joan Dye Gussow, an NOSB member and nutritional scientist from Columbia University, "It was very clear to me that you could not regulate anything like organic in that heavy-handed way. You just couldn't. It was a belief system. It was a way people wanted to work with nature."[45] Organic farmers approached the farm as a holistic system. They did not want to see organic rules reduced to a materials list. Yet given the need for explicit regulatory codes, much of which was supported by consumer advocates, organic standards increasingly concentrated on allowable and nonallowable materials. Within this limited arena, consumers pressed for strict limitations. This frequently conflicted with organic farmers' needs and the growing practices they had used for decades.

The NOSB rejected some of the more extreme demands made by consumer advocates, such as proposals that organic products be tested for pesticide residues annually or that nonorganic compost be banned from use. While there were some debates between farmers and consumers about materials allowed for use in farming, most report that this was more of a process of education than outright conflict. Far more significant controversies would emerge later around synthetic material use in the processing of foods, not growing them.

Perspectives on the use of synthetic materials in processing shifted over the years as organic rules were developed. Some synthetic ingredients, such as baking powder, had long been used and were considered benign within the organic community, and allowances for certain synthetics were built into many organic standards. Coody was among the first to try to systematically distinguish acceptable ingredients beyond the basic synthetic ban when she was developing standards for the state of Oregon.

We were the first law in the nation not to go on the synthetic and natural designation. We used a list. We had, actually, a positive list, and we said, "If it's not on this list, you can't use it." And we just listed everything that was okay.... I have a master's degree in systematic ecology, so I studied quite a bit of chemistry. We realized we really were using some synthetic materials. So I made the case that we should not just totally prohibit synthetic materials. We should allow some specific ones. I coined the term "compatible synthetics" to describe those things. And by that I meant compatible to the principles of organic. I proceeded to write ... this research paper about ... evaluating materials based on principles that were gleaned from the organic principles.[46]

Although there were exceptions made for some synthetics in existing organic standards, at the time that the federal law was being considered, there were few processed organic goods and, in general, synthetics were still viewed as inconsistent with basic organic principles. Even leading processors at the time spoke out against allowing the use of synthetic materials in food production. Hence, the OFPA included language that prohibited the use of synthetic materials in processing.

When drafting the language for the legislation, Merrigan had her doubts about an outright ban on synthetics in processing, but the dominant view, even among some leading processors, favored a strict stance on the issue. Given this, the law was written prohibiting the addition of "any synthetic ingredient during the processing or any post harvest handling of the product."[47] This line in the legislation would serve as the basis for some of the most heated controversies in organic history. While the greatest conflicts would not emerge until after the NOP took effect, once again, the seeds of dissension and the ultimate trajectory of the organic industry were planted during this period when federal rules were being crafted.

Despite the apparent ban written into the OFPA, the NOSB still took up the issue soon after it first convened in 1992. There was recognition among several board members, not just processors themselves, that the outright prohibition of synthetics in processing may have been poorly thought out and "it may be necessary to recommend certain exemptions" regarding this provision.[48]

Some opposed exemptions on the grounds that foods that required complex processing should not qualify as organic in any case. Longtime organic adherents recognized that there were some contradictions in regard to synthetics, but that allowances should be limited to those materials that one would traditionally use in their kitchen at home. Exemptions should not include the chemicals that were pervasive in conventional processed foods.

Gussow, who served on the NOSB as a consumer representative, was skeptical about allowing synthetic ingredients in food labeled organic.[49] Referencing the iconic manifestation of industrial food manufacturing, Gussow posed the question of whether there could or should ever be an "organic Twinkie." This question drove home the issue of whether there was something more to organic food than a product that meets the ingredient criteria. Giving voice to the tiller perspective, Gussow felt that organic had to mean something more, that it should assure consumers that organic's implicit characteristics—"appropriate scale, localness, community control, personal knowledge, good nutrition, social justice, broad citizen participation, close grower/eater relationships and farmer connections with schools and communities—were embedded in what we ate. When a certified Organic Twinkie or its equivalent turns up in the supermarket it will be a signal that organic no longer carries such assurances."[50]

While many share Gussow's view in retrospect, at the time organic advocates struggled with the question, and the issue of synthetics in processed food would be hotly debated within the NOSB. Some argued that foods that contained any synthetics should only qualify for the "made with organic ingredients" designation allowed by law. Products labeled in this way had to include at least 70 percent organic ingredients, while the remaining 30 percent was allowed to be nonorganic. This is in contrast to the 95 percent organic threshold required to bear the official USDA organic label. But the question of allowing synthetic ingredients to be among the 5 percent in a certified organic product was troubling to organic purists.

Gussow herself went back and forth about what rule would be most reasonable given the diverse interests within the organic community. A rigid restriction would have significant implications for the future of the organic industry, likely limiting it to a niche market for a long time, if not forever. Allowances for synthetics, on the other hand, could strip organic of its essence.

Processors knew early on that the ban included in the OFPA was a mistake that needed to be rectified. DiMatteo recognized the error and implications for the organic market if exemptions weren't made.

We didn't even think about it at the time [of the OPFA's passage].... [Without exemptions for synthetics] we're never going to expand the industry. We're never going to expand farming opportunities, if we can't offer people a multitude of products.... If people aren't eating raw soybeans, you have to make tofu out of them, or soymilk, or rice milk. You've got to make rice cakes and you've got to have these other things.... Cereals. Nobody eats raw wheat.... [P]eople like cookies. They like cakes. We're no longer a society ... where we make everything at home.[51]

Given modern eating habits, DiMatteo reasoned, without some allowance for synthetics in processing, organic would forever remain a small fraction of the food market. Producers would have no incentive to seek organic ingredients for products that would never qualify for the label given the necessary inclusion of banned substances.

This debate carried on among NOSB members and the broader organic community. Sligh explains how those in favor of allowances ultimately won the day.

The farmers put a few categories of exemptions for synthetics used in growing in the legislation—things like pheromone traps and mechanical pesticides. Processors used that argument to get their own list, too. If the farmers were given a short list of synthetics, people felt it was only fair to give the processors one, too.... A minority on the board thought we should not even vote on it as it was not in the law. But the USDA said they would take it up with their counsel and they thought there was a way to resolve it, and wanted us to make recommendations. A majority of the board wanted to do that, and the disagreements were papered over.[52]

While remaining a synthetic ingredient skeptic, Gussow proposed a compromise rule that would allow for the use of nonorganic materials under limited circumstances, and after careful review that would take into consideration health and environmental impacts, the purpose of the ingredient, and whether there was an organic substitute. With these conditions, in 1999, the NOSB unanimously passed the rule allowing for synthetic material exemptions.[53] As specified in the original law, the NOSB would retain control over which materials could be placed on that list, and each substance would have to be reviewed every five years with the expectation that materials would be removed as organic alternatives were developed. This would prove to be a false hope. Once again, compromises among organic's diverse community would set the stage for a different outcome from what many organic proponents had envisioned.

The issue of materials revealed multiple cleavages within the organic community. Consumers had their notions about the purity of organic products, and at times both processors and farmers would advocate for practices that conflicted with consumers' understandings of organic. As the organic industry grew, including the entrance of larger enterprises and those that specialized in processed foods, the issue of synthetic materials would create a still-larger divide. Major conflicts between industry groups and most of the rest of the organic community would eventually take center stage. But the negotiated agreements reached among members of the organic community in the course of the initial NOSB process would provide the basis for the battles to come.

### The Role of Farmers and Demand for Independent Oversight

Aside from questions about the definition of organic, a second way in which consumer interests parted from those of organic farmers was in regard to how regulation should be conducted. For much of its history, certification was carried out by organic farmers' associations. Yet this approach was challenged when consumers mobilized around the creation of a federal system. The original draft of what would become the Organic Foods Production Act was written by legislative staffers before significant input from farmers was directly solicited at the Leavenworth meeting of organic farmers' associations. As is common and desirable from the perspective of consumer advocates, in that initial draft, public officials were placed in charge of organic oversight. Although public agencies are not always funded adequately and the risk of capture by private interests is always possible, consumer groups generally support direct government regulation since government regulators are considered to have the necessary independence to regulate and enforce rules effectively. To the extent that there are problems in a public regulatory system, officials are ultimately accountable to the public, which can press for improvements.

This stands in contrast to private entities, which are often vulnerable to interference due to their need for private funding. In instances where private oversight bodies are dependent on funding from the entities they are supposed to regulate, the risk of corruption is obvious. And if the membership of the organizations charged with oversight is composed of the regulated actors themselves, then a clear conflict of interest exists. Such was the nature of the organic system prior to the federal program. Organic farmers were members of the very association certifying their practices. As one critic put it, the regulated were in a position to fire the regulator.

According to Rangan, it is a fundamental principle that actors within any given industry should not be trusted to police themselves. She expressed the consumer's position on the structure of certification systems. "Organic standards being independent from the people being certified is really ... important. That's not what was going on before. Farmers were part of creating those standards. And while I think farmers have a role to play in terms of the input they have, it needs to be an independent body that's ultimately making their decisions."[54]

The initial draft legislation reflected consumers' interest in having direct government oversight of the organic industry. This was met with objections from farmers. Kirschenmann describes the farmers' reaction on learning about this aspect of the proposed law: "[Merrigan's] first version of the law really took the private certifiers and all those other private parts of the organic sector, shelved them, and put everything in the hands of the government. We were unhappy about that."[55]

The initial legislation gave certification authority to the USDA. Organic farming associations, the primary organizational vehicles for representing farmers in the organic industry, would have been completely stripped of their certification role based on what was originally proposed. This was one of the primary objections raised by the farmers and certifying organizations represented at the Leavenworth meeting. They wanted to ensure that they would maintain control over the meaning and enforcement of organic standards.

When organic association members assembled at Leavenworth to hammer out their position on the federal law, Henderson of NOFA New York had proposed that peer review be built into the legislation. According to her,

That went over like a lead balloon.... Kathleen Merrigan and Senator Leahy said that the government wouldn't stand for anything that was as participatory as that. And we didn't really get any support from anyone outside the very small organic farming groups that saw our logic. The consumer organizations felt that government intervention was important to control industry, that nobody with a financial interest of any kind, however small, was trustworthy.[56]

Many farmers were already going along with the federal initiative with trepidation. The elimination of the organic associations from the organic system was too much to bear. It violated the farmers' traditional orientation toward private control and self-regulation. To the degree that they were compelled to turn to the federal government for support, they wanted to retain as much control as possible. In the final version of the organic legislation, Merrigan reintroduced a role for private certifiers, but in order to satisfy consumer demands, the certifying function of these groups had to be severed from the advocacy and educational component. Many well-established organic farmers' associations reorganized into two separate sister organizations—one acting as a private certifier, and the other carrying on the educational and advocacy aspects of the work. The USDA would accredit the private certifiers, thereby keeping this function in private hands while also ensuring independent public oversight of the system.

While this approach was designed to satisfy all the relevant constituencies, it had significant consequences for the membership makeup and level of participation in these groups by organic farmers. Lewin explains how this has affected his chapters in California.

Basically we've removed the chapter participants from the certification process. Certification is done in a more centralized, more professional paid level, and we're no longer using volunteers to meet at, like, a pizza joint to read the files and recommend certification from a home office. We're not doing that. But the removal of that role has taken away some of the reason for the chapters to exist. They haven't rallied around new causes, by and large. Some of them have, but others are adrift or just don't have a reason to meet now that they are just a group of growers.... They are increasingly saying, "Well, certification is done by professionals. Why do I need to go to a chapter meeting?" ... We're just trying to figure out where this organization is going. How do we keep people involved? Do we need to keep people involved?[57]

Hoodes recognizes the problematic outcome of this change to the organic system.

When the law came in, certifiers were required to get the farmers off their board because it had to be independent. That presented a problem. So farmers have fewer places to come to the table. Although many organizations have split, as NOFA has, they have their certifier and then they have their other organization. But it's been harder for farmers to have their voice.[58]

The removal of the certification function from grassroots farmers' organizations is just one way in which the NOP's structure further disempowered the farmers who had once served as the foundation of the movement. And this did not occur because of big agribusiness; it was largely a result of consumer pressure, perhaps the most powerful allies that organic farmers had throughout this process. This again demonstrates how the NOP and the contemporary state of the organic industry need to be seen as the product of complex interactions between the many groups that make up the broader organic community. Decisions that had a specific intent during the crafting of the law and subsequent debates about particular provisions of the policy had long-lasting consequences for the movement along with the general direction in which the organic industry would develop.

### The Organic Standard: Ceiling or Floor?

Another significant issue that divided the organic community was whether the federal organic standard would serve as a ceiling or floor for organic practices. Should it be a single unitary standard, thus providing the benefits of consistency and simplicity? Or should the federal government just set a minimum standard and then allow certifiers to independently require more stringent measures as they saw fit? The unitary standard, which critics considered a ceiling, would inhibit individual certifiers from adopting an organic standard above those set by the USDA. No one would be able to claim that they were "more organic" than any other certified entity by adopting additional measures that they might consider superior or more consistent with traditional organic principles. Yet at the same time, the unitary standard offered clarity and consistency, and would avoid potentially destructive competition between certifiers who might confuse consumers with claims of being more organic or the "real organic."

During the initial standards development process, before the rules were established and the NOP went into effect, there were not yet many large organic firms, and conventional operators had not entered the organic sector in any significant way. But disagreement on this issue foreshadowed the greatest division that would come to characterize the organic field: that between organic processors and large firms, on the one side, and small farmers and traditional organic activists, on the other. During this battle, the processors won, with profound consequences for the movement and industry.

Organic processors and larger organic industry actors had an obvious stake in having a single universal rule, and they took the lead in pressing for the unitary standard. Yet to some degree consumers also stood to benefit from having a single label with a clear, consistent meaning. In contrast, many small organic farmers and certifying organizations opposed the organic ceiling. They thought that the USDA standard should be the floor above which certifiers would be allowed to add additional criteria if they so chose. This would give certifiers, several of which were still affiliates of organic farmers' associations, the ability to exercise greater control. All certifiers would be required to meet the USDA minimum standard, but some could choose to apply extra requirements and therefore establish themselves as a higher caliber of certification. This might appeal to certain consumers, and as such, farmers or the makers of organic products would have an incentive to seek that certification if they wanted to reach that market. In the meantime, the presence of such an organic "gold standard" would create pressure for others to improve or for the USDA standards themselves to be raised. If the USDA set the floor, the minimum standards needed to qualify to use the term organic would solve one of the major problems recognized by all. It would give the term real meaning and would prevent unscrupulous actors from applying their own disingenuous definition of organic—an issue that was emerging under the private system.

One of the main reasons that organic farmers' associations wanted the federal standards to serve as a floor above which certifiers should be free to rise is that they believed that this would ensure continuous improvement to farming methods and help push the market toward the movement's ultimate goal of a truly sustainable system of agriculture. Since the inception of organic standards, farmers never considered those in use at any given moment to be the end point of their efforts. From the first attempts by organic farmers to collectively define what made their practices unique, organic was always a work in progress, a reflection of the best agreed-on methods based on current knowledge, but always subject to revision and improvement. Going all the way back to the 1940s, the pages of Rodale's *Organic Farming and Gardening* were full of discussions and debates about organic practices. The magazine served as much as a forum for defining and improving growing methods as it did an instructional resource for those wishing to adopt organic practices.

Later, when farmers created formal organizations, the intent behind establishing a clear system of standards was not simply to codify what they were doing but also to share ideas and improve their methods. This commitment to improvement was expected even at the individual farm level. The "farm plan," required in virtually all organic standards, not only concerned how a farmer intended to meet organic standards; it supplied a framework for future planning and improvement.

An opportunity to institutionalize this drive for improvement through the federal system was lost when a unitary standard was adopted for the NOP, thereby creating a ceiling for organic above which there was little incentive for growers to rise. The alternative of establishing a federal floor, while allowing states or certifying organizations to apply their own requirements beyond that minimum, would have altered the incentive structure for farmers and certifiers, likely yielding improvements to organic methods over time.[59]

Hoodes describes how organic standards spiraled upward under the prior certification system:

Previous to the federal involvement, standards in organic agriculture were raised constantly. Farmers found better ways of innovating upward. When I first started raising sheep, I couldn't figure out a way to not use parasiticides.[60] And in the ten years after I stopped raising sheep, farmers have found that by planting their fields with certain plants and with homeopathic things, they don't need parasiticides. So parasiticide use has dropped, and standards are innovating upward in farmers' minds, although not on the federal level. But in the past, it would have been that in NOFA New York, if New York farmers found a way, they would say, "Let's ban parasiticide use." And then another [certifier] would do it. And then, as the technology grew, it would be more widespread.[61]

Advocates of the floor approach cite the way that competition would have driven standards upward as certifiers would seek to raise their own requirements to match industry leaders. Had the standards been established as that floor, however, some of the problems facing consumers and processors would have remained. From the processor's end, organic businesses would have found themselves pressured to raise their own standards to secure certification from those organizations with a reputation for applying the most rigorous criteria. Processors using multiple ingredients would have found themselves in a situation similar to the one they experienced when multiple standard regimes were operating simultaneously. If certain certifiers came to be recognized as superior because of the more rigorous standards they applied, then processors would again face difficulty assembling all the ingredients necessary to obtain that higher level of certification. If a processor needed to source ten ingredients for their product, but could find only nine that were certified to the higher standard, then the final product would not qualify as certifiable at the higher level. They could only utilize the services of a certifier that applied a weaker standard, and then only if the more lenient certifier did not have any provisions of their own that exceeded the standards used by others. Certifiers that only adhered to the minimum USDA criteria may have developed a reputation as inferior to others, thus lowering the price premium such products could command.

In at least some ways, smaller operations would have been advantaged by allowing certifiers to impose their own, higher standards. In some cases, smaller operators are more able to utilize certain sustainable practices. Small farms, for example, frequently have more diversified crops than their larger counterparts. If a certifier independently adopted a provision that disallowed large monocultures, big operators would have been unable to secure such certification. Indeed, a certifier could add requirements regarding the size of the farm or the ownership structure, thereby offering a standard that better reflected common understandings in which organic goods are associated with small, family-operated farms. Had that been allowed, the USDA organic seal alone might have fallen into disfavor, at least among a more committed and conscious segment of the consumer population. This has occurred to some degree with the addition of new labels and certification systems—an issue that will be explored in the coming chapters. But allowing such diversity under the organic label itself was something that processors fought hard to prevent.

The kind of certifier competition and upward pressure on standards that farmers' associations had hoped for was inhibited by the adoption of the minimum floor. The uniform standard provides big players with an advantage in the organic market. Large operators can capitalize on economies of scale while achieving the same level of recognition as smaller actors utilizing more ecologically sound practices. Under the uniform system, certified is certified. Everyone receives the same USDA organic seal, and there is nothing to differentiate products beyond that.[62] Organic certifying organizations have no means to demand more than what the federal standards call for, and farmers and others in the organic market have little incentive to do anything but the required minimum. In fact, for private, for-profit certifying organizations, there are incentives to do the opposite. As with all regulatory enforcement, inspectors exercise some independent judgment and have some discretion. Being more lenient in the enforcement of the federal standards is a way to attract the business of less committed growers and processors who want the organic seal at the lowest possible cost.[63]

Processors pressed for a uniform standard as this issue was debated within the NOSB. DiMatteo explains the position of the Organic Trade Association as well as the processors who dominated the organization.

[Having a uniform standard] was big for the OTA, and it caused some friction with other groups. We wanted to ensure that the federal law, and the standard and the regulations, would be the *only* organic standard, that certifiers couldn't have additional criteria, and be able to market either themselves as certifiers having a higher standard or, in the marketplace, to say that there are "levels of organic." The OTA was very much committed to the fact that the industry, the community, would never get traction if there was multiple sets of organic standards, and the competition was about who was more organic, and which certifier could add additional requirements or more organic requirements.... We would constantly be competing with each other, as opposed to competing with the nonorganic market. [With the uniform standard] your market identity wasn't tied to the certifier, it was tied to the USDA seal. So if you changed certifiers, you wouldn't have to start marketing all over again and have someone say, "Why did you change certifiers? Weren't you good enough for that other certification organization?" ... We really felt it would limit our market opportunity.... [W]e'd be exactly where we were before, which was making very little headway in terms of converting land to organic production.[64]

In some ways, consumers would have benefited from the minimum standard approach supported by small farmers given the potential for continuously improving standards and the options made available by certifiers whose standards exceed the federal minimum. But in other regards, this would have perpetuated the confusion that consumers experienced when multiple certification systems were in effect. Consumer advocates tended to favor clear and consistent information, and a single seal with a single meaning serves that function. As will be discussed in the following chapters, some consumer advocates now support the creation of additional certifications designed to address different criteria "beyond organic." Ironically, this may cause even more consumer confusion as new labels proliferate.[65] In the case of organic itself, though, a uniform standard did much to reduce confusion for consumers. At the same time, it altered the incentive structure for organic businesses in ways that placed small organic farmers at a competitive disadvantage while also inhibiting the upward spiral in standards that had previously characterized the organic field.

## Environmental Advocates and Organic Farmers

Yet another important fissure in the organic community can be found between environmental organizations and organic farmers. By all counts, environmental groups had not made organic agriculture a priority through the 1980s.[66] To the extent that environmentalists paid attention to agricultural issues, their sentiments were against pesticides, not necessarily in favor of organic agriculture as practiced by farmers at the time. Organic farming was still marginal, and remained a politically sensitive issue among policy makers and conventional agriculture industry actors. Feldman explains environmentalists' reluctance to embrace organic.

People in the environmental community weren't there because to utter the O word in the agriculture community meant that you risk the loss of your credibility, because it really was so nonviable. It didn't have commercial viability, certainly. And the kind of people who were practicing this form of agriculture really weren't reputable people. You still have the people who thought of organic farmers as hippies.... [S]o it was like, "Don't mention that word." We never testified that organic is a solution. You didn't hear that. In many cases, what the environmental community was talking about is how we can mitigate risks. We can take this chemical off the market and restrict it this way, that way, the other way. It was not "organic is a solution." Today, yes, you hear more of that. But then, ... organic [was] viewed as the fringe, and [there was] no ... heavy environmental community involvement in ... organic.[67]

Despite avoiding the O word for many years, environmental organizations were drawn into the legislative fray by advocates who wanted to see an organic law passed. And those environmental groups would prove to be decisive when it came to convincing lawmakers to get on board with the OFPA. According to Merrigan, "We would not have had ... success in the 1990 Farm Bill if the Environmental Working Group and a couple of other key big environmental groups hadn't come in and helped us work the floor vote in the house."[68] But once on board with the organic program, environmental organizations were, for the most part, of the spreader mind frame. Environmental advocates had long sought to restrict toxic pesticide use due to the threat such chemicals posed to the water, air, and wildlife. From the perspective of most in the environmental community, the more rapidly that acreage could be brought under organic cultivation, the better.

Like consumer groups, environmental advocates tended to favor strict government oversight and a uniform organic standard. In some instances, they favored controls that were unrealistic and unworkable from a farmer's perspective. This was evidenced in a dispute about the exemption from formal certification procedures for very small organic farmers. Organic practitioners wanted to exempt any grower making five thousand dollars or less in sales annually—a small operator for who even minimal additional paperwork or cost would likely drive them away from organic certification of any kind. Some even argued for a higher sales threshold exemption since small farmers barely make ends meet even with gross sales over twenty-five thousand dollars. Environmentalists wanted to limit the exemption to one thousand dollars. While farmers were successful at getting the threshold exemption of five thousand dollars, the dispute is indicative of the differences in interest and ideology that separated these groups.

In general, because most environmental advocates had little involvement with agriculture and even less with organic agriculture, there was a steep learning curve. Environmentalists wanted to be as strict as possible and get the greatest environmental return from any policy measure. One farmer representative on the NOSB complained about the approach taken by environmentalists on the board.

The person who represented the environmental community ... wanted much tighter regulations, and tighter regulations in a rather dysfunctional way.... [T]here was that segment in the environmental community that thought it could solve all the problems by simply [establishing] strict rules and regulations ... and apply[ing] uniform regulations to local situations, which in the real world generally doesn't work.[69]

Other than having representation on the NOSB, by most accounts, environmental organizations have only been intermittently involved in the organic cause. They have played a decisive role in some of the major struggles, but few have made advancing organic agriculture a central component of their work. Organic advocates have reached out to environmental groups, and have done much to educate them about both the ecological benefits associated with the organic approach as well as the practical limitations. Nonetheless, most environmental groups remain on the sidelines of organic struggles.

## Identity, Policy, and Efficacy: From Community to Coalition

It is crucial to note that despite these divisions within the organic movement, disagreements rarely yielded open conflict or hostility between subgroups. On the contrary, according to most participants and observers, representatives of the different constituencies got along well, and went to great lengths to maintain a sense of cohesion and community. Yet in some sense, this culture of conflict avoidance may have actually hampered the organic movement's ability to press strongly for its objectives in the policy arena.

Merrigan studied the sustainable agriculture community during this period, when efforts were being made to bring unity to the movement and effectively influence policy development. What she found was that participants expended great effort to foster a sense of harmony among themselves to the point of papering over differences in ways that inhibited the development of real policy goals. The very notion of an organic community, referenced frequently by virtually everyone affiliated with the cause, masks difference and may project a false sense of unity.

According to Merrigan, organic advocates were hampered in their ability to operate effectively in the policy arena not by divisions within the sustainable agriculture community but instead by their failure to acknowledge those divisions. In her analysis, this is a product of the type of community that developed among organic agriculture proponents over many years of marginalization and ridicule from establishment institutions. This sense of mistreatment isolated them from conventional agriculture industry actors, policy makers, and others with whom the group would eventually have to work in order to advance its interests. Merrigan characterizes the movement not as a traditional policy advocacy coalition but rather as more of an "identity group" in which participants greatly elevated the value of internal bonds with their organic peers. "Because identity masks differences, it is more difficult for members of the group to develop a policy agenda than if those differences were recognized and efforts were made to address them."[70] In other words, organic proponents were reluctant to openly address disagreements out of fear of offending their brethren in the organic community. By not doing so, they were less able to debate and negotiate to come to a clear, unified policy stance.

The concept of identity groups was first applied in the 1980s to politically organized racial and ethnic groups, religious groups, feminists, gays and lesbians, and others who appeared to depart from traditional pluralist interest group politics. As opposed to the pluralist framework in which individuals act politically based on a range of different interests intersecting and overlapping with diverse groups in complex ways, those engaged in "identity politics" define themselves by a singular group membership and identify all interests in association with that group. Critics charge that this deeply felt sense of exclusive group membership inhibits shared understandings and the ability to recognize common interests with those outside the group.[71]

Merrigan observed many conferences, retreats, and other meetings, including those of the NCSA, the largest umbrella group for organic advocates. At such meetings, Merrigan noted that sustainable agriculture proponents carry out hours of self-affirming discussion, but achieve little in terms of identifying clear policy positions, let alone a concrete agenda for moving their cause forward. She concluded that having to assume policy positions would have forced groups to confront real differences between them—differences for which compromises could have been achieved had such groups been open about their particular interests and the ways in which they disagreed with some others.

Hoodes's portrait of how the group operated is consistent with Merrigan's assessment. Hoodes notes that the coalition was united on central values, but that there was no transparent process for negotiating unified positions on specific policy matters. "We have some general principles ... [but] we don't have a great process of how we come about positions.... We're such a loose network that we don't have any votes." The NCSA would make unified statements when consensus was achieved around general principles, but when there were differences, the "National Campaign stays out ... and everybody else takes whatever position they want."[72] The NCSA had roughly 150 affiliated organizations, perhaps 40 of which would be present at any given meeting. Meetings primarily served as information-sharing venues as well as community-building or community-maintaining exercises. In Merrigan's view, hammering out differences and adopting unified policy positions would have enabled the coalition to operate more effectively in the policy arena.

Merrigan also found that there was an insular quality among sustainable agriculture groups. While happy to engage with others within their identity community, these groups were closed to relations with conventional agriculture organizations or other outsiders perceived to differ from their shared worldview. In Merrigan's assessment, this limited understanding of the needs as well as interests of other groups prevented sustainable agriculture advocates from utilizing knowledge and discoveries relevant to their cause. Harkening back to early disputes about the role of science and controlled experimentation, knowledge from outside, including scientific findings, was discounted in favor of beliefs espoused by organic practitioners within the identity group.

Corresponding with this propensity toward internal conflict avoidance, organic advocates also steered clear of formal coalition affiliation and the establishment of any unified leadership structure. The NCSA was always defined as a network as opposed to an actual coalition. Coalitions typically come together around specific policy goals. But NCSA affiliates remained steadfast in their desire to retain their autonomy and not commit themselves to particular policy positions. This was apparent over the many hours of observation of NCSA meetings that I conducted. Discussions were informative, and members benefited from learning about the work and perspectives of the affiliated groups, but no attempt was made to seek a united position when differences emerged. Instead, ongoing meetings and conversations served to reinforce the sense of collective identity without yielding practical actionable positions.

Lacking a unified voice and a defined coalition structure with leaders who could speak for the entire group, sustainable agriculture advocates, as a collective, remained relatively weak. Yet policy development moved forward, and the interests of individual segments of the organic community, each pursuing its own goals, would make their way into the final policy outcomes.

Although many, like Merrigan, charge that this focus on identity maintenance is damaging in the context of concrete policy development within a pluralistic governing system, in the case of the organic movement, the emphasis placed on community building and organic identity can be seen as a reflection of the movement's historical adherence to a prefigurative political strategy. Organic activists see themselves as change agents and as part of a movement designed to transform the agriculture system, if not society as a whole. But historically their plans for doing so never involved policy or Washington politics. They saw themselves as developing a parallel system—one that would expand as the broader public learned about different ways of living. The strategy revolved around education and lifestyle change, not tinkering with policy within a fundamentally flawed industrial agriculture system.

In this context, maintaining clear identity distinctions is of fundamental strategic importance. The movement needed to demonstrate that another system is possible, and to make that apparent, an alternative system had to be vividly distinct. It did not see organic production as an adjunct to the conventional agriculture system; the movement wanted the system to be functionally whole by itself. Compromises that might blur the difference between the practices of the organic community and those of conventional agriculture players could be detrimental, and as such, sustaining distinctions was paramount. This is why Merrigan, observing the organic movement during the 1990s, saw a dysfunctional group—one that had such an exclusive and rigid sense of self that it was incapable of working with others in the policy world of DC bargaining.

The passage of the OFPA and the process of creating federal standards compelled organic advocates to reorient their political work. This was the shift from a prefigurative strategy to one of conventional interest group politics. Organic proponents would be forced to look beyond the "deep core" principles that united them as a movement and begin to confront the policy specifics on which they would differ.[73] As the process progressed, the movement adopted tactics more common to conventional policy advocacy groups. They created formal coalition organizations and hired full-time lobbyists as they sought to develop a more coherent policy focus.

This transition did not come quickly, nor was the transformation ever complete. According to Hoodes, the NCSA, the main umbrella group for organic advocacy organizations from every sector, never ceased to operate as a network as opposed to a formal coalition. She observed that the movement's transition forced it to more deeply engage in the policy arena. "Sometimes we're a coalition and sometimes we're a network. When there's a particular issue that everyone is in agreement on ... it comes together as a coalition, but sometimes it slacks off into just networking and keeping everyone involved."[74] Some groups would eventually see the shortcomings of the NCSA structure, or lack thereof. In 2003, several key organizations within the campaign broke off to form a separate organization, the NOC, a more conventional coalition in which members develop a common platform and engage in collective lobbying.[75]

Although some segments of the organic movement continue to take on the characteristics of traditional advocacy coalitions, the emphasis placed on preserving that deeply felt but ultimately vague sense of identity was not always a liability. At times, the strong sense of collective identity that Merrigan depicted as problematic actually aided the organic movement. Organizers working with such a group can easily draw on that shared feeling of identity to activate members of the community, especially when confronted with threats or an attack from "outsiders" like the federal government or agribusiness corporations. Although effective work on day-to-day policy issues may have been lacking, there is no doubt that the strong sense of identity among members of the organic community inspired them to act forcefully when they felt that their core principles were being challenged.

### The Big Three: Mobilization and Unity

The organic community's ability to act as a movement and cohere around common goals was demonstrated most spectacularly when the USDA announced the first proposed organic rule in 1997. After three years of work, the NOSB had submitted its recommendations to the USDA in 1995. The USDA hired Gershuny, who had been involved in organic issues for many years and had helped to establish NOFA's organic certification system, to turn the NOSB recommendations into formal regulations. The organic program was still severely underfunded, and Gershuny found herself with little support as she undertook the daunting process. She insists that she remained true to the NOSB recommendations as she did her work, clarifying seeming contradictions and translating the proposal into language that would pass muster with regulatory agents. But the rules she crafted were eventually subject to review at several levels, including at the Office of Management and Budget (OMB), an executive branch office that acts to ensure that all agency actions are consistent with the president's priorities. Gershuny says that much mischief was done at this level.

OMB officials demanded fundamental changes in the rules that Gershuny presented, ignoring crucial recommendations painstakingly developed over several years by the NOSB. Gershuny reported that she knew that the revised rules violated basic organic principles, but felt confident that corrections would be made following the public comment process. While the draft rules released by the USDA in 1997 altered many provisions included in the original NOSB recommendations, organic advocates were particularly enraged by three measures, which came to be known as the "Big Three." The USDA proposal would have permitted the use of irradiation on certified products, it allowed for the use of sewage sludge as a fertilizer, and it did not include a ban on GMOs.

The organic movement rallied against the Big Three like never before. The mobilization around the original passage of the OFPA was largely one of a small group of organic insiders and their allies from the professional environmental and consumer nonprofit sector in Washington, DC. By 1997, the grassroots organic base had grown much bigger and opposition to the Big Three was unprecedented in organic history. The USDA received 275,603 comments from opponents of this gross redefinition of organic—the largest number of comments ever to be received by the government agency.[76] In the face of this firestorm of opposition, the USDA reversed course and withdrew the draft.

Had the Big Three become part of official organic policy, it is likely that organic would have disappeared altogether. True organic adherents would have abandoned the term, rejected products with the bogus organic label, and developed alternatives based on more traditional organic philosophy. While organic proponents may have been an isolated fringe group for most of its history, the Big Three mobilization demonstrated that the community was capable of decisive collective action and that there was widespread public support for the cause.

Some of the details behind the USDA's 1997 proposal remain murky. Gershuny says that conventional agribusiness interests did not seek to meddle in the standards development process, at least not in any direct way that she experienced. It was her sense that those conventional players who were interested in organic supported rigorous standards based on the belief that they would have the resources to comply with any such regulations. These players sought to keep standards high, such that even if the organic sector remained small, organic products would command a significant price premium. Those conventional players that Gershuny had contact with even supported the ban on GMOs.[77]

The biotechnology provisions inserted by the OMB appear not to have been the work of conventional food industry actors seeking to water down organic standards but instead were motivated by larger trade issues of concern to the Clinton administration. While undoubtedly influenced by corporate actors with broader interests at stake, the perversion of organic rules was not actually about organic per se. *Mother Jones* magazine uncovered an internal USDA memo that linked the allowance for genetic modification in organic goods to trade negotiations taking place at the time. The United States, a leader in biotechnology, was attempting to gain access to European markets for the export of genetically engineered crops. Europe by and large shunned such technologies. According to the USDA memo, "The Animal and Plant Health Inspection Service and the Foreign Agricultural Service are concerned that our trading partners will point to a USDA organic standard that excludes GMOs as evidence of the Department's concern about the safety of bioengineered commodities."[78] While wanting to maintain a consistent stand in defense of the safety of biotechnologies, USDA officials were also concerned about activist groups that had pledged to "wage war" against the USDA should GMOs be allowed within organic rules. USDA officials attempted to obscure the issue by classifying GMOs as synthetic substances that are typically, but not always, prohibited under organic rules. But the failure to include an outright ban set off the mass protest, and in effect, unified and strengthened the organic movement. For her part, Gershuny feels that she was made the scapegoat in the political fiasco that ensued.

Ironically, the battle that united the organic movement also set the stage for new divisions. As noted earlier, Merrigan asserted that the organic movement had a culture of conflict avoidance for much of its history, especially when it came to internal differences.[79] The Big Three struggle gave rise to brash new organizations whose leaders did not feel this constraint, and who saw aggressive action and open criticism as essential tactics in the fight for organic integrity.

## The Debate on Tactics

The organizations that make up a movement coalition may encounter internal disagreements due to the varying positions that groups within the coalition occupy. As discussed above, organic farmers may differ with consumers, who differ from animal welfare advocates, who differ from environmentalists, and so forth. This can be understood on the basis of the distinct goals that such organizations have within our highly fractured interest group system.[80] In addition, groups may disagree on tactical or strategic matters—divides that have also characterized the organic movement.

The main tactical divisions within the organic community lie in messaging and the types of pressure that groups seek to bring to bear through publicity as well as grassroots mobilization. Tension emerges when some groups publicly identify those they consider to be bad organic actors, or when they convey to their members and the media that USDA policies represent the loss of organic integrity. Some of the more militant groups have called for boycotts of businesses within the organic industry thought to be undermining standards in some way. These groups have filed lawsuits, called for institutional disinvestment from bad organic actors, and carried out corporate campaigns designed to apply pressure on targets by generating bad publicity for their companies. This causes dissension within the organic community because some believe that the organic industry is still too young to withstand such internal criticism. These actors fear that just as the general public is being exposed to organic and being convinced of its benefits, groups calling the organic system into question will deter consumers from ever coming on board.

Two organizations in particular are commonly identified (and self-identified) as those that utilized more controversial tactics. The first is the OCA. This Internet-based group was born out of the Big Three controversy following the USDA's first proposed organic rule in 1997. Tens of thousands of organic consumers, previously uninvolved in organic politics, were mobilized during the Big Three struggle. Outreach through food co-ops, independent grocers, and other venues drew thousands of organic consumers into the movement. As petitions were signed and names were collected, these individuals were brought together to form a significant new grassroots force. Those newcomers, along with thousands of seasoned activists, were primarily organized into the OCA. Today, the group claims 850,000 affiliates, and bills itself as "the only organization in the US focused exclusively on promoting the views and interests of the nation's estimated 50 million organic and socially responsible consumers."[81]

Cummins describes the group's beginnings during the Big Three controversy:

We had inside sources, and we knew eight months previous that [the USDA] was going to suggest that genetic engineering, food irradiation, and that using toxic municipal sewage sludge would be OK as well as some other basic degradations of traditional organic standards. So we launched this campaign the day they put out the proposals, December 16, 1997. Immediately it was a firestorm of resistance to this on the part of organic consumers.... [T]he Organic Trade Association was basically waffling, ... and the sustainable ag groups across the country were pitifully weak in terms of having very small mailing lists, an inability to react quickly, and a sort of politeness that kept people from just stating the obvious, which is, basically they're trying to take over organic standards. Our phones were ringing off the wall. We could barely send out leaflets and petitions to natural food stores as quickly as they were requesting them. We only had a few campaign staff people. We realized in January that we needed to form an organization that represented organic consumers.... There were organizations representing the industry. There were statewide organizations representing organic farmers. But there was nothing to represent the consumers.[82]

With some seed money from the Consumers' Union, the OCA was launched.

Cummins, with roots in the movements of the 1960s, proudly acknowledges the group's reputation as "the radical wing of the organic movement," and defends OCA's campaigns to protect organic integrity and directly criticize those who undermine it. He views other organizations as limited in their ability to be openly critical of problems in the organic industry. "Most of them are constrained from telling it like it is because they're dependent on foundations for their money. We are not. We get our money from the grass roots, and we can say any damn thing we want, and no one can shut us up, because we're too big."[83]

He ties the ability to act independently to the group's financial independence as well as the ideological orientation of its directors.

We basically created an organization of people who share the same vision for radical transformation.... [O]ther organizations ... are hampered in taking controversial stands. We don't have that problem and we never will.... You don't want your board of directors being a bunch of rich dilettantes, which most organizations are tempted to do, because that's how you raise money. Our people are hard-core committed activists who are willing to take the heat.[84]

The Cornucopia Institute, a Midwest-based small farmer advocacy group, is the other organization commonly seen as controversial within the organic community. According to cofounder William Fantle,

Our presence has created some tension, primarily because we are willing to name names and to identify some of the bad actors. Some elements of the organic community, particularly as you get closer to the corporate level, are concerned that we may be giving a black eye to organic. That has created some tension. There are people and organizations in the organic community that think we should just sit down, shut up, and clap louder, and that's not going to cut it. At some point, consumers are going to learn if there is a fraud taking place, and they may become totally disillusioned with organics and abandon it. We think the time is right now to weed the garden, to prune, to remove some of the bad actors, and to show that the system works, and that they can enforce these standards when there is somebody violating it and make those rulings that will protect the integrity of organics. We don't expect everybody to adopt our modes, and our tactics are very aggressive.[85]

As might be expected, the Organic Trade Association and the organic companies targeted by these groups are those most opposed to these types of tactics. Cummins reports that he is regularly approached by industry representatives at conferences and elsewhere who say, "You've got to stop talking about organic standards and fighting to preserve organic standards. The mass media is starting to convince people that you can't trust the organic label anymore." Fantle says that the Organic Trade Association has put direct pressure on Cornucopia board members not to adopt measures that show any organic producers in a bad light. Industry representatives are clearly concerned about any threat to organic's public image, but criticism of these tactics also comes from some farmers' associations and certifying groups.

Wonnacott was among those who expressed concern that harsh criticisms foster division and weaken the larger effort to promote organic. She also ties this to the divisions between farmers and consumers discussed earlier. Her doubts about the role that some consumer groups have played in organic politics go back to the mass mobilization in 1997 that gave rise to the OCA.

I think right off the bat, these consumer groups, while playing an important watchdog role, really elevated the mistrust, and that has persisted.... [T]hey could rev up all of these masses of people that knew nothing [about the realities of on-farm agricultural production] and get them to write in letters, ... making a spectacle of the whole thing, ... making it the most commented-on draft that the federal government has ever gotten feedback on. It really elevated the discussion, but is it a good thing for organic? I don't really know.... [I]t made a lot of consumers mistrust ... federal involvement in national certification.[86]

A representative from another certifying organization was more forth-right in his assessment of these groups.

Basically what the problem is, is that the Cornucopia Institute and the Organic Consumers Association have ... been extreme in their methods and disingenuous in their messaging.... Their rallying cry is, "It's an attack on organic standards, they're lowering organic standards!" ... They are the single-biggest problem within organic, where the people who hate us can go fight another fight, because we are tearing ourselves apart on the inside. They are the primary causes of that.... They care; they legitimately care. They are good people who care. It's just that their methods are really questionable."

Unable to appreciate the basic strategic disagreement about how best to defend and promote organic principles, this critic charged that Cornucopia and the OCA "exaggerate for [the purpose of] fund-raising."[87]

Charges and countercharges that groups are acting to protect their organizational and financial interests, either by moderating one's message to appease wealthy funders or making overstated claims to "rev up" grassroots supporters, are not uncommon among groups that differ on tactics. Some social movement scholars have also sought to understand organizational behavior in this way. Movement organizations have an interest in achieving the goals for which they are designed, but they also have an interest in protecting the organization that provides for the livelihood of leaders and staff, and is seen as an essential vehicle for advancing ongoing reform. Much has been written about the way in which, once established, the survival of an organization can become a goal in itself, especially among those who are employed by the organization or hold influential positions within it.

According to classic political theorist Robert Michels, control over organizations, even those with democratic structures, eventually falls into the hands of a few.[88] Personal power and material interests then drive those in charge of organizations to redirect action in ways that protect their interests and control. The so-called iron law of oligarchy diverts organizations from their original purpose and makes organizational survival an end in itself. This could motivate decision makers to take actions not based on careful analysis of how to achieve political goals but rather for the purpose of organizational maintenance and survival. Movement organizations depend on the support of members and funders, and part of what these backers look for is evidence that the organization is relevant. Ideally this is achieved through the regular work of the organization, with the evidence of effectiveness being the desired outcomes. But organizations may also be tempted to take bold actions, albeit not necessarily effective ones, simply to draw attention to their group by making headlines and getting media coverage.

One need not adopt Michels's cynical perspective to understand strategic disagreements within a movement. After all, it would be difficult to explain why individuals motivated by power or material self-interest would choose to pursue a relatively underpaid career in what was, for a long time at least, a marginal movement. Other than those directly tied to organic business enterprises, it would be hard to find many professional organic advocates who earn salaries commensurate with their skills and education level.

A better explanation for why organizations adopt conflicting strategies is not that they're seeking to garner attention and make a name for themselves but instead that movement leaders simply have different strategic sensibilities owing to personal history or other anecdotal factors. The fact that such actors would make tactical choices that would also serve to maintain organizational survival is not surprising given that most movement activists recognize organizational capacity as necessary for any effective social change effort.[89] In other words, despite what movement actors might think of one another, there is little reason to believe that organic advocates are intentionally adopting tactics that they believe are harmful to the cause just to bolster their organizational profile.

Regardless of the source of divergent tactics, these differences can serve to create animosity and division within a movement. The consequences of this are unclear. Some social movement scholars have theorized that strategic and tactical diversity may not be harmful to a movement's success. The so-called radical flank effect—that is, having more militant factions that can make other actors appear reasonable and moderate—can potentially yield benefits unavailable to a more strategically unified movement.[90] Yet even if this is the case, it is rarely understood as such by movement participants, and many organic advocates expressed frustration with those who they perceived as misguided or self-serving for adopting approaches out of line with their own.

## Conclusion

Participants in the organic movement regularly refer to the organic community, but the notion of a unified group was perhaps never accurate. Prior to the late 1980s, organic proponents were separated geographically, with little contact between them except on the local and regional levels. There was some common affinity for organic principles generally, yet different regional associations were divided on several of the specifics, and a sort of rivalry existed among these associations. During the mobilization for the OFPA, many new groups, such as those made up of consumers and environmentalists, were brought into the organic circle. Despite the unity implied by the continued reference to the organic community, the movement was in fact composed of a diverse array of constituents who differed on a number of issues.

Open conflict within this broad organic community was rare. Disagreements would be hashed out among the participants in the NOSB, and in the public eye, the movement appeared unified and stronger than ever. This is exemplified by the successful mass mobilization around the Big Three. The USDA backed down and withdrew its 1997 draft rule proposal, and later issued a new set of rules much more in keeping with the NOSB's recommendations. In December 2000, the new rules were entered in the *Federal Register*, and the NOP was officially launched in 2002. For the first time, products began to appear on the shelf bearing the official USDA organic seal.

Yet the compromises reached during the negotiations leading up to the official rules set the stage for even more dramatic changes—and not the kinds that many had been fighting for decades to achieve. The rules laid out in the NOP gave rise to the wholesale incursion of multinational agribusiness and food enterprises into the organic market. In previous decades, there was not much that could be considered an organic industry. Some organic enterprises were growing into fairly large and successful businesses. Companies like Muir Glen and Cascadian Farm had become multimillion-dollar enterprises by the 1990s.[91] But they had started as small businesses operated by organic activists, who still headed them, and they were few in number. Some disagreements were beginning to emerge between these entrepreneurs and others within the organic community during the 1990s as the NOSB was hammering out the regulations. These differences were no more significant than those that arose between consumers, farmers, environmentalists, and others. Nevertheless, following the creation of the NOP along with the entry of large conventional food and agriculture players, many would come to perceive a fundamental divide between Big Organic and the rest of the organic community.

Even today, however, it is an oversimplification to say that these battle lines are clear. There still exists an ongoing strategic and ideological
division between spreaders and tillers embedded throughout the farmer,
consumer, environmental, and other organizational manifestations of the
movement. Some side with the organic industry, cheering the astronomi-
cal growth of the organic market and its reach into previously untouched
customer populations. They value small organic farms serving local communities as a welcome supplement to the broader food industry, but see
large national food producers, processors, and distributors as essential players in the modern world economy. Others lament what they consider to be
the loss of organic integrity in a market that in many ways reproduces the
industrial food system they were fighting against from the start. In the next
chapter we will consider the basis for these disagreements by assessing the
current state of the organic industry.

# 5  Are We Better Off? Movement Achievements and the Threat from Big Organic

In my opinion, "organic" is now dead as a meaningful synonym for the highest quality food.

—Eliot Coleman, "The Benefits of Growing Organic Food," December 2001

The U.S. organic market is experiencing strong expansion, with organic food and farming continuing to gain in popularity. Consumers are making the correlation between what we eat and our health, and that knowledge is spurring heightened consumer interest in organic products.

—Laura Batcha, executive director and CEO of the Organic Trade Association, press release, May 2014

The organic movement has been through a lot since 1990, when the OFPA was passed. The passage of the act itself came rather abruptly. The movement had been primarily focused on its market-based strategy of spreading organic through consumer education and grassroots outreach to farmers and gardeners. It had successfully built an organic community, or rather, many local and regionally based organic communities, each of which had painstakingly developed organic standards and procedures suited to their needs. But internal problems with the private organic certification system coupled with external threats to their cause thrust activists into the world of public policy and traditional politics. Grassroots activists, used to coming together to discuss rules for their self-governed systems, now found themselves dealing with lawmakers, lobbyists, and bureaucrats in an effort to shape a new system that would be centrally administered, and ultimately, controlled by individuals and agencies with no prior connection to their familiar community. The process was tumultuous. It took years to shape these diverse farmer-derived standards into a uniform set of rules and procedures suitable for federal adoption. There were many hard-fought struggles along the way that tested the strength and fortitude of this increasingly complex movement. Eventually, though, the activists got it done.

In the time since the OFPA was passed and the NOP went into effect, the organic world has been profoundly transformed. Organic is no longer a fringe practice espoused by radical critics of the industrial food system; it has been embraced by virtually every establishment sector from the federal government to research universities to multinational corporations. Food and agriculture firms whose leaders once denounced organic as a fraud have entered the market with force, dramatically changing the way organic goods are produced and distributed. The market for these products has grown enormously. A majority of consumers now purchase at least some organic goods.[1] Consumers are drawn to the promise of organic's health and environmental benefits—at least some of which have by now been verified through scientific research.

The development of the organic market over those years has provided movement leaders and activists with an opportunity to assess how far they have come as well as where they want to go from here. There are certainly impressive achievements, and thus organic activists can be proud. But some continue to feel that in the process of establishing organic as an accepted, recognized agricultural practice and growing the market, perhaps beyond anyone's wildest expectations, the soul of organic has been lost.

Many feared such an outcome even as the process was just beginning in the late 1980s. Others held out hope then, and today still feel that the movement has achieved great things despite some shortcomings. This chapter will consider the state of the organic industry, including the assessments of those who have fought for decades to bring us to this point. In what ways can activists claim victory? To what extent have weaknesses within the program and developments in the market nullified some of the gains that have been made?

### The Critics, the Evidence, and Organic's Credibility

Among the most significant changes of the last few decades are the credibility that organic agriculture has achieved and its acceptance within mainstream institutions. These shifts can be attributed, at least in part, to the fact that organic has been subject to greater scientific scrutiny and there is now ample empirical support for at least some of the claims long made by organic proponents. It is beyond this book's scope to offer a comprehensive assessment of the ecological advantages of organic agriculture or the health benefits gained from an organic diet. But it is worth reviewing how organic is currently viewed within the scientific community, and how health and environmental claims have been received by pundits, policy makers, and the general public.

In the early years of the organic movement, proponents were ridiculed by the leaders of most mainstream institutions as hucksters or deluded mystics opposed to the advance of science. Criticisms can still be heard today, although the more common charge is one of elitism.[2] The loudest denunciations come from conservative pundits and right-wing think tanks supported by industry funders with a vested interest in the conventional system along with committed ideologues who oppose all things perceived to be liberal.

Despite mainstream embrace and corporate involvement in the industry, conservative critics still consider organic philosophy to be subversive and in some way counter to free market principles. Groups like the Center for Consumer Freedom attack the "holier-than-thou, organic-only political movement" in the name of consumer choice. Dennis Avery of the conservative Hudson Institute's Center for Global Food Issues is among the most outspoken contemporary critics of organic. Avery refers to organic as "the deadliest food choice," and regularly writes columns denouncing organic health and environmental claims.[3] An avid proponent of biotechnology and conventional agriculture, Avery even repeats the classic threat-of-starvation assertions touted decades earlier by Nixon era officials.

Not surprisingly, both the Center for Consumer Freedom and Hudson Institute receive support from a host of agribusiness and biotechnology corporations with an interest in keeping organic proponents in check. Although they make questionable use of science, Avery and others have been able to perpetuate doubts about the benefits of organic through conservative channels and mainstream media outlets always seeking to cover both "sides" of an issue, no matter how clear the evidence.[4]

Unlike the early years of organic in the United States, the movement is now much more prepared to counter attacks from the likes of the Center for Consumer Freedom, the Hudson Institute, the American Enterprise Institute, or other detractors. The Rodale Institute, which used to be virtually alone in addressing the charges of organic critics, continues to conduct extensive scientific research on organic issues. Other organic research, education, and advocacy organizations have emerged to support that effort, such as the Organic Farming Research Foundation, founded in 1992. The Organic Center, a research and education group affiliated with the Organic Trade Association, also plays a role in compiling data and publicizing the benefits of organic practices.[5] Corporate involvement in the organic sector, for better or worse, has brought well-funded interests willing to act in defense of the now-lucrative organic label. The Organic Trade Association regularly stands alongside the host of advocacy organizations that work to educate the public about organic's many benefits and to counter its critics.

While organic's advocates have a better organizational footing to fight back against detractors in the media and across the Web, they also have been able to draw on the rapidly growing body of scientific literature on organic issues. After years of underfunded grassroots study, federal and state governments have made more resources available to empirically examine organic methods.[6] Federal support for research on sustainable agriculture practices has increased dramatically since funds were first made available in the late 1980s. The 2002 Farm Bill created an organic-specific grant program known as the Organic Agriculture Research and Extension Initiative, and allocated fifteen million dollars for organic research and education.[7] The allotment was increased to seventy-eight million dollars in the 2008 bill.[8]

The number of universities with programs dedicated to sustainable agriculture research has also increased substantially. Early on, the University of California at Santa Cruz had a program dedicated to organic farming. The Center for Agroecology and Sustainable Food Systems traces its origins back to the 1970s, making Santa Cruz one of the first universities to give serious consideration to alternative farming practices. Support for research on sustainable agriculture was expanded throughout the University of California system in the mid-1980s.[9] The Leopold Center for Sustainable Agriculture at Iowa State University was founded around this time, too. Washington State University started a program in 1991—the same year that the University of Minnesota launched its Institute for Sustainable Agriculture. Almost all universities with agriculture programs do at least some work on alternative farming practices, and this has generated a significant body of research on organic matters. The number of published reports on the nutritional quality of organic foods alone has increased fivefold since the 1980s.[10]

Abundant new data have altered the battlefield in the war of ideas about organic. Despite the fact that much more empirical evidence is available to support or undermine each side, the science is far from settled given the relative newness of this research. Nonetheless, the growing body of empirical research has provided a somewhat-clearer picture of what organic practices actually do and do not do. In some instances, organic methods appear to fulfill many of the promises long touted by their proponents, while in other regards the evidence is less compelling. Of the two main areas of interest to organic consumers and researchers—the environment and health—the evidence about the environmental benefits of organic production is more firmly established.

**Environmental Benefits**

The ecological advantages of organic methods are many.[11] Among the most pressing issue facing agriculture in the world today is the degradation and loss of soil—and organic agriculture can play a vital role here.[12] The United States has seen considerable soil degradation due to overgrazing and conventional agricultural practices. Intensive conventional agriculture tends to deplete soils. Repeatedly growing one or a few plant species in the same fields drains the soil of the nutrients necessary for those crops. Without the return of organic matter to sustain the nutrient cycle, off-farm synthetic inputs and a host of other technological fixes are necessary to maintain productivity. But given the intensive methods typically deployed in conventional agriculture, these fixes do not reverse the overall decline in soil quality.

Organic farming practices, especially when fully deployed, can protect and even build the soil. This should come as no surprise. Building and maintaining soil health is a central goal of the organic approach. Traditional elements of organic practice include the use of compost composed of animal and green manures, cover cropping, crop rotation, and other soil-building methods. Indeed, an emphasis on the health of the soil—as opposed to the health of the plant—represents the essence of organic philosophy.

This holistic approach stands in contrast to the reductionist NKP mentality, which seeks only to get essential chemical nutrients (nitrogen, phosphorus, and potassium in particular) into the plant, most often through the use of synthetic off-farm inputs. The problems of conventional chemical-intensive agriculture have been obvious for decades, although the bulk of the funding directed toward agricultural research continues to flow toward developing technological fixes to the problems of soil degradation. The evidence is now clear that organic practices offer a true low-tech solution for saving soil and farmland.

Organic's ecological benefits go well beyond soil health. Research has also confirmed that organic practices protect groundwater and surface water.[13] The extensive use of pesticides in conventional agriculture has resulted in considerable water contamination, and some of these toxins have made their way into drinking water supplies. Studies have linked this contamination to a myriad of environmental and health problems.[14] Industrial-scale livestock production also poses a dire threat to water supplies. Concentrated animal feeding operations (CAFOs), in which thousands of animals are densely packed for feeding, accelerated growth, and efficient production, create severe ecological hazards due to the waste they generate.[15] Organic methods, in which the pasturing of animals is required, and where there is less reliance on synthetic pesticides and fertilizers, act to better protect water resources. Given diminishing supplies of freshwater throughout the world, organic agriculture, in combination with water conservation and improved irrigation techniques, offers a viable solution to one of the most serious global environmental problems.

Organic methods can also help to address what many consider the most pressing environmental threat of our time: climate change. Agriculture accounts for between 10 and 12 percent of total greenhouse gas emissions.[16] Research has demonstrated that organic practices reduce greenhouse gas releases by using less energy and fewer petrochemicals, while also helping to sequester carbon in the soil.[17] Relative to conventional production, organic methods have been found to use 45 percent less energy while generating 40 percent less greenhouse gas.

Maintaining biodiversity is another area in which organic methods have proven superior. While all agriculture is, in a sense, an effort to limit the biodiversity found in nature for the purpose of growing particular desired species, organic techniques better allow for the sustenance of biologically diverse environments. Conventional industrial agriculture is characterized by massive monocultures as the drive for efficiency leads farmers to specialize in one or a few crops over a vast area. In 1900, farmers typically grew five or more different commodities; today, most grow only one or two.[18] Monocultures, by their nature, greatly diminish the number of species inhabiting farmland and its surroundings.[19]

In a tangle of pathology, the measures that farmers must take to maintain this uniformity ultimately have ecological repercussions beyond the farm.[20] Large monocultures render crops more vulnerable to pests, thereby necessitating the use of synthetic pesticides. These chemicals tend to diminish in their effectiveness over time as pests develop resistance to the treatments.[21] Farmers find themselves on a "pesticide treadmill" in which stronger pesticides or new technologies are required to achieve the same results previously attained through more benign chemicals.[22] Stronger toxins in turn pose greater risks to human health and other species.

Genetic modification is the latest innovation in the battle to impose biological homogeneity on a planet characterized by vast diversity. Touted by manufacturers as a means to reduce the use of dangerous pesticides and herbicides, bioengineering gives rise to a whole new set of risks to both health and the environment. Monsanto is the world's leading producer of bioengineered seed. Its products have come to dominate the corn and soybean market through the use of its genetically engineered Roundup Ready seed varieties. These seeds are engineered to withstand treatment with Monsanto's corresponding herbicide, Roundup, which kills plant life lacking the engineered genetic traits, thus leaving only the genetically modified crop standing. This has allowed farmers to control weeds without mechanical processes or more targeted herbicide treatments, while intensifying production and further diminishing biodiversity.

Evidence is already emerging that some "superweeds" are developing immunity to Roundup, thereby rendering the process less effective and necessitating still-newer technological innovations to stay on pace.[23] Similar resistance has been found to develop in pests after prolonged exposure to Monsanto's insect-resistant bioengineered corn, in which the production of a toxic bacteria is built into the plant's genetic code.[24] Critics contend that introducing these novel organisms into the open environment could disrupt ecosystems in fundamental ways. In addition, many fear that foods produced with these engineered traits pose threats to human health.[25]

In the meantime, organic farming techniques address pest problems without the use of genetic engineering or most synthetic pesticides and herbicides. This better maintains the biodiversity that more closely reflects the rest of the natural world. Although some big organic growers do not maximize the full potential of these methods, those who practice organic farming based on the NOP rules still outperform conventional agriculture.

Perhaps the most significant finding is that all the environmental benefits associated with organic production can be achieved while still generating yields that match those of conventional chemical-intensive approaches. A few still cite the starvation myth, a notion propagated by biotech firms that present genetic engineering as the world's only solution to hunger and malnutrition.[26] Yet a growing body of evidence indicates that organic yields are equivalent to those of conventionally grown crops and even exceed conventional yields under certain circumstances.[27] The Farming System Trial, conducted at the Rodale Institute, a thirty-year side-by-side comparison of organic and conventional growing, has demonstrated that organic matches conventional yields during normal years and outperforms conventional methods during times of drought. In short, on the ecological question, research indicates that organic does indeed deliver on its many promises.

### Health Issues

Although there is now an abundance of empirical data proving the environmental advantages of organic production, thus far the findings on health outcomes are mixed. There are essentially two questions to consider regarding the health effects of eating organic: Do organic foods have greater nutritional benefits, and do they help to prevent or avoid disease? One could also add the issue of farmworker health—a serious concern for those with the most intense exposure to chemicals used in growing and an important social justice issue. The lack of definitive conclusions on these matters is in part due to the fact that the widespread conscious consumption of organic is relatively recent, and hence analyses of the long-term health consequences for consumers are difficult to assess. Nonetheless, researchers have deployed other methodologies to try to get at the question of whether or not organic lives up to the health claims.

One study that received a great deal of media attention in 2012 showed no tangible health benefits to eating organic. The meta-analysis conducted by researchers at Stanford University looked at findings from over two hundred prior studies.[28] In regard to nutrition, the Stanford researchers found no significant differences between organic and conventionally produced goods. This is consistent with another major metastudy published in the *American Journal of Clinical Nutrition* in 2009.[29] But the Stanford results did not go unchallenged by organic proponents.[30] Metastudies like that conducted by the Stanford researchers compile data from a large number of prior research projects. Such studies are subject to criticism based on the particular methodologies used to combine data from different sources.[31] Depending on how the data from multiple studies are selected, weighted, and merged, one can reach different conclusions.

Although nothing has garnered media attention like the Stanford study, a number of other analyses have yielded directly opposite conclusions.[32] For example, a British meta-analysis published the year before found that organic methods increased the nutritional quality of fruits and vegetables relative to conventionally grown produce, and that such benefits have measurable implications for human health and longevity.[33]

Assessing the effects of growing methods on the nutritional quality of food can be difficult. Many variables can affect the nutrients found in fruits or vegetables, including variety, weather conditions, climate, soil quality, and time of harvest. Side-by-side conventional and organic growing, where these variables can be controlled, may be the best means of verifying the effects of farming practices on the nutritional quality of the produce. Some such studies have demonstrated that organic methods do in fact yield more nutritious products than those grown conventionally in most cases.[34] Even the Stanford study concluded that organically grown food contains higher levels of certain compounds thought to prevent cancer and fight disease. On the matter of nutritional quality, therefore, there are indications that organic is superior in some respects, yet further research is necessary before scientists can say with confidence that eating organic leads to better health.

In terms of exposure to potential toxins, the evidence is fairly clear. Those who eat organic are consuming fewer pesticide residues.[35] Although organic goods were never purported to be entirely free of synthetics, due mostly to the pervasiveness of chemicals in the environment and contamination from conventional crop spraying, organic fruits and vegetables have far fewer synthetic chemical residues than conventional products. Of course, there is debate about the effects of exposure to these materials on human health. Organic advocates cite the prevalence of cancer and other diseases associated with such chemical exposure. Others, such as the Stanford researchers, counter that since the amount of chemical residue found on conventional foods is well below federal standards, there is no reason to conclude that such exposure has real health consequences. Organic proponents respond by citing the frequency of instances in which conventional foods do contain residues above those considered acceptable. Once these materials are in use, it is impossible to fully ensure that levels will always remain within required limits.

One must also consider that there is significant variation in exposure levels and vulnerability to the effects of such chemicals. Agricultural workers and even those living in the vicinity of agricultural operations, for instance, are exposed to far higher levels than those found on the finished product. Vulnerable populations such as children and pregnant women face risks even at levels of exposure that may be considered safe for others.[36] Plus, as history often shows, substances once thought to be harmless can prove otherwise in the long term. Organic advocates point out that numerous synthetic substances in widespread usage in conventional agriculture have not been thoroughly tested, and the long-term effects of technologies like genetic modification cannot be known with any certainty, since they are relatively new.

Overall, the weight of the evidence suggests that organic methods do fulfill their claims of ecological benefits and at least some health advantages. Given the growth of organic production, the environmental gains are already substantial and the success of the organic approach holds promise for achieving a far more sustainable agricultural system. And since millions of consumers are now partaking of organic foods, there are likely to be significant long-term health benefits.

A handful of conservative critics persist in challenging the advantages of organic, and scientists continue to investigate such assertions, but in official circles, many important players have sided with organic proponents. Although in the United States, official policy still only recognizes organic as an environmental marketing term with no implied endorsement of health advantages, organic agriculture has nevertheless been embraced at the highest levels.

First Lady Michelle Obama made headlines when she introduced the organic garden she would be planting at the White House to highlight her campaign for healthy eating and an end to childhood obesity.[37] Officials in the executive branch have also changed their tune on organic. At the USDA, organic has gained considerable credibility since the days of Earl Butz. As described above, substantially more federal resources are being dedicated to organic farming research, and the NOP received more staff and funding to support its operations. Organic advocates were concerned when President Barack Obama appointed former Iowa governor and biotech champion Tom Vilsack as USDA secretary. But the appointment of organic proponent Merrigan to the position of deputy secretary of agriculture guaranteed that organic farming would get far more attention from the USDA.

Organic has achieved even greater official acceptance elsewhere in the world. While the United States and Europe continue to be the biggest consumers of organic goods, sustainable farming methods offer great promise for less developed nations, many of which produce organic goods for export. But as analysts note, in addition to presenting economic opportunities, organic agriculture can play a role in providing food security in developing nations. In 2002, the UN Food and Agriculture Organization made organic agriculture a priority area for research and investment, citing not only health and environmental benefits, but also this vital food security role.[38]

Much has been written about the effects of the global industrial agriculture system on peasants and the poor in less developed countries.[39] In contrast to the claims that biotechnologies can "feed the world" and help fight hunger, these technologies have disrupted traditional farming communities and thus left millions more vulnerable.[40] Those capable of growing food for themselves without relying on expensive technology controlled by distant foreign actors are more secure in many ways than those drawn or forced into the global industrial food system.[41] In addition to the endorsements of the United Nations and numerous nongovernmental organizations, many governments across Europe and elsewhere have embraced organic practices, and many offer subsidies to encourage farmer conversion to organic methods.[42]

## A Booming Organic Market

In terms of winning scientific credibility and the support of government officials, organic proponents can claim many victories. This kind of elite backing can be crucial for any movement. But the core strategy of the organic movement does not depend on passing laws or securing other forms of state support. It is based in large part on market success, and as such, winning the hearts and minds of consumers along with the public at large is most important. Here, too, organic advocates can claim considerable achievements.

Abundant market data provide evidence of organic's mainstream acceptance. Organic has gone from a marginal, largely unknown exotic practice to one that is recognized and positively considered by a majority of Americans. Seventy-five percent of US consumers report buying organic products, and most of them are convinced of organic's health and environmental benefits.[43]

The organic market achieved growth rates of over 20 percent for several years following the establishment of the NOP.[44] Organic sales of food and beverages in the United States jumped from $1 billion in 1990 to $26.7 billion in 2010. While these increases have slowed since then, sales growth in the organic sector continues to outpace the food industry as a whole and those of most other segments of the economy.[45]

With the rapid rise in consumer support for organic products has come a major increase in the land under organic cultivation. Total organic acreage went from less than one million acres in 1992 to well over five million acres in 2013. The most dramatic increase in organic acreage came in the years immediately following the NOP's establishment in 2002, when organic farmland increased by 123 percent in just five years. During that time, the number of certified organic operations nearly doubled from just over seven thousand to nearly thirteen thousand.[46] Even with this rapid growth in organic production, supplies could not keep up with skyrocketing demand, drawing still more acreage into organic cultivation.[47]

By many measures, these figures signify great success for the organic movement. While still a relatively small segment of the overall food market (roughly 4 percent in 2010), organic sales and acreage continue to trend upward. The current levels of consumer interest and official support were unimaginable just a couple of decades ago. The fact that a relatively small group of marginalized alternative agriculture advocates could achieve broad popular support, win over powerful advocates right up to the White House, and build a multibillion-dollar industry is impressive. Today, Americans are far more conscious about the food they eat, and where and how it is grown. Organic movement activists deserve a great deal of credit for this profound transformation. While there is still a long way to go and many structural barriers to overcome, based on the gains made in recent years, activists can at least envision a day when their dream will be realized—a day when all farming will be organic.

Yet many die-hard activists are not pleased with the trends they are witnessing. Organic sales and acreage are growing, providing inspiration to the spreaders within the movement. But those who favor a slower, *deeper* transformation of the food and agriculture system are troubled. Since the OFPA's passage, the organic industry itself has been transformed from top to bottom. How organic crops are grown, by who, and how they get to the consumer have all been dramatically altered, leaving many to feel that in the face of all these achievements, something important has been lost.

### Failure in Success?

Much of the change in the organic market can be attributed to the NOP's creation. Although a segment of the food-buying public was ahead of the curve, there is little doubt that official embrace of organic as a legitimate and recognized farming practice through the NOP boosted consumer awareness as well as mainstream acceptance.[48] In a mutually reinforcing spiral, big agribusiness and conventional food industry retailers bought into organic once the official federal system was in place.[49] The coherence that federal oversight brought to the market provided these firms with the certainty they needed to venture into the previously chaotic organic world. Once committed, conventional food companies, utilizing their vast resources, extensive distribution networks, and mainstream retail outlets, gave consumers more information about as well as access to organic goods than was ever possible through food co-ops, farmers' markets, or other alternative venues. Bringing organic to consumers through the places they most commonly use to buy their food contributed significantly to this dramatic growth.

### Big Organic in the Market

It is not necessarily the organic market's size that troubles the tillers of the movement; it is the nature of the contemporary organic industry. Through the 1980s and into the 1990s, one finds few references to the organic industry at all. Organic was primarily of interest only to those engaged in it: farmers and a small segment of consumers. Active participants in this cause only referred to the organic movement or, more commonly, the organic community. As described in previous chapters, the community was diffused through a hodgepodge of independent farmer and consumer organizations around the country. Despite this lack of coordination, most of those involved shared a common vision. They imagined a day when all food would be grown by farmers committed to working with nature by mimicking its processes, not conquering it with industrial technology. The composition of the organic community was complicated by increased involvement from environmental groups and traditional consumer organizations during the struggle to pass the OFPA, but even into the 1990s, the whole organic sector was by and large a grassroots, activist-led effort. This is no longer the case.

While still retaining the notion of the organic community, Sligh offers the following portrait of this transition and the new institutional environment in which activists now find themselves:

We're still actively involved in trying to preserve the integrity of the organic claim, and trying to get the actual role of government to better perfect it. The difference is that there is now, clearly, an organic industry. There is an organic civil society, and there is a big role by the government, ... and there is an organic industry. So you have these three different groups that make up the organic community, so to speak.[50]

The strain to conceive of government and industry as residing within the cherished organic community is evident in Sligh's assessment. When he began his work, there were obviously farmers and businesses with economic interests in organic. But there were few operations of any size, and even those were run by organic advocates who had successfully grown small enterprises. There were a few millionaire entrepreneurs among them, yet the organic community was still fairly unified, and the whole effort was taking place largely outside the state. Today, the presence of both a movement and industry is clearly evident, and political institutions, not just farmland and farm stands, serve as the arena through which these actors seek to advance their goals.

When Sligh refers to the organic industry, he means Big Organic, those large organic firms or conventional companies that have entered the organic arena in recent years. It is these firms that have come to dominate the organic market and cause the most consternation among many long-time organic advocates.[51] They meet organic standards as established by the NOP. Yet in terms of scale, ownership, the types of products they make, and the methods of growing, processing, distribution, and sale, their practices mirror those of the mainstream food and agriculture system.[52] As introduced earlier, some refer to this transition as the conventionalization of the organic industry.[53] Evidence of this conversion can be found everywhere.

The Hartman Group's research clearly demonstrates that it was the sale of organic goods in conventional retail outlets that led to the explosive growth in the organic market. By and large, these are not whole foods grown by small organic farmers. A short time ago one would have to visit the food co-op or an alternative grocer to find organic products. Today such goods can be found on the shelves of nearly any conventional supermarket, and even in many big-box stores like Walmart and Target. The percentage of organic sales occurring at conventional retailers such as supermarkets and big-box stores rose to 54 percent in 2010. Natural foods retailers, which include large national chains like Whole Foods, claimed 39 percent of organic sales. The remaining 7 percent was divided between online sales, boutique stores, and direct sales by farmers to the consumer.[54] The fact that farmers' markets and other direct sellers, once at the core of organic distribution, now make up such a tiny percentage of the market is indicative of the profound transformation in the organic sector.

It is not simply that organic farmers are utilizing different distribution mechanisms. The types of products on the market have also changed with a significant shift toward processed organic goods.[55] Such goods are primarily produced by conventional food and agriculture firms. According to a report by the Hartman Group, "Major corporations seized on the opportunities created by smaller pioneering brands and drove explosive growth into the organic category."[56] The strategy used by most conventional businesses to enter the organic market has been to purchase successful independent organic enterprises. Philip Howard, a researcher at Michigan State University, has documented this wave of corporate acquisitions in his ongoing analysis of "who owns organic."[57]

Howard sees 1997, the year of the USDA's initial draft rule proposal, as a watershed for the incursion of conventional agribusiness into the organic market. After decades of decentralized movement control and years of program development under the authority of the USDA, the release of an actual set of national organic standards (however flawed and controversial) assured conventional enterprises that a viable business opportunity was in the making. In the ten years that followed, one-third of the top thirty North American food processors acquired organic brands.[58] In 1999,

General Mills purchased Cascadian Farm, maker of organic cereal and gra-
nola in addition to frozen fruits and vegetables, along with Muir Glen, a
major producer of organic tomato sauces, soups, and salsas. Dean Foods
bought Horizon and Alta Dena, two big organic dairy product companies.
Through these acquisitions, Dean would eventually claim two-thirds of
supermarket organic milk sales. Dean also came to dominate the soymilk
market with its acquisition of WhiteWave Foods. Kellogg bought Kashi,
an organic company specializing in cereals and snack bars, in addition to
MorningStar Farms, a producer of a variety of soy-based meat substitutes.
Howard found that during this period, corporate giants purchased virtually
all the most successful organic enterprises founded by activists from within
the organic community.[59]

To many in the organic food-buying public, this shift in control of the
industry was hidden from sight. Corporations practice "stealth ownership,"
concealing their corporate identify behind the brand names and logos traditionally used by the original organic companies, and thus hiding any
connection with the parent corporation.[60] Marketers commonly perpetuate
this illusion of independence through bucolic images on labels and marketing materials that give consumers the impression that small farmers are
supplying independent companies. Perhaps more cynically, some companies have subtly removed organic content or lines from some of these well-
known products, such as Odwalla beverages or Silk soymilk, now owned
by Coca-Cola and Dean, respectively.[61] Brand loyalty among those who do
not carefully reread food labels helps to sustain the success of these now-
conventional products.

Conventional and natural foods retailers have also claimed a stake in the
organic market through the creation of their own private label lines. The
Safeway supermarket chain, for example, introduced its O Organics brand
in 2005. Trader Joe's, a supermarket chain that features many natural and
organic products, has its own organic goods. Walmart launched its organic
milk brand in 2006, and Target introduced its own organic line, Simply
Balanced, in 2013.[62] The appearance of an organic market still in the hands
of small independent producers to some degree remains, but in reality any
given organic product found on supermarket shelves today is likely produced by two or three large firms.[63]

Beyond the shift in the processing, distribution, and sale of organic
goods, the farmers supplying these companies are of a different nature.

Although many predicted that a national program would draw conventional companies into the organic world, few anticipated how quickly this would occur or the speed with which the organic market would expand around the time of the NOP's introduction. Small farmers were not capable of supplying the large firms that wanted organic materials for mass production. The high prices being paid for organic materials in limited supply enticed many conventional farmers to try their hands at organic growing. Many converted a portion of their holdings to organic, thereby accounting for much of the growth in organic acreage that followed the creation of the NOP.

As with the conventional corporations that stormed into the organic market, activists and traditional small organic farmers are wary of these newcomers to organic farming. These actors were not drawn into the organic field out of ideological commitment, as had been the case during the decades when the movement and its practitioners were scraping by.[64] A rush of new participants, even those motivated by financial reward, would not necessarily constitute a problem if the organic system that movement participants helped to construct ensured that all the promise of organic agriculture would be fulfilled merely by complying with its rules, or even if there were readily available safeguards to correct shortcomings once they became evident. Unfortunately, at least some activists have concluded that their standards and rules were not designed in a way to ensure the achievement of their hopes and dreams. What's more, they have found that the system is not responsive enough to their demands when serious deficiencies are identified. Many have felt their influence diminish with the incursion of these powerful new actors in the organic world.

## The Conventionalization of Organic

Many longtime movement participants are clearly troubled by the changes they have witnessed within the organic market. These concerns are laid out in the *National Organic Action Plan* (*NOAP*), a manifesto of sorts for the activist wing of the organic movement.[65] The plan grew out of a series of discussions held among movement leaders concerned about the direction of organic since the NOP's implementation. Beginning in 2005, under the leadership of RAFI-USA, individuals from a broad spectrum of advocacy organizations, including many who had been involved with the organic movement for decades, conducted a number of dialogue meetings. Five years later RAFI-USA published the *NOAP*, which delineates a number of concerns and has served as a guidepost for many organizations as they seek to improve the organic system.

The heart of many perceived problems with the new organic world order is that the new entrants to the system are, first and foremost, profit-driven entities. Given that these actors are motivated more by economic interest than ideological commitment, it should come as no surprise that when economic gain conflicts with ecological practices, profit wins out. Thus, critics note "the fear of organic becoming an 'input substitution' approach, where farmers and processors receive certification to the lowest enforceable standard."[66] This fact is a central component of the conventionalization of organic. New entrants at all levels adhere to the letter of the organic law, while otherwise superimposing traditional conventional practices on the system. For instance, rather than having a diversified farm where on-site material is composted for fertilizer in a holistically integrated system, large growers rely on purchasing organic inputs from off the farm. Instead of using extensive crop rotation or interspersing crops for the purposes of soil management and pest control, the new large organic operators tend to specialize in just a few crops, and utilize permissible organic fertilizers and pesticides. While only materials meeting organic standards are used, these operations are far from what activists envisioned when they sought to transform the world of food and agriculture.

Livestock handing provides a vivid example of what critics most fear about conventionalization. Both the rules in the OFPA and the standards developed for the NOP were particularly vague given that livestock was not a significant part of the organic market at the time. The standards that were implemented required that animals be given access to pasture as is consistent with their natural diets and living conditions. Fantle of the Cornucopia Institute describes what developed in the organic dairy industry once large players moved in.

What we've ... seen has been the rise of factory farms in organic dairy. They are doing confinement agriculture. They've modeled their practices after conventional agriculture, where feedlots are the rule. You get an economy of scale with your production; you are able to milk your cows three times a day rather than two times a day.... Aurora dairy ... is the biggest organic dairy in the country.... This company is the largest provider of privately labeled milk. If you go into Walmart and look for the Walmart-brand milk, if you go to Target, if you go to Costco, if you go to Safeway, this is the milk brand.... We don't necessarily call it a farm, but a factory farm. They've got fifteen thousand animals on five different farms, largely confinement conditions.... [The animals] can see pasture over a fence, but they have no gate to get at it.[67]

As was feared when NOSB members were debating whether federal organic standards would serve as a ceiling versus floor, some organic proponents see a system that is at best stagnant and at worst deteriorating. Many believe that the fundamental principle that organic farming practices should continuously improve is falling by the wayside. As discussed in chapter 4, the private certification system that was in place prior to the federal program allowed for that improvement. Farmer innovations would make their way into the standards manuals of private certifiers and eventually come to be adopted by others. Many organic activists worry that just the opposite is occurring under the NOP. According to the *NOAP*, "The lack of continuous quality improvements in organic standards and the difficulties in tightening federal organic regulations remain common concerns."[68]

Although the pasture rule abuses were particularly egregious (and have since been addressed to at least some degree), in general activists are dismayed that organic producers are finding ways to fully mimic conventional industrial production by simply substituting organic materials in place of the synthetic ones typically used in growing or food processing.

The significance of this shift in practices for advancing the organic cause depends on who you ask. Spreaders and tillers clearly have different perspectives. Given the growth that the newcomers have brought to the organic market, one could celebrate the fact that millions of additional pounds of synthetic substances have been kept off of our food, out of the fields, and out of our water. As confirmed by the studies cited earlier, soil is healthier, water is more protected, and biodiversity is greater relative to land on which conventional agriculture is still practiced. Yet a large organic monocrop of corn or soybeans being processed into frozen dinner entrées to be shipped across the continent does not yield nearly the environmental benefits as that of a small, diversified organic farm distributing fresh produce to a local market. Health benefits are likewise diminished and the potential for real social transformation is lost if a wholly organic food system simply means different labels on the products that consumers grab from the supermarket shelf.

### The Battle for Control of the Organic System: The NOSB, the USDA, and Congress

The architects of the federal organic system from within the organic community worried that all would not be perfect once the program was in place. Now they are striving to implement changes or at least prevent any additional degradation. Yet some fear that they have lost control over the organic system.

Riddle, an organic proponent who did much to professionalize the role of organic inspectors by helping found the International Organic Inspectors Association, is among those who were surprised by the political challenges that came with the NOP's creation. He favored having a federal system, and was optimistic about its prospects based on his experience of working with government agencies on organic policy in Minnesota. On reflection, Riddle concludes,

I think I was naïve. Having worked with a state system, there just is a whole lot less corruption at the state level than there is at the USDA. The influence of big business and the Cattleman's Association, all of the commodity groups, Monsanto, and the agribusiness lobby is just phenomenal at the USDA, and the revolving door where people are appointed from those industries into government and back out again, I wasn't really cognizant of that and how that could impact the failure to enforce the act once we had an act in place.[69]

The grass roots was obviously in charge of all things organic prior to the federal government's involvement. Farmers' associations developed the organic standards, and most business was conducted among a network of farmers, small producers, co-ops, and alternative grocers. Even in places in which state government played a substantial role in the organic system, advocates like Riddle felt that the organic community had ample voice. In the years that followed the passage of the federal legislation, grassroots members of the organic community continued to exercise a great deal of control through the NOSB, organic farmers' associations, and the organic industry, which until the late 1990s, included few conventional enterprises or large organic companies.[70] Now large corporate players exercise extensive influence over the organic system at every level, from the NOSB to the NOP to the halls of Congress.

The NOSB is still heralded as a revolutionary regulatory mechanism since it gives actual statutory power to a citizens' board that includes representatives of all stakeholders in the organic community. But organic activists, including some who have served on the board, charge that in practice, the NOSB does not have the control granted to it through the OFPA. According to Scowcroft of the Organic Farming Research Foundation, "We really thought that we had a public-private partnership in place with the USDA that would work. We were wrong. The USDA's Agricultural Marketing Service has utterly failed to build a cooperative relationship with [the NOSB] and the larger organics community."[71]

One shortcoming in the system has to do with the actual board appointments, which are made by the secretary of agriculture. Some have charged that the appointed individuals do not always reflect the interests of the constituencies that they are intended to represent, and that there is a bias in favor of large players. The initial board included Kahn of Cascadian Farm and Craig Weakley of Muir Glen, two successful organic entrepreneurs. Both individuals had been involved with the movement for many years and they founded small enterprises, but they were certainly not representative of organic processors at the time—a sector that mostly consisted of much smaller operations.

Some see the justification for the appointment of successful organic entrepreneurs to the NOSB, even if such individuals are uncharacteristic of the broader sector. After all, people like Kahn and Weakley worked their way through the organic maze that was the system prior to the NOP. They would be expected to have valuable insights into building a more coherent system, and demonstrated their commitment through their years of effort in the organic world.

It is the appointment of conventional corporate representatives to the board that organic advocates find most troubling. Alongside Kahn and Weakley on the original NOSB sat an executive of Beech-Nut, a food industry giant that just a few years prior had been found guilty of fraudulently selling artificially flavored sugar water as apple juice—hardly a sound representative for advancing organic integrity in the new system.[72]

The Cornucopia Institute is an outspoken critic of the agriculture secretary's questionable appointments to the NOSB. In 2012, it released a report titled "The Organic Watergate" in which it identified eighteen agribusiness representatives who have served on the board, holding not just slots reserved for processors, but also positions designated for environmentalists, scientists, farmers, and consumer advocates. The list includes employees of General Mills, Dean Foods, PurePak, and other agribusiness giants.

Perhaps the most egregious case of corporate board stacking was when George W. Bush's agriculture secretary appointed Katrina Heinze, a General Mills employee, to a *consumer* advocate seat on the board. Movement backlash led Heinze to decline the appointment as a consumer representative, only to be reinstalled to the scientist seat on the NOSB. The Cornucopia report concludes that "the corporate stacking of the Board leads to the erosion of the integrity of the organic label, with real repercussions for organic consumers, farmers, and the public interest."[73] Cornucopia and other activist critics charge that these actors are not capable of offering objective opinions on organic rules given their ties to large corporations more rooted in conventional food businesses.[74]

In addition to the questionable board appointments, grassroots groups have also expressed occasional concerns about NOSB procedures. In recent years, activists have complained that the NOSB no longer provides adequate opportunity for the public to participate in the process of developing recommendations.[75] Unlike the early years of the NOSB, votes by the board at general meetings, where movement representatives are in attendance, are frequently conducted with little discussion. Debate about measures takes place prior to NOSB meetings at smaller committee gatherings that few members of the public attend. Thus, activists are not able to adequately strategize or mobilize allies when measures are coming up for a vote.

Although questions are sometimes raised about the NOSB, this body remains the most accessible to organic activists and most responsive to their demands. Moving up the organizational structure, greater problems can be found. For years critics complained that USDA and NOP officials often fail to act on the NOSB's recommendations. Organic proponents have long sought the development of an accreditation manual to provide organic certifiers with clear and consistent guidelines for the interpretation of regulations, along with other mechanisms that would bring the program into compliance with internationally recognized accreditation standards. This is just one indicator of the lack of serious attention given to the program by USDA officials. The *NOAP* cites the "underfunded and under-staffed NOP, its poor enforcement record, and the lack of clarity in standards development" as issues of central concern.[76]

While some have charged mismanagement and incompetence on the part of those appointed to head the NOP, for at least part of the program's history, this appears to be failure by design. There have been modest improvements in recent years, but for a long time the NOP was starved for funds. Sligh, the first NOSB chair, recalls that the board was provided little financial support, leaving them to "panhandle … to get money to have meetings."[77] This did not improve quickly. In 2007, five years after the program went into effect and when the organic industry had reached $14 billion in sales, the NOP had a staff of just nine people and a budget of $1.5 million. For the sake of comparison, the *New York Times* reported that in the same year, the $83 million dry pea industry received $47 million in support from the USDA.[78]

It is important to recall that USDA officials had steadfastly opposed organic for decades and representatives had actually testified against the OFPA, which situated the organic program within its department.[79] Sligh points out the irony of the situation: "There they were, in charge of implementing a program that was diametrically opposed to the big business, biotech approach they supported."[80] Henderson insists that at least during the Bush years, "anyone who got a job at the National Organic Program who turned out to be too friendly to organic got transferred to some other division."[81]

Under these circumstances, it is not surprising that the NOP was not given adequate staffing and support within the USDA, or that NOP staff members would not always heed the citizens' board appointed to advise them. There have not been incidents of USDA subversion of organic as extreme as the 1997 draft rule debacle and subsequent Big Three fight that occurred under the Clinton administration, but the Bush White House did its part to hinder the organic cause. Inadequate funding, the appointment of conventional food industry representatives to the NOSB, and general neglect of the NOP left advocates frustrated during the program's crucial early years. There have been some improvements under the Obama administration. Organic activists heralded Merrigan's appointment as deputy secretary of agriculture in 2009.[82] Advocates also successfully lobbied to have the NOP's funding increased. Staffing and financial support for the program has doubled since Obama took office.[83] The organic cause suffered again in 2012 when funding for key programs was cut under the temporary extension of the Farm Bill, passed when Congress could not reach full agreement on agriculture policy. Hence, despite some gains, critics maintain that the USDA itself has been insufficiently supportive of the NOSB and the organic cause, even with a more sympathetic president in the White House. The

NOP is severely underfunded in general proportionate to the size of the organic industry. This, in combination with the facts that corporate actors exercise undue influence over the program and USDA bureaucrats harbor residual resistance, leaves many organic proponents disgruntled.

Influence over NOSB appointments along with the actions of USDA officials and those at the NOP are all ultimately tied to political influence more generally—another area where advocacy groups feel outgunned by corporate giants with deep pockets, teams of lobbyists and lawyers, and ready access to elected officials. Critics contend that big players have abused their influence in Congress and at the White House to the detriment of the organic system, and that the grassroots organic community is left out of important decisions.

The risk of corrupt political influence was made clear soon after the NOP went into effect. In 2003, Georgia Republican Nathan Deal, at the behest of poultry producer Fairfield Farms, managed to slip a one-sentence clause into a three-thousand-page omnibus spending bill that would have severely undermined organic standards for livestock. The measure would have nullified organic feed requirements for organic livestock if the price of organic feed was too high. The entire organic community spoke out against this measure, including the Organic Trade Association and a number of big corporate organic companies. In this instance allies in Congress led by Senator Leahy successfully passed legislation to overturn the measure.[84]

Other controversies about the interpretation and enforcement of organic standards would plague the system for years to come, but this one was an early sign to organic proponents that industry actors could wield considerable political influence. Organic proponents would not only have to contend with battles at the NOSB, NOP, and USDA levels; they would need to develop their own political capacity and closely monitor threats to organic in the legislative arena. As will be discussed in the next chapter, organic activists have continued to step up their political game, but whether they have the strength to defend the integrity of organic standards from threats outside and even within the organic community is a matter of debate.

### The Threat to Standards

The sense that traditional organic advocates have been marginalized and Big Organic now has undue control is a concern only because of what these large actors do with their influence. As described above, critics already assert that profit-driven new entrants to the organic world, be they multinational agribusiness firms or even just conventional farmers, are doing the minimum necessary according to organic rules and otherwise introducing conventional industrial practices to the organic industry. But there is greater concern that these powerful players are now using their influence to further undermine the system's integrity.

Oftentimes this takes the form of seeking allowances for additional synthetic materials to be used in processed organic foods. Organic activists find themselves having to fight rule by rule, ingredient by ingredient, to prevent revisions that they feel violate organic principles. As noted in chapter 3, the initial organic standard included allowances for certain synthetic materials for use in growing and processing. These are included on the national list of allowable and prohibited substances as recommended by the NOSB and approved by USDA officials. Goods can qualify for the USDA organic label provided they are made up of 95 percent organic ingredients (measured by weight, and not including water or salt).[85] The other 5 percent can be composed of nonorganic materials or synthetic ingredients included on the national list.

Advocates, even most grassroots organic ones, long considered certain exceptions for nonorganic substances as essential. Proponents also recognized that certain ingredients might not be available in organic form and that nonorganic ingredients should be allowed, at least in the short term, until organic materials could be found to use in their place. This is why the national list was considered necessary. But these exceptions were intended to be rare and temporary. According to the law, ingredients deemed allowable on the national list must be reviewed every five years. The aim was one of continuous improvement: that synthetics or other nonorganic ingredients considered allowable by necessity at one point in time, would eventually be removed as substitute materials were discovered or organic versions became available. The problem is that few materials have been removed from the list, and in fact, just the opposite has occurred. The list of 77 ingredients included on the list at the program's start in 2002 had grown to 250 by 2012.[86]

Many such additions are easily justifiable from the spreader perspective. Each allowable new ingredient is seen as an opportunity to launch new types of organic products that would not be possible to make otherwise.

This expansion of the organic market will create more demand for the production of additional organic crops. Allowing some tiny amount of synthetic material in a finished product could foster significant expansion in the production of organic raw materials that would, by law, compose the bulk of any product labeled organic.

Carrageenan is a prime example. A synthetic material derived from seaweed, it is commonly used in dairy products, such as ice cream and yogurt, as a thickener and stabilizer. It gives foods a smooth creamy texture and prevents mixed ingredients from separating out. Carrageenan is also commonly used in things like jelly and pie filling. Spreaders argue that consumers who expect jelly to have a smooth consistency may never purchase the organic version if it is watery and separated. It is claimed that by allowing organic producers to use just a small amount of this substance, a great expansion in organic berry production for the purposes of organic jelly manufacturing is made possible. From the spreader perspective, this conversion of farmland to organic would never take place otherwise, and thus this minor synthetic ingredient exception moves us closer to the day when all agriculture will be organic. An exception for a synthetic ingredient that will compose just a tiny percentage of a product seems a small price to pay from this vantage point.

In contrast to this encouraging image, tillers see organic dying by a thousand cuts through this process. Carrageenan in particular has been associated with intestinal health problems. Many did not believe it should be included on the national list, and spoke out when it came up for reconsideration. From the perspective of tillers, the inclusion of synthetic materials with potential health implications undermines the integrity of the organic label and poses a fundamental threat to the success of organic as an alternative food system. Many consumers associate organic with purity and health. As they learn that organic products actually include synthetic materials, especially those with health ramifications, there is the risk that they will lose faith in the system and abandon organic altogether. Rather than seek a big, rapid expansion of organic that would largely be captured by large producers anyway, from the tiller viewpoint it is best to maintain high standards, and continue to seek ways to improve growing and production methods, even if that means slower organic market growth. Opponents of the loosening of organic rules have been successful at preventing many materials from being added to the list of allowable ingredients, but dozens of other substances have been approved. In 2012, the NOSB voted ten to five in favor of keeping carrageenan on the allowable list.[87]

The war over the soul of organic is primarily fought through battles of this sort. To the extent that activists feel that the organic label is losing its integrity, it is happening one ingredient at a time, although occasionally more sweeping issues emerge. For example, right around the NOP's launch in 2002, the USDA made a blanket policy decision that "contact substances," those materials used in food processing, but that are not actual ingredients in the finished product, would be exempt from the restriction on synthetics. This frustrated organic activists given that the NOSB had declared many such materials prohibited ingredients. It added to the concern that USDA officials would be unresponsive to the NOSB and grassroots organic activists. Yet this fight paled in comparison to a later epic struggle over synthetics—one that laid bare major divisions within the organic community. While it is sometimes characterized as a battle between the organic movement and the organic industry, it is best considered a case of the ideological and strategic divide that separates spreaders and tillers.

After just a few years of struggling back and forth about particular ingredients, the fight over synthetics came to a head in 2004. This is when some segments of the organic community seized an opportunity to try to wipe the slate clean and institute an even stricter standard than that used prior to the NOP. This stemmed from crucial wording included in the OFPA itself: the provision that stated that handlers "shall not, with respect to any agricultural product covered by this chapter, add any synthetic ingredient during the processing or any post harvest handling of the product."[88]

When the rule was originally drafted, some had predicted that this language—indicating an outright ban on synthetics in processing—would eventually lead to a lawsuit, and indeed it did.[89] Arthur Harvey, a seventy-two-year-old organic blueberry farmer from Maine, long opposed what he considered the corruption of the organic ideal through several provisions of the NOP. He filed a lawsuit against Secretary of Agriculture Ann Veneman just a few days after the NOP took effect in 2002. Harvey sued on nine counts, claiming a discrepancy between the rules established by the USDA and the actual organic law—the most critical of which was contradictory statements about the use of synthetic ingredients in processing.[90]

The case sharply divided the organic community. While virtually everyone recognized the need for some synthetic materials, many organic advocates including RAFI-USA, the Organic Consumers Association, the Sierra Club, Beyond Pesticides, the Center for Food Safety, and the National Cooperative Grocers Association, among others, viewed the case as a means to halt what they considered to be USDA abuses whereby synthetic ingredients were being permitted without full NOSB analysis and approval.

In 2005, an appellate court in Boston ruled in Harvey's favor on three counts. It eliminated certain nonorganic material exemptions based on the lack of availability of organic ingredients and a provision regarding transitioning livestock to organic, but most significantly, it ruled that products made with synthetic ingredients could not bear the organic label. The court ordered the USDA to rewrite the rules to bring them into alignment with the law. Products that did not meet the legal criteria within two years could no longer be sold as organic.

This ruling sent shock waves through the organic world. Many advocacy groups saw the Harvey case as a means to reassess and rebalance the power of determining which, and under what circumstances, synthetic materials could be used. Even Harvey supporters assumed that after some negotiations and reforms, allowances for certain synthetic ingredients would be reestablished. But the Organic Trade Association went into crisis mode. Processed organic goods were by this point a multibillion-dollar industry, and in the association's view, the lawsuit threatened the entire organic market. Industry-affiliated organic advocates such as DiMatteo felt betrayed by those who sided with Harvey. The matter of synthetics had been thoroughly debated during the original rule-making process within the NOSB, and the final decision agreed to, even by the majority of grassroots organic activists, was that some synthetics would be allowed. In DiMatteo's view and that of Big Organic as a whole, Harvey's supporters had violated the agreement.

In a move that would solidify the already-obvious division between many organic advocates and the organic industry, the Organic Trade Association maneuvered to thwart the Harvey decision. It used its influence in Congress to have a provision quietly inserted into an agriculture appropriations bill that essentially nullified the Harvey ruling and reestablished the allowance of synthetic ingredients.[91] According to DiMatteo, "Without these amendments, there would have been great disruption for organic businesses and farmers that would result in the loss of markets, reduction of organic acres, and reduced choice for the organic consumer."[92] DiMatteo maintains that Organic Trade Association lobbyists had saved the organic industry and reestablished the arrangement that had been agreed on since the NOP's founding. From the standpoint of many organic advocacy groups, the Organic Trade Association could no longer be considered an ally and the industry-advocacy division was firmly established. Organic proponent Emily Brown Rosen warned organic activists in an article in the *Natural Farmer*; following its behind-the-scenes maneuvering on Harvey, "We can't assume that the Organic Trade Association represents all interests in the organic community."[93]

As with the Harvey lawsuit, many organic advocates believe that even when they win, they lose. They often feel thwarted by the system within which they must work. As Brown Rosen put it, "It seems like there is always a new 'crossroads' being encountered in the recent history of organic regulations and growth.... Sometimes it seems like we must be turning in the same direction each time ... and ending in the same place."[94]

In some instances, organic activists have been successful at defeating measures they consider to be violations of organic principles. They achieved significant changes to the pasture rule that had been abused by some industrial livestock operations, as described earlier. But again, some feel that the case provides another example of this defeat within victory. After years of pressuring for a clarification of the pasture rules that they hoped would bring an end to the abuses being carried out by large confinement livestock operations, organic activists finally achieved this in 2010, when the USDA issued new guidelines largely in keeping with the movement's demands.[95] Under the new rule, animals must be allowed to graze for the full length of the season, as determined by organic certifiers based on local conditions. This must be a minimum of 120 days, although in many parts of the country the grazing season would be considerably longer. The rule also stipulates that animals must be getting at least 30 percent of their food from the pasture during that season.[96]

While many in the organic activist community heralded the new rule as a victory, once again, the challenges of working within a rule-bound system outside the direct control of committed organic practitioners made for a bittersweet win. According to Hoodes, "We won that one, but we got it wrong."[97]

The final rule fostered a compliance evaluation that focused on certain technical measures. Adequate conditions for livestock were based on such factors as whether or not the ribs of the animal are visible and how clean they are. While these factors can suggest mistreatment, they do not neces-sarily indicate that animals are being treated poorly. The visibility of an animal's ribs is natural at some life stages and certain times of year, and it is common for a pastured animal to get muddy following a rain. Thus, the rule that activists had sought for so long created potential hardships for the small organic farmers they sought to protect.

These seemingly inappropriate measures stem from the issue that has plagued organic practitioners since certification began. While there are forms of evaluation that utilize a systems approach, taking into consid-eration many factors associated with the particular context, the tendency in any certification regime is to develop standardized, easily measurable empirical markers of compliance. Much as schools have come to rely on standardized tests for children to indicate educational achievement, the measurement tool is not always the best means of actually capturing what one seeks to know.

This is the same issue that organic farmers confronted when they first set about creating formal standards for organic production. The earliest versions were more like philosophical statements, but these quickly devel-oped into more technical requirements with empirical measures as organic farming associations sought to have clear criteria for certification purposes. Organic rules still include a requirement to maintain a farm plan, which is used to describe how farmers are meeting organic standards for inspec-tion purposes. But the plan was originally designed to encourage farmers to think of their operations holistically and consider not only how they were going to meet organic standards but also how they would improve over time. With USDA-accredited inspectors concentrating on materials and technical measures, this aspect of the farm plan has largely been lost. As such, the question raised by organic practitioners decades ago is still relevant in contemporary organic politics: Is organic an abstract philosoph-ical approach to agriculture or an empirically measurable set of farming practices?

Despite the fact that the NOP has been in effect for a number of years, issues that proponents feel threaten the integrity of the entire system con-tinue to crop up. This occurred again in 2013 when activists learned that apple and pear growers in the Pacific Northwest were using antibiotics on their orchards to prevent fire blight, a disease that can cause exten-sive damage to orchards.[98] Antibiotics are a hot-button issue for many consumers, and organic proponents had long claimed that the organic sector was wholly antibiotic free. It came as a surprise to many, even to some veteran organic activists, that the materials designated as allowable on the national list for the prevention of fire blight were actually antibiotics.

Intensive negotiations got under way about how to quickly and quietly dispense with the use of these particular materials. As has been the case from the start of the federal program, different stakeholders had conflicting perspectives on how to handle the situation. Consumer groups wanted an immediate end to the use of the practice as soon as the materials were due to sunset from the allowable list. The growers who relied on this safeguard as well as the industry groups that depended on their products sought the common five-year extension. The NOC found its own members divided on the matter. In this instance, leaders advanced a compromise two-year extension with firm commitments that the practice would be discontinued at the end of that period.

## Industry over Movement?

Both the industry and the movement have been profoundly changed since the NOP's development. The way in which corporate actors have moved into the organic sector has led some to suggest that the movement has been overtaken by the industry—that in essence, organic as a movement is no more.[99] Yet there are a number of challenges to such a dire conclusion, the first of which is how we draw the line between movement and industry given the market-based approach that has been the centerpiece of organic strategy from the start.

In any movement based at least in part on expanding markets for a type of product, corporate actors participating in the industry cannot be cleanly separated out.[100] After all, they are producing goods in accordance with the rules that activists developed. They are educating consumers and raising consciousness about the importance of organic production. The line between pursuing profits and recruiting adherents to a set of beliefs and a way of life—that is, building a movement—can be hazy for a cause focused more on personal lifestyle choices than state action.

Lifelong organic activists can, in some instances, still be found in these corporate boardrooms, and many continue to see themselves as advancing a mission, not just running a business. Gary Hirshberg is the president and CEO of Stonyfield Farm, the third-largest yogurt company in the United States. He campaigns around the world for causes such as preventing climate change and advancing agricultural reform. The tops of Stonyfield yogurt containers bear political messages on a wide range of environmental and social issues, and 10 percent of the profits from Hirshberg's company are donated to environmental causes.[101] In these senses, Hirshberg is an activist advancing a cause. But he also runs a multimillion-dollar company now owned by international food giant Dannon. Walmart is one of Hirshberg's biggest accounts, and he sources ingredients for his yogurt from all over the world. Hirshberg is not the typical organic activist, but neither is he the typical corporate CEO.

People like Hirshberg are clearly part of Big Organic, yet as an individual, Hirshberg still may be considered part of the movement. He started off as an activist. For several years he ran an alternative living research center, but later adopted a market approach as a means to spread his message and advance his ideals. "I used to run non-profits and I had zero economic power. In my 20s I became aware of my impotence as a change agent. I realized then—if you want to change the way the world operates, you need to marshal your economic power."[102] Although he began Stonyfield, Hirshberg likens building his customer base to building a movement. As in any movement, success is measured partly by the number of adherents brought to the cause. Expanding production and growing his company therefore is a means to reach more people and spread the organic gospel.

Hirshberg and similar people complicate any movement that seeks to advance change using market mechanisms. It is difficult for organic proponents to address actors who successfully utilize the tools that they themselves adopted as their primary means to reform the system, even when the outcomes undermine fundamental elements of their vision. While many activists wince at multimillion-dollar organic enterprises, Hirshberg is not uncomfortable with the size to which some organic enterprises have grown. In fact, he celebrates it. When imagining a sustainable, wholly organic food system, Hirshberg insists, "All these dreams are made possible by scale."[103] He views his company as a social change vehicle—one that can only succeed by adopting the methods of other successful businesses. That includes not just having a good product but also producing it in a way that makes it affordable and desirable for consumers. From this perspective, the success of the movement is measured, at least in part, by the profitability of Hirshberg's company.

Attempting to separate large producers such as Hirshberg from the movement is compounded by efforts to separate small movement farmers from the organic industry. Even the most ideologically committed small farmer still has economic interests that they seek to protect. Some small and midsize growers even supply some of the agribusiness operators, or distribute at least some of their goods through national natural foods chains such as Whole Foods. Hirshberg boasts of the family farms that survive based on their relationship supplying Stonyfield. Other large corporations like to tout their support for the small farmers from whom they source a small percentage of their ingredients. While some of these big players undoubtedly exaggerate their reliance on small organic growers, some farmers have benefited from and come to rely on these relationships. These small actors are still more likely to have an ideological affinity for the goals of the organic movement, even if they are not engaged with organic issues beyond selling organic goods. But the market-based nature of the organic movement's core strategy in which the central players in the movement for social change are also businesspeople blurs the distinction between movement and industry.

Although some operations now straddle the divide between movement and industry, analysts point out that there is also a fairly apparent bifurcation within the organic sector.[104] While many tend to emphasize the size and influence of Big Organic, there has also been substantial growth in farmers' markets and CSA programs alongside the explosive growth in the conventionalized organic sector.[105] Even if one draws a rigid line between the movement and the industry, by many measures the movement is growing and is stronger than ever.

## Conclusion

The organic movement has clearly been transformed since the OFPA's passage and even more so since the start of the NOP. The creation of the national program signified and amplified the credibility that organic farming had achieved. The NOP's benefits were recognized and verified by university researchers as well as government leaders. Consumers flooded into the organic market, as did conventional agribusiness enterprises seeking to capture those customers. The acreage under organic cultivation grew dramatically along with sales of organic products. Many within the organic community celebrate these achievements, but others do not.

The growth came largely from the entry of big new players in the organic sector. These primarily conventional firms usually followed the letter of the law, but the rules that organic movement participants had devised were not designed in ways that would prevent abuses. Battles ensued. While organic activists were at times successful in defending their standards or clarifying vagaries to establish requirements as they had intended, in other instances they lost. Large food and agriculture firms were able to practice organic in ways never imagined. Many felt that their control over the cause they had fought so hard to promote was slipping from their hands.

Different segments of the organic movement adopted different strategies in response to these developments. Some are largely pleased with the direction of the organic industry. They remain steadfast in their support of the NOP and continue to fight when they feel that the standards are under threat. Others are disillusioned with the NOP, and have sought new ways to transform the food and agriculture industry. The paths chosen by these activists and the state of the movement will be explored in chapter 7. Before analyzing these strategic variations, though, we should consider the issue that lies at the core of much discontent within the activist base of the organic movement: the question of social justice.

# 6 Searching for Social Justice

The better earning class of the public will pay a high price if they can be shown its value.
—Jerome Irving Rodale, *Organic Farming and Gardening*, May 1942

The organic movement has gone through many phases, and different values and goals have taken priority among the various segments of the movement at different times. As discussed in chapter 2, health was an important consideration for some early US proponents such as Rodale. Many consumers today are still drawn to organic food out of health concerns. Environmental issues gained prominence in the 1970s, and currently remain a major motivator for numerous organic advocates and consumers.

On the health and even more so environmental fronts, there are ample benefits to organic agriculture. Its spread and mainstream acceptance signify crucial gains in these areas. Even many of those most concerned about the entry of large conventional corporate enterprises still recognize that organic methods yield significant environmental and health advantages. Rules would have to be much more seriously degraded before these returns are lost completely relative to equivalent conventional growing methods and products.

Despite consensus within the organic community about these benefits, the tillers within the movement, those who hoped for a more fundamental transformation of the food and agriculture system through the spread of organic, have serious grievances. Although Big Organic may have limited environmental benefits by introducing some conventional practices to organic farming, and despite diminished health gains through the incorporation of more synthetic materials in its overprocessed products, the *social* dimension lies at the heart of much discontent with the organic industry's

transformation. The social goals of the organic movement can be difficult
to disentangle from the myriad of other objectives and values found among
organic proponents. What is it, specifically, about organic agriculture that
suggests it can advance social justice? How is social justice being conceived?
This chapter explores the question of social justice within the organic system, especially as it pertains to family farmers, low-income consumers, and
farmworkers. It examines the competing needs of these groups and dilemmas they pose for organic advocates seeking to address issues of social justice within the context of their movement strategy.

Many are drawn to an idyllic agrarian image of a nation composed of
self-sufficient independent farmers meeting all the food needs of each local
community through ecologically sustainable practices and fair exchange
with consumers.[1] This is a vision that many hold about the United States'
past, despite the fact that such an agrarian ideal was never truly manifest. While the independent self-sufficient farmer could be found in certain places at certain times, this portrait ignores the history of slavery,
indentured servitude, the exploitation of immigrants and migrant labor,
and domination by railroad robber barons and big landowners that more
commonly characterize US agricultural history. To the extent that organic
proponents are inspired by this agrarian ideal, however, there is a sense of
justness in such a system—one in which hardworking families can make a
simple living by mixing their own labor with the land and providing their
communities with needed wholesome nourishment.

Yet most recognize that this picture leaves out other factors that mark
contemporary US society. Even if small farmers are able to make a living,
few are able to do so without at least some hired help. It is important to
consider who these hired workers are and how they are treated. There is also
the matter of low-income consumers, particularly communities of color in
urban areas, and their ability to access and partake in the farmers' bounty.
These are matters of social justice, and how the organic system addresses
the needs of each of these populations requires attention.[2]

The centrality of social justice and fairness to the organic vision should
not come as a surprise. After all, the contemporary organic movement finds
its roots in the 1960s' counterculture movement in which social justice was
the primary focus. Ecological sustainability was recognized as critical, but it
was only one component of a wholly reformed social order in which human
needs were to be met in a way that provided fairness and dignity for all.

The 1970s and 1980s saw a splintering of the broader, more encompassing movements of the prior decade. The civil rights and black power movements were joined by a host of other racially and ethnically based empowerment struggles, in addition to the growing environmental and antinuclear movements, feminism, gay rights, and a plethora of other causes. Back-to-the-landers who carried on with organic agriculture became a cause unto themselves. Although organic farmers remained isolated from the wider farming community due to their unconventional ways and counterculture roots, their common vocation would eventually bring them into closer contact with other groups seeking to protect family farming as a whole. Some viewed the development of organic agriculture as a means by which family farmers, battered by market fluctuations, unbearable debt, and federal policies that favored the big players, could once again receive a fair return for their labor.

## Justice for Family Farmers

Conventional family farms had been in crisis for decades by the time the organic movement was on the rise during the 1970s and 1980s. "Get big or get out" farm policies established following World War II, together with technological developments, led to concentration in the agricultural sector. Small farmers were forced to either invest in costly technology, specialize and expand their operations, or get caught in the squeeze with shrinking prices paid by an increasingly concentrated food industry sector with ever-growing supplies available.[3] All these options spelled hard times for farmers who were either driven out of business or into debt to finance the equipment, seed, and other costly inputs purchased from off the farm. Downward price swings would take their toll on farmers who assumed too much debt in their efforts to grow their farms into prosperity. The number of farms in the United States went from 6.8 million in 1935 to fewer than 2.2 million today.[4] Many of the surviving family farmers have been forced to seek off-farm supplemental income. Fewer than half of all farm operators now make their living exclusively from farming, and small farmers are even more likely to need paid work off the farm.[5]

The plight of the family farmer is also evident in the declining percent of the retail cost of goods returned to the original producer. This proportion has fallen from about 40 percent in 1910 to 15.8 percent today.[6] Although family farms were being lost throughout the century, the 1980s' farm crisis brought on by a sharp decline in prices had the effect of heightening public consciousness about the problem. Farm Aid benefit concerts were organized to raise funds and rally support for ailing family farmers, but few saw any solution forthcoming that would reverse the trend of family farm decline.

It was in this context that some began to see the organic agriculture sector as offering a possible refuge for family farmers. Small organic farmers were able to survive the crises that were afflicting others in the conventional market. Organic farmers did not need to purchase expensive new technologies or off-farm inputs, and so avoided a great deal of debt. And organic goods brought a price premium, giving farmers better return on their products. Expanding the organic market appeared to provide an opportunity to allow family farms to survive. Thus, in large part the social justice sentiments within the organic movement started to concentrate primarily on justice *for farmers*. How can the hardworking people who feed the nation, those being squeezed by agribusiness and banks, receive fair compensation for their efforts?

The hope that the organic market offered also captured the imaginations of many because it reflected a return to the agrarian ideal.[7] Many organic farmers were still selling closer, if not directly, to consumers. Massive scale and corporate involvement was largely absent from the organic sector in the 1980s, when conventional family farmers were being hard hit. Although it was midsize farms that suffered the most during the decades of farm crises, organic supplied a way to save family farming through an approach that many believed could not be replicated on an industrial scale. This is why many organic activists are so troubled by the changes in the organic industry. Large players were ultimately able to work their way into the organic sector. The rules that activists had helped to develop were not crafted in a way that would keep agribusiness out and keep organic production limited in scale and centered on whole foods.

There were some attempts made along the way to create a stronger bulwark against corporate involvement and protect small farmers. During the development of the California organic law as well as discussions of the OFPA and NOP, consideration was given to rules about scale and ownership—provisions that were ultimately rejected.[8] Despite the associated values of those in the organic community, organic rules had always focused on growing methods. Many were reluctant to try to add disparate new provisions as rules were being transformed into state policy. Besides, some believed that the use of proper organic methods would preclude industrial-scale production, thus there would be no need to append other restrictions. Others saw size or ownership restrictions as unnecessary as well as undesirable. Spreaders within the organic community even welcomed the idea of organic methods being adopted by large players. DiMatteo, giving voice to the spreader perspective, sees no problem with large-scale organic production and does not think the organic movement should concentrate on issues of size.

"I believe that there should be opportunities for all scale farmers," says DiMatteo. "I don't like a lot of the government policies that have … made it very difficult to survive as a family farm. I think those are all important things to work on.… Does organic have to deliver that? The answer for me is, 'no.' Organic doesn't have to deliver all those things." DiMatteo insists that scale of production or independent ownership was never directly tied to organic, and adding such considerations would hamper the movement. "It wasn't the Organic Foods Production Act [that eliminated scale or ownership considerations]; even when independent certifiers and the regional groups … [and] the private sector efforts were done, … what came forward in all the standards was that it's about how farming is done.… [W]hat do you do for soil health, and what do you do for animal health, what do you do for the environment?" DiMatteo maintains that incorporating other considerations into organic rules would have impeded the movement. "You have to stay focused.… If you've got a million different things you're trying to accomplish, you're not going to get any of them done. So even before OFPA came in, the standards had to narrow themselves to … our core of farming," she asserts. "If we didn't, we'd be still so much smaller than we are."[9]

By emphasizing farming techniques and materials alone, organic rules would not be sufficient to advance the social goals that motivated many activists. The organic market would not necessarily provide family farmers with a comparative advantage allowing for the rebuilding of a food and agriculture system based on the agrarian ideal. But it is not simply that the organic circle would widen to include large corporate players that depart from the traditional understandings of organic farming; even more concerning was that the infiltration of such actors would harm those who *do* embody that ideal. Not only would the new organic system fail to grow the number of small and midsize farms, it threatened to harm those who had been able to survive in the organic niche. Just as occurred in the conventional foods market, the entry of larger players has created a price squeeze for organic farmers. This can be seen when examining the market for many types of organic products.

Mixed salad was among the first organic products to enter the mainstream market, and this industry has been used by some to exemplify the impact of market development on small growers.[10] Mixed salad or "mesclun" was initially promoted by alternative restaurants catering to a more upscale clientele. These venues were typically supplied by small growers, some of who had exclusive relationships with restaurants that featured their close ties to farmers.[11] But entrepreneurs eventually brought bagged, premixed organic salad to a mass market.

Earthbound Farm was a leader in this industry. It was a pioneer in expanding the organic market, but in doing so introduced a conventional industry approach to the organic lettuce sector. In order to produce for a mass market, Earthbound and other big producers intensified lettuce production. They contracted with multiple growers who would produce several crops per year of the fast-growing baby greens. Some took full advantage of organic rules that allowed for the use of controversial inputs such as Chilean nitrate, a fertilizer banned in organic production in other countries, yet allowable under US rules. Processing technologies for cleaning and packing products allowed for production on a large scale.

Samuel Fromartz, in his book *Organic, Inc.*, describes the market logic driving companies like Earthbound—the same competitive pressure that characterizes market systems generally.

While Earthbound Farm had "industrialized" over the course of the 1990s—getting bigger, driving down costs, and eliminating competition each step of the way—it had done so to compete against Dole, Fresh Express, and ReadyPac, the big conventional companies in mainstream markets. Earthbound's goal was to be as efficient in harvesting, processing, bagging, marketing, and distribution as conventional produce companies, narrowing the price differential between organic and conventional lettuce and prompting more consumers to try their product.[12]

The Earthbound story is a good one if the only goal is to grow the organic market, and achieve at least some of the associated environmental and health benefits. The problem as some see it is that Earthbound proceeded to drive long-term, ideologically committed, small and midsize organic growers out of the market. The social justice, the fair return to farmers that many organic advocates had been seeking, was being undermined through the very success of their movement and the program they helped to create.

Fantle explains how he watched the transition of the organic dairy industry.

[Factory farms] can flood the market with milk and squeeze out all the smaller competitors, which has been what has happened. I'm from Wisconsin and I have watched it happen for the last thirty years [in the conventional dairy industry]. The smaller farmers are being crushed and moved out of conventional markets because of the incredible growth and consolidation of power that has taken place in the farming sector with the rise of larger and larger operations.

Producing premium-priced organic milk supplied a viable opportunity for those small producers, but Fantle complains that now industrial-scale organic producers like Aurora Organic Dairy are "generating a lower price … and undercutting brands that are produced by ethical family farmers around the country. Most consumers don't know the story about organic milk and where it's being sourced."[13]

Organic consumers do *not* know the story about organic milk because they see the same USDA organic seal on *all* organic products. Clever marketing and bucolic images on packaging further obscure the industrial-organic reality. While some may imagine that buying organic is helping to support small family farms, there is no such requirement in the organic standards. In that sense, the quest for justice for farmers in the form of fair prices is not necessarily achieved through the organic program. Like conventional family farmers, organic growers are subject to the same price squeeze that happens when big players use market power and economies of scale to capture market share.

New distribution systems that have helped to propel the growth in the organic market have also placed many small and midsize growers at a disadvantage. Prior to the NOP and the development of the industry, retailers and restaurants that carried organic goods had to use small regional distributors or purchase directly from growers. More direct linkages and shorter distances meant that small farmers could capture a higher percentage of the food dollar. With the development of the industry, national distributors moved in, displacing these local and regional networks.

One such distributor is United Natural Foods International (UNFI). It was formed in 1996 with the merger of two major regional distributors. The national merger was followed by a series of acquisitions, including a number of co-ops and smaller independent operations.[14] UNFI is now the largest wholesale organic foods distributor in the United States with clients that include the natural food retail giant Whole Foods, among many others. The availability of organic goods through distributors such as UNFI along with the consistency in products and year-round availability that they can bring has undermined the access that small organic farmers used to have to some markets.

Coody, who now works as a consultant on organic issues, notes the added distribution challenges brought by the industry's development. "Access to markets has changed quite a bit for small farmers. It's really hard. We used to be able to run around and sell things at the backdoor of restaurants, and to retailers directly, and now that doesn't happen."[15] While national distribution has certainly helped to grow the organic market and provided some advantages to consumers, many small and midsize organic farmers have suffered as a result.

This is not to suggest that no small or midsize organic farmers have benefited from the organic expansion into the mainstream market. As will be discussed in the next chapter, some activists and farmers have developed creative ways to remain economically viable in the new organic environment, usually by focusing on specialty crops or utilizing direct-to-consumer marketing. But beyond the survival (and even growth) of the small farm sector that relies on selling through innovative alternative avenues, some other small and midsize farmers have found ways to benefit from organic's spread into the conventional retail market.

First, as mentioned in the previous chapter, some large firms do source some products from smaller growers. Stonyfield claims to support thousands of family farmers through its organic yogurt operation. Even companies like Walmart purchase some percentage of their organic products from family farms—a fact that they boast about profusely.[16] Although the percentage of the retail sales cost retained by these organic farmers is small compared to direct sellers, working with Big Organic has proven a viable strategy for some family farms. How this will play out in the long term is yet to be seen. Large retailers typically seek to drive down the prices they pay for goods. Walmart in particular is notorious for squeezing suppliers. Whether any organic farmers beyond extremely large producers will be able to survive such pressure is doubtful. Even at this point it is reasonable to suspect that corporate giants keep small farm suppliers on for symbolic purposes only.

Perhaps more important is the means by which some organic family farmers have organized to capture more of the price that consumers pay for their goods. The Coulee Region Organic Producers Pool (CROPP) is an example of how some small and midsize dairy farmers have benefited from the new organic market. The CROPP cooperative was founded in southwestern Wisconsin in 1988. Originally focusing on organic vegetables, it has since grown into a major national dairy producer with a range of products sold under its Organic Valley brand name. The cooperative is made up of over 1,800 family farmers. These co-op members elect the organization's board of directors and collectively set their pay price. Although this does not fully protect them from the vicissitudes of the dairy market, the cooperative's size and democratic decision-making process provides these farmers with a measure of security and control over their economic destinies.

While some CROPP members are fairly large (one member has a herd of over 1,500, which is still a fraction of the size of the typical conventional herd), most are small or midsize. Over 80 percent have fewer than 100 cows, and only 6 have over 750.[17] Co-op membership rules require that farms be family owned and operated. The co-op also maintains standards that go beyond the USDA organic requirements, and it strives to distribute its products as close to the point of production as possible.[18] Even though CROPP is a cooperative, its size, with its national brand, may not mesh with the agrarian ideal envisioned by some organic proponents. Still, it does offer an illustration of how family farmers can achieve fair returns and remain economically viable within the new organic marketplace.

The growth of international trade in organic goods has also provided challenges and opportunities for small farmers. Organic agribusiness operations have been established in many countries. These operations then export goods into the United States, which further exacerbates the price squeeze experienced by US growers. But some small growers and peasant farmers in the global South have benefited from the organic market development. Some small farmers have organized themselves into "grower groups" in order to access the US market. Grower groups are essentially farmer cooperatives where independent farmers growing the same crops using the same methods obtain organic certification collectively. USDA-accredited external inspectors visit each member farm on initial certification, but thereafter only a percentage of the individual operations need to be inspected in order to secure certification for the entire group. Meanwhile, internal controls are used to ensure that all the individual sites are in compliance with NOP rules.

It is this element of internal oversight that at times raises concerns among consumer groups, which instead want independent inspectors to assure compliance. Although consumers were not pressing for change, NOP officials reinterpreted the rules in 2007 to require regular third-party inspections of *every* operation within a grower group—a move that would have made certification cost prohibitive for these small farmers. Organic advocates of all types mobilized to oppose the change in order to protect peasant growers, submitting a petition signed by over five hundred organizations from twenty different countries. The NOP reversed course and reestablished group certification procedures, thus allowing these small farmers to benefit from the growing international trade in organic goods.[19]

Seeking fair prices for farmers is the social justice issue that has received the most attention within the organic movement, but farmers are only one part of the social justice equation. There are also the concerns of social justice for farmworkers and ability of low-income people to access healthy food. These are vexing issues for the organic movement, in part because there is a tension between addressing the needs of these populations and the goal of ensuring that family farmers receive adequate financial return.

### Organic for All?

That the organic movement should struggle with matters of poverty and workers' rights is ironic given that the contemporary organic movement was originally propelled by radical activists who sought broad social transformation explicitly revolving around issues of inequality and social justice. Some of the earliest countercultural food activists got their start distributing food for free and seeking to ensure access to fresh, healthy, nutritious food for all.[20] The fact that organic food commands a price premium that many consider to be unaffordable is a bitter irony, although not one that should come as a surprise to activists utilizing a market-based approach to social change. As the quote at the beginning of this chapter indicates, Rodale himself saw hope for organic's growth premised on meeting the needs of "the better earning class" capable of paying a higher price for quality goods.

Organic farming was viewed as an opportunity to secure fair prices for small farmers because organic goods command a price premium in the market. The markup on organic goods relative to their conventional counterparts varies, but it can be substantial. The USDA's Economic Research Service found that the price premium for organic milk over conventional to be between 60 and 109 percent, although premiums for other items such as fresh produce were usually under 30 percent.[21] Given that food prices in general are relatively low in the United States, a 30 percent premium may not seem significant, yet it can serve as a barrier to low-income consumers and others used to prices in our cheap food system. Although most consumers indicate a willingness to pay more for organic, price is the most commonly cited barrier for those who do not purchase organic goods.[22]

Even though organic premiums may not be great in some instances, perceptions of organic food as being the expensive province of the wealthy can be difficult to overcome. Critics of organic "yuppie chow" often exaggerate the cost differential and propagate the stereotype of organic as "an elitist ... indulgence" for those immersed in their "affluent narcissism."[23] Even if just a few organic items are priced far above conventional products, it propagates the idea that all organic goods are out of reach, and deters price-conscious consumers from even considering items with an organic label or venturing down the natural foods aisle in the supermarket. Expensive specialty restaurants and upscale natural foods vendors such as Whole Foods also reinforce the stereotype through their high-markup organic goods catered to elite shoppers.

Providing low-income people with access to healthy food is obviously an issue that extends well beyond the organic sector. Numerous studies have documented the lack of access to fresh, healthy foods in impoverished communities. The history of racism and racist policies in the United States make this problem especially acute for many communities of color.[24] Redlining, racialized urban planning, and the citing of industrial development has in many instances created isolated Africa American, Latino, and immigrant communities where access to nutritious food is severely limited.[25] The USDA identifies the prevalence of "food deserts," those "urban neighborhoods and rural towns without ready access to fresh, healthy, and affordable food." These communities often lack supermarkets and grocery stores, and residents are forced to rely on fast-food restaurants or convenience stores that rarely stock fresh produce or other healthy options, let alone organic goods. The USDA estimates that 23.5 million people live in such areas in the United States.[26] Thus, access to organic food for lower-income families is extremely limited due to both physical proximity to organic food markets and the additional cost associated with most of these products.

Efforts have been made by a number of organizations seeking to increase access to healthy food in low-income areas, but these are typically carried on outside the traditional organic community. Robert Gottlieb and Anupama Joshi document the work of a wide range of organizations seeking to achieve this aspect of "food justice."[27] Nuestras Raices is one such organization. It has been operating in Holyoke, Massachusetts, for over a decade. This depressed postindustrial town is predominantly Puerto Rican, and has long suffered from a number of food security issues. Nuestras Raices started as a community garden project. Community gardens, or "urban agriculture" as those with greater ambitions refer to them, provide underserved urban communities with a way to supplement food needs, while also imparting skills and facilitating self-sufficiency.[28] They re-create the link between consumers and the production of food—a lost connection in many places, but one that is almost completely lacking in many lower-income urban areas.[29]

The community garden started in Holyoke spread into multiple gardens, then a greenhouse, restaurant, and shared communal kitchen. Eventually, Nuestras Raices developed an extensive farmer-training program and food business incubator.[30] This program has served as a model for how disadvantaged people in neglected communities can begin to meet their needs in terms of fresh, nutritious food outside the global industrial food system. While some see the turn to self-help programs as the abandonment of hope for achieving policy change at the structural level, others believe that community-based food production can be a part of a broader agricultural reform strategy.[31]

Growing Power provides another example of grassroots food justice activism. This organization, founded in Milwaukee by former basketball star Will Allen, offers education and training in urban agriculture in low-income communities. It uses profits from sales at upscale farmers' markets to subsidize food banks and CSAs supplying food-insecure populations with access to healthy fresh foods. Just Food, another food justice group, has an expansive program in New York City in which underserved communities are given access to healthy food through CSAs and farmers' markets.

It offers training in urban agriculture, healthy cooking, and food-business support.

While many of these groups utilize sustainable growing methods in their training programs and provide access to organic foods, there is little connection between these groups and the traditional organic movement. The Web site for Growing Power states that although its operates at or above USDA standards, organic certification is "not a priority.... We would all much rather be in the fields than filling out lots of paper work for the government."[32]

The disconnect between urban food justice groups and organic advocacy organizations is indicative of the fragmentation that occurred within the more encompassing social change movements of the 1960s. Just as with the organic movement, many contemporary food justice efforts can trace their ideological roots back to antipoverty and racial justice campaigns from decades before.[33] But the activists involved are rarely the same ones who dedicated themselves to defending family farms and who sought to develop sustainable agriculture practices and organic certification systems. Food justice groups more often trace their lineage back to the environmental justice campaigns that emerged in the 1980s—a movement equally distinct from the predominantly white segments of the environmental movement.[34]

Given the structure of the food market, including that for organics, the kinds of projects that food justice organizations have undertaken may offer low-income populations the only hope for access to fresh, nutritious, sustainably grown goods. Without fundamental structural change, market mechanisms would otherwise place organic food out of reach for poor communities. Meanwhile, the progressive inclinations of many activists within the organic movement fosters ongoing angst about inaccessibility and the air of elitism that surrounds an organic system that relies on higher-than-average prices for food.

In many instances, organic activists have advocated for programs designed to provide low-income people with greater access to these goods. Since 1992, the federal government has funded a farmers' market program in association with the Special Supplemental Nutrition Program for Women, Infants, and Children (WIC). This supplies a small amount of dedicated additional funding (up to thirty dollars annually beyond the standard benefits) that can be used by WIC recipients at participating farmers' markets.[35] Organic activists have long advocated support for this program.

In another attempt to expand access, many organic CSA farmers use a sliding scale for the purchase of farm shares. Some have a percentage of shares dedicated to low-income consumers, or donate surpluses to food pantries or soup kitchens.[36] Such efforts reflect organic activists' concerns regarding the larger issue of food justice, but these ad hoc measures do not address the fundamental tension that exists between seeking to support farmers by bolstering their incomes from the sale of goods and the inability of low-income people to afford those goods due to their lack of funds.

Food justice is not an issue that can be addressed by the organic movement itself. Critics are correct to point out that the organic movement has in many cases failed to give ample attention to matters of class and race. But addressing economic inequality and racial injustice is particularly challenging given the fact that organic price premiums are what allow small organic farmers to earn a living in the face of competition from low-cost conventional goods. If anything, the decline in the organic price premium advanced by large producers capitalizing on economies of scale and market power poses the greatest threat to organic movement farmers. This creates a difficult dilemma for those who aim to address the economic concerns of those on both the consuming and producing ends of the organic market relationship.

The matter of organic food is just one dimension of the politics of food justice. There are internal tensions even among those whose primary goal is to address food and nutrition issues for the poor. In the 1960s and 1970s, federal antihunger measures were wedded to agricultural policies through the Farm Bill. As a result, food aid and school lunch programs designed to serve disadvantaged communities are linked to agricultural subsidies, which primarily benefit large commodity-crop producers. Lawmakers representing urban areas with few or no ties to agribusiness interests lend support for the commodity programs, largely because such projects are bundled with those that provide food aid to their constituents. Some antihunger groups that have long been supported by nutrition programs tied to commodity-crop production are even at times reluctant to back new food justice approaches. They are concerned that funding may be siphoned off for the food aid programs on which many communities and schools have come to rely.

Between urban poverty groups and big agribusiness, the coalition of supporters with a vested interest in the industrial food system is significant.[37] This situation may change in the years to come. Conservative lawmakers have sought to sever agricultural supports from poverty programs, maintaining the former while cutting support for the latter. Although problematic in the short term, this may create new opportunities should these programs be permanently separated. In the meantime, food justice organizations face a difficult challenge when seeking support for empowerment programs that depart from the cheap food model backed by a diverse constellation of constituencies.

The organic movement has lent its voice to the food justice cause, but rarely has this issue been a top priority. The sense among many is that simply defending organic standards and trying to protect small farmers is more than organic activists can handle. As much as these proponents may want to make organic food available to all, the market-based strategy inhibits the movement's ability to advance food justice for low-income consumers.

## Laboring on the Organic Farm

The matter of social justice within the organic industry also extends to farmworkers. Early on, communes and independent family-operated farms only had to deal with issues of fairness and equity within their small operations (and scholars of the movements of that era have noted that equity was still a major problem within the activist community, including issues of race and gender). But even many family farms require additional hired labor. Fair wages within the agricultural sector are difficult to find anywhere.[38]

Agricultural workers have always been among the most exploited populations in the United States, starting with slavery and the indentured servitude of indebted European immigrants through the mid-nineteenth century, to Chinese and Japanese immigrants put to work in the fields of California's rapidly industrializing agriculture sector. These immigrant populations, along with Mexicans native to the region, were eventually supplemented by desperately impoverished victims of the Great Depression from the Midwest, dramatically portrayed in John Steinbeck's *The Grapes of Wrath*.

The Bracero Program, established during World War II, institutionalized the regular migrations from Mexico in an effort to address labor shortages caused by the war. That program of allowing the temporary migration of Mexican workers was extended through 1964 at the behest of agricultural capital, which insisted that these workers were needed to address ongoing labor shortages after the war. Mexicans, more recently joined by migrants from Central American nations and the Caribbean, continue to make up the bulk of the agricultural workforce and even much of the labor for other food industry sectors, from meatpacking to retail outlets and restaurants.

This history is characterized by hyperexploitation along with periods of intense struggle and conflict. Farmworker strikes occurred as early as 1902.[39] In 1930s' California, labor conflict intensified. The Congress of Industrial Organizations and other radical groups sought to organize farmworkers, but the Bracero Program undermined the position of workers and inhibited efforts to unionize.[40] Although labor struggles took place throughout this history (including collective actions by Bracero Program participants), it was not until the mid-1960s that farmworkers began to make real progress.

The UFW union under Cesar Chavez's leadership successfully utilized consumer boycotts and public education campaigns to pressure employers and secure labor rights. The UFW's work on pesticide issues bolstered the claims of organic proponents and did much to raise consumer consciousness about the hazards of the toxic chemicals used in conventional agriculture.[41] While claiming victory in many organizing campaigns throughout the 1960s and early 1970s, the UFW found it difficult to sustain a union composed of a largely immigrant or migrant population. During the 1980s, under Reagan, a stridently antiunion president, the UFW, not unlike the rest of the labor movement, saw significant declines from which the union has never recovered.

Federal policy has always served to place farmworkers at a disadvantage. Agricultural workers were explicitly left out of the National Labor Relations Act, the federal legislation that guarantees the rights of workers to form unions and afforded a number of employment protections. Although California, the biggest agricultural producer, passed a law in 1975 giving farmworkers the right to organize, migrant workers from Mexico or elsewhere are always vulnerable due to the threat of deportation, inhibiting the ability of many to take an open stand against abusive employers.

A labor-contracting system also makes it difficult for farmworkers to exercise any voice. Independent farm labor contractors collect fees from farm owners to provide workers, yet the contractors in turn exact a payment from the farmworkers. The contractor creates a buffer between workers and the farm owner, but the system is rife with abuse, including the withholding of wages, undercounting piecemeal production, and price gouging for housing and other supplies needed by the workers.[42]

The organic community has long discussed farmworker issues, yet explicit enforceable requirements regarding wages or the treatment of farmworkers are absent from organic rules. Along with size and ownership provisions, labor standards were among the ideals rejected for inclusion in California's organic standards as they were being formally developed.[43] OFPANA originally included some social criteria when the trade association was attempting to create national organic guidelines, but scrapped those items soon after, concluding that "the organic label could not be used to redress every problem in the food system."[44] Not surprisingly, federal organic standards contain no labor criteria.

Only very general principles regarding respect for workers' rights have ever been included in formal organic standards. For example, IFOAM has provisions that address matters of social justice. Chapter 8 of the IFOAM standards includes general principles related to social rights along with some more specific recommendations about farmworker access to health care, water, and fair wages, among other basic concerns. The standards identified in the document include globally recognized rights such as prohibitions on forced labor and the right of workers to form unions.[45] Despite including these standards, there is no established provision for inspecting growers for compliance with social issues.

Although many organic growers support farmworkers' rights in principle, there appears to be little will to include specific social requirements in federal organic standards among farm operators. A study of organic farmers in California found that most oppose adding such criteria to organic rules and feel that the financial burden of requirements such as health insurance would be too great for them to bear.[46] As such, despite the fact that organic farmworkers escape exposure to some of the toxic chemicals used in conventional production, conditions are otherwise not much better for organic farmworkers than those in the conventional agriculture industry. Wages are low and benefits such as health insurance are rare across the entire agricultural sector.

Some have found that farmworkers employed by small, ideologically committed, all-organic growers fare somewhat better than those working in mixed conventional-organic operations.[47] Small organic-only farms tend to provide more year-round employment, as opposed to the highly exploitative subcontracted seasonal employment arrangements that characterize most conventional farm labor. But even many small and midsize organic farms run by ideologically committed individuals rely on low-wage workers or young low- or no-pay "interns" who carry out farm labor in exchange for room and board along with the learning experience.

Unionization is rare across the entire farm labor sector given the lack of protections afforded under federal law and the challenges of organizing an often-transient population. Workers on one organic farm, Swanton Berry Farm in Davenport, California, are organized under the UFW, but this is exceptional. The farm's owner, Jim Cochran, essentially invited the workers to unionize. He saw the shortcomings of organic standards when it came to issues of social justice, and bemoaned the lack of consciousness about the issue among organic consumers. "Everybody cares about how the bugs are treated, but nobody cares about how the workers are treated," Cochran told one reporter.[48]

Workers at Swanton receive a living wage, benefits, access to low-cost housing, and vacation time. Cochran sees the union contract as codifying fair treatment in ways that are not accounted for in organic standards alone. He highlights the fact that his employees are unionized, and uses it as a means to raise awareness about issues regarding workers' rights.[49] "The public has no consciousness about this," he said. "It's important to raise the issue in the public domain."[50]

Remaining economically viable is a challenge for farmers who seek to provide better treatment and fair wages for their workers. Competition from large growers who drive down wages and costs whenever possible tightens margins and creates downward wage pressure across the industry. Recognizing the scope of the issue, the authors of the California organic farm labor study concluded,

To create production conditions that are favorable to a broader conception of social justice, change is needed in the entire food system, not just at the point of production. Indeed, to move beyond the silence about labor within the sustainable agriculture and organic communities, we must situate these issues in the context of the entire food chain (production, processing, distribution and consumption). Only then can we hope to envision and create agriculture that is characterized by a truly comprehensive definition of sustainability: ecologically sound, economically viable, and socially responsible.[51]

Yet even this more encompassing perspective on the agricultural system may be unrealistic. There is little reason to believe that economic justice could be established within the agricultural sector without policy reforms that would affect labor across the entire economy. Worker justice remains a vexing issue for the organic movement. Once again, the quest to secure a fair income for farmer-employers comes into conflict with paying a living wage and fostering social justice for others connected to the food and agriculture system. As with issues of poverty and access to healthy food, worker justice is not something that can be addressed in isolation within the organic agriculture sector.

## Conclusion

Organic was to provide an opportunity for small farmers to escape the wrath of the industrial agriculture system and secure a fair return for their efforts by getting a good price for a good product that could not be produced in an industrial fashion. The entry of large conventional corporate players into the organic market undermined that strategy. By introducing conventional practices, these businesses drove down prices and put many smaller organic producers in a squeeze. Some small growers still managed to survive and even thrive in the new organic market, and as will be explored in the next chapter, more and more small farmers are succeeding through innovative marketing techniques.

Still, economic survival and justice for small farmers is just one goal. Making organic food available to low-income consumers represents another challenge. Aside from some piecemeal efforts by individual farmers to get a portion of their products to needy consumers and organizational support for federal programs that provide broader access to organic goods (or at least to healthier food), organic movement activists have not found it easy to address this concern. The goal of good prices for farmers runs up against the aspiration of getting organic food to disadvantaged communities where people simply can't afford to pay a good price.

Additional contradictions emerge when it comes to farm labor. Since worker standards are not a part of the NOP, conventional players in the organic market have employed the same practices they use for all their farm laborers. Even though some ideologically driven organic growers may treat their workers better than is typical, there is no guarantee implied in the organic label. Consumers may imagine that anyone producing an organic product also has worker justice in mind, but this is not the case. Even many committed organic growers find that they cannot pay workers a living wage while also successfully competing in the marketplace.

On the whole, the cheap food model that has characterized the US food system for decades creates an insurmountable structural barrier for organic activists to overcome issues of social and economic injustice. When the goals of the movement were narrowed to agricultural issues from the grand social transformation envisioned by activists in the 1960s and 1970s, matters of social justice became that much more elusive. Organic proponents recognize this shortcoming. In a summation of the many unresolved issues of social justice, the *NOAP* states,

Concerns about farmer and farmworker rights, migrant labor, changing organic contracts, and farmer and worker wages and benefits are now being challenged. The need to find ways to institutionalize fair prices, wages and benefits; to build bridges with the worker community, and to address scale, ownership, and control of the organic sector are all viewed as critical to the long-term success and sustainability of organic agriculture.[52]

Organic agriculture in itself is likely to endure in any case. It is another question altogether as to whether the movement is capable of advancing the social values that many believed to be in some way linked to organic agriculture. What's more, *how* those values can be advanced, and whether or not that effort will move forward under the banner of organic agriculture, are even bigger questions. In light of all of these challenges, the diverse strategies adopted by different segments of the sustainable agriculture movement will be considered in the next chapter.

# 7 Strategic Innovation: The Three Trajectories of the Organic Movement

A market and a sentiment are not quite the same thing as a political movement.
—Michael Pollan, "Vote for the Dinner Party"

The activists of the organic movement laid the groundwork for the organic system that transformed the organic industry. Developments in the organic industry have in turn transformed the organic movement. I examined the movement's evolution since it was founded. It went from a following of scattered individuals drawn to the writings and speeches of a few national proponents of alternative agricultural practices to an appendage of a mass movement designed to foster broad social transformation into a movement in its own right with an initially loose collection of regionally based organizations that developed increasing focus and coordination. The political struggle for a national organic program fostered more organizational formalization and attracted allies to the cause from various other movements with a stake in agricultural reform.

The NOP's creation and the changes to the market it spawned have now yielded more new developments within the movement.[1] There are now three discernible segments of the contemporary organic movement. The food justice movement could be thought of as a fourth, except that its efforts are separate enough that it should be considered a movement in its own right. The other three branches might rightly be termed the "sustainable agriculture movement," or as some say at this point in time, the "food movement." Not all formally place organic at the center of their work. Yet all three emerged from the same organic root, and making sharp demarcations between them is difficult in any case. Rather, the distinctions can be seen in the kinds of strategies that different groups and individuals within the movement tend to emphasize. To analyze the current state of the movement, I'll return to my hometown and the three individuals intro-duced at the opening of the book—individuals who embody the directions that the movement has taken.

I live in New Paltz, New York, a small town in the Hudson River valley about eighty miles north of New York City. It is home to one of the col-leges in the state university system. Tourism is also a significant part of the economy. The Shawangunk Ridge (the "Gunks") lures rock climbers, hikers, and other outdoor enthusiasts, and fall foliage and farms draw city folks looking for a bucolic getaway. New Paltz is a predominantly white com-munity, although the college attracts some students of color and there are more diverse populations nearby, so it is not entirely homogeneous.

New Paltz is a progressive community founded in 1678 by French Hugue-nots fleeing religious persecution in Europe. A number of stone houses built by the founders still stand, allowing town boosters to boast of being home to the "oldest street in America." A more recent claim to fame occurred in 2004, when the Green Party mayor began marrying same-sex couples before it was legal to do so in New York or anywhere else in the country, gar-nering worldwide attention through this act of civil disobedience. Another notable anecdote is that New Paltz was the first community to defeat the siting of a Walmart—a struggle documented in the book *Megamall on the Hudson: Planning, Walmart, and Grassroots Resistance.*[2] Although Walmart was turned away, like the rest of the region, New Paltz is threatened by sprawl creeping up the Hudson River corridor from New York City. None-theless, being on the far reaches of a reasonable commute to the city, New Paltz is still situated in a largely agricultural area. The county is among the largest apple producers in the country, and in addition to the orchards, a number of small farms are still in operation.

In many ways the town can be seen as a microcosm of the world of organic agriculture. Alongside two conventional supermarkets, each of which has the obligatory organic and natural aisle, there are two alternative markets that carry organic produce and other products. Their names reflect the two primary sentiments that motivate organic consumers: personal health and environmental stewardship. The more upscale of the two is called Health and Nutrition. It stocks a wide array of nutritional supplements along with organic produce, dairy, and processed goods in its tidy, brightly lit aisles. While much smaller than a Whole Foods, it serves a similarly affluent clien-tele. The other is called Earthgoods. It originally opened in 1977, at a time when co-ops and other alternative food venues were establishing a footing by catering to post-1960s' skeptics of "the establishment," including the conventional food industry. It, too, carries some health-related products, but they are a small percentage of the mishmash of organic goods, produce, and other items interspersed with amateur paintings, Buddhist statues, and environmental books scattered throughout the cluttered aisles. It is situated among the yoga studios, tie-dye clothing stores, vegetarian restaurants, and coffee shops that serve the downtown alternative crowd. Even more true to the organic movement's origins is the High Falls Food Co-op a few miles away. Founded in 1976, it's one of the oldest continuously operating food co-ops in the country.

The region has also seen significant growth in farmers' markets and CSA programs. While there is no scientific proof to support the claim given the geographic units on which official statistics are gathered, some insist that the New Paltz area has among the highest per capita CSA membership in the country. New wholesale food distributors specializing in local and organic goods are also cropping up alongside conventional distributors, which in turn are adding local and organic features.

To some degree, the developments witnessed in New Paltz are evident across the United States. This is not only true of food-buying habits; it also reflects trends that have emerged within the organic movement in the last couple of decades. Those individuals introduced at the opening of the book—Liana Hoodes, Ron Khosla, and Dan Guenther—each represent the different strategies that segments of the organic movement have adopted.

There is no rigid demarcation that locks organizations and individuals into distinct strategic approaches. Hoodes, Khosla, and Guenther certainly agree much more than they differ on issues related to sustainable agriculture. But like the organizations and individuals with whom they are affiliated, these three direct their efforts to advance the cause in different ways. One continues to fight in the trenches of Washington, DC, battling it out with policy bureaucrats and conventional corporate food interlopers to defend the integrity of the official organic label. Another sees little hope in reforming the federal program. Instead, effort is directed toward creating an alternative label and certification system to replace or augment the NOP. From this perspective, what is needed are systems that encompass all that organic was meant to be, small and locally focused, and that return control of certification to the farmers and activists who have always stood at the heart of the movement. A third path is one that largely forsakes labels and certification schemes altogether. To these sustainable agriculture proponents, government policy and certification systems are secondary. The political action necessary to transform agriculture is to simply know your farmer and buy your food locally.

### Defending Organic Integrity in the NOP with Liana Hoodes

Not unlike the back-to-the-land visionaries from decades before, Hoodes and the organic advocates who she works with seek to transform the food and agricultural system in the United States. To quote from the *NOAP*, the guiding document for the policy arm of the organic movement that Hoodes coauthored, the goal is to "to establish organic as the foundation for food and agricultural production systems across the United States."[3] But unlike some other sustainable agriculture proponents, Hoodes believes that the path to get there runs through the federal government and the Washington DC policy system. Much of her work involves analyzing law and policy, organizing coalition partners, and mobilizing grassroots activists to pressure lawmakers and USDA officials to create policy aimed at advancing the organic agenda. The *NOAP* makes policy recommendations on everything from incentives for transitioning farms to organic to increased federal funding for organic research to stricter enforcement of organic standards.

Over the course of decades, many activists like Hoodes went from being organic farmers working the fields to policy experts working the hallways of DC office buildings. For years the diverse array of organizations that make up the NCSA, and now the NOC, have gathered in the days just before NOSB meetings to share ideas and plot strategy for defending organic standards. A lot of time has been spent scrutinizing policy details and ingredient lists, and arguing against proposals put forth by conventional agribusiness actors to loosen organic standards in ways that would allow them to more easily extend their reach into the lucrative organic market.

Cummins is also directly engaged in the policy world. He acknowledges that the NOP isn't all that he would like it to be. Similar to Hoodes and many others in the policy arena, his ideal is one in which small and midsize farms are the basis of our agriculture system, and he recognizes that the federal program does not deliver that. Yet he asks, "What do you do? Do you junk the USDA organic standards and come up with some other ones, or do you keep fighting to keep the standards high? We believe it's strategically best to keep fighting for high standards." When asked to reflect on his success in the policy fight he says "We're basically holding our own. For years they've tried to lower the standards, and most of the time we've been able to beat them back."[4]

Indeed, through a combination of professional lobbying and grassroots pressure, these groups have thwarted many attempts to water down organic standards, from the epic battle to defeat the Big Three to improvements in livestock handling to preventing numerous synthetic substances from making their way on to the allowable list. Grassroots support lies at the heart of any effective challenge to established moneyed interests, but having experts closely monitoring policy developments in DC is essential to the success of any national movement. These professionals inform and help to mobilize the activist base.

As described previously, the evolution of grassroots groups into professional policy-oriented entities is common among social movements of all types.[5] Virtually all movements ultimately turn to the state to institutionalize their goals. Although organic advocates were slow to develop their own policy organizations, the years spent creating the federal standards fostered new organizational forms and provided many with the skills necessary to engage in this arena. They were joined and aided by more established consumer and environmental groups that socialized organic activists into Washington's professional policy culture.

It is typical for movements in the early stages of development to be staffed by activists who rose up from the grass roots. Many of those playing key roles in organic advocacy at the federal level got their start as farmers or activists working with organic farming associations. It stands to reason that these individuals and organizations would work to defend the NOP. Many are seasoned veterans who have dedicated themselves to the cause for many years. They witnessed the growing threats to the movement as less committed actors began infiltrating the market. They helped usher in the federal legislation intended to protect their cause and then battled USDA officials who threatened to fundamentally redefine the meaning of these rules. They spent years hammering out definitions and poring over standards and ingredient lists in order to create a system they felt not only had integrity but also embodied the ideals that had motivated them for decades. It would be difficult for anyone so invested to abandon the enterprise.

Many organizations include the defense of organic standards in their mission. Organic farmers' associations such as Oregon Tilth, the Northeast Organic Dairy Producers Alliance, CCOF, and the NOFAs all seek to defend the NOP. Several of the independent organic farmers' associations that served as the primary organizational vehicles for the organic movement have created branches that now act as USDA-approved certifying agents, thereby tying them to the NOP. Other industry groups like the National Cooperative Grocers Association along with public interest organizations such as Beyond Pesticides, the Union of Concerned Scientists, and Consumers Union are also committed to defending and strengthening the NOP. Many of those striving to have a hand in national policy do so in conjunction with coalitions that work in Washington, DC, such as the National Sustainable Agriculture Coalition or the NOC.

The organic movement's DC presence is still relatively small, but it plays a vital role. Like many professionalized movements, a good deal of these groups' work involves a combination of grassroots mobilization, lobbying, and legal action. Affiliates of the NOC and similar organizations receive regular action alerts from Hoodes calling on them to sign petitions, make phone calls, or submit public comments. Most of these actions target lawmakers or the USDA directly. Professional staff members will testify at NOSB hearings, file legal complaints, and lobby elected officials to maintain or strengthen organic rules. Some of the more confrontational groups, like the Cornucopia Institute and the Organic Consumers Association, also utilize boycotts against companies that they believe fail to live up to organic principles. As discussed previously, this type of action is controversial since many professionals within the movement believe that it may damage the public image of organic and ultimately threatens to turn consumers away from certified organic products altogether. Despite some variation in tactics, though, a central intent of all these groups is to protect the integrity of the USDA organic label. The professional policy-oriented groups that have resources, expertise, funding sources, offices, and organizational structures serve as a vital link to individuals concerned about organic integrity.

Although lifelong activists lead many organic advocacy organizations, as they transition out, professionals specifically trained in policy will likely assume these leadership positions. Some view the transition of movements from grassroots mobilization to professional advocacy with skepticism, fearing that organizational staff members and leaders will become entrenched and co-opted, and more interested in maintaining their organizations and positions than advancing the original cause. But the fact is that these professionals perform a crucial role defending and advancing the hard-fought gains made by the movement. There is no telling where the organic industry would be or even what organic would mean if these groups had not joined the policy fray on behalf of organic farmers and consumers. Right now, the makeup of the movement's policy wing is in transition. Some policy professionals are in key positions, yet there are still many advocates who got their start as farmers.

The organic advocacy coalition is by this time well established, and the need to maintain focus on the national system is not a question for the segment of the organic movement that has chosen this path. People like Hoodes are not ready to give up on the NOP and the hard-won gains that have been made since the OFPA's passage. Without clear universal standards and systematic enforcement, there are no guarantees in the marketplace, and as far as the policy advocates are concerned, only the NOP can serve that function effectively. According to Joe Mendelson, Hoodes's ally and the legal director at the DC-based Center for Food Safety, "One of the things that the National Organic Program does is that it says, 'You've got to meet these standards. You're getting inspected by a certifier, and here are the standards for everyone to see.' ... That's important. There's documentation that you're meeting that standard."[6] Many in the organic movement have been occupied for almost a quarter century attempting to get that system enshrined in law and enforced by the federal government. Whether that approach will ever bring to fruition the dream of an entire agriculture system based on organic production is another question. It is one that has inspired others to seek alternative means of advancing the cause.

## Bringing Farmers Back in with Ron Khosla

The moment that the wheels were set in motion to develop a national organic program, proponents knew that if the system was not designed appropriately, or if the integrity of the organic standards was not maintained, people would begin casting about for an alternative. Most of those involved in setting the standards and developing the program feel that in spite some administrative flaws, the standards and system are themselves sound. Most of the concerns expressed by those engaged in the policy arena stem from a lack of adherence to or enforcement of NOSB recommendations, or inadequate funding. Obviously those immersed in the policy arena, like Hoodes, believe that the NOP is worth fighting for. They have dedicated themselves to defending organic integrity against corporate organic players and government officials not fully committed to the organic mission.

But inevitably, the NOP has its critics, and as expected, some of them have sought to develop alternatives. Some are highly critical and feel that the battle for organic, at least under the guise of the NOP, has essentially been lost. Organic farmer and advocate Eliot Coleman denounced the organic system upon its implementation. He wrote in *Mother Earth News* that "the U.S. Department of Agriculture–controlled national definition of 'organic' is tailored to meet the marketing needs of organizations that have no connection to the agricultural integrity organic once represented." In Coleman's view, "'organic' is dead."[7] The fact that farmers can't even use the term unless they have USDA permission demonstrates that the federal government now "owns" organic, and it is no longer worth fighting for the term or engaging with the Washington bureaucrats, politicians, and their corporate paymasters.

Khosla would not go so far as to say that the term organic is meaningless or that the NOP is a total failure; he sees it as a useful tool for supermarket shoppers seeking to avoid synthetic materials in their food. But he does not believe that the system is suitable for small farmers, and his main intent is to recapture the essence of the original movement by advancing an alternative farmer-controlled label and certification system that he feels truly reflects the agricultural and social principles once represented under the organic banner: small, sustainable, locally oriented farms.

The diminished place of farmers in the certification process is among the most significant changes in how the organic system operates. Although farmers are generally considered to be the originators of the movement and the true stewards of organic ideals, ever since the movement shifted from a grassroots-controlled alternative food project to one that engaged in state and federal politics, farmers have seen their role curtailed. The severing of organic farmer-led associations from the standards development and certification process undermined the primary organizational base that farmers had within the system.

To quote from the *NOAP*,

An absence of an organized political base for organic farmers is a growing concern. … Many view [the] USDA's regulation that prohibits organic farmers from sitting on their own organic certification boards as having "decapitated" organic farmers and stifled their meaningful participation in the organic regulatory process.[8]

It is important to recall that the removal of farmers from the certification associations was not just a measure imposed by the USDA. Consumers were a strong voice in favor of separating the role of regulators from the regulated. As with many of the perceived problems with the current organic system, the seeds were planted early on through the dynamic process of program development carried out by the diverse constituencies that make up the organic community. Khosla is among those who believe that farmers need to have a greater role. That's why he founded an alternative certification system outside the NOP in which farmers work together to ensure sustainable farming practices. As introduced in the opening, Khosla is the founder of CNG, "a Grassroots Alternative to the USDA's National Organic Program."

According to its Web site,

The USDA program requires an enormous amount of record keeping for diversified farms—a paper trail of everything that happens from seed to sale. This is appropriate for large farms that grow only a few types of vegetables and sell in bulk to chain stores or big processing plants. But it has proven excessive and largely irrelevant for diversified small farms that may grow upwards of 200 varieties of vegetables, herbs, flowers and fruits and then sell them directly to consumers right in their own neighborhood.[9]

The CNG requires adherence to the NOP growing standards, but it is a fully independent, non-USDA-accredited entity. There are no USDA-accredited inspectors, and CNG participants cannot call themselves organic or use the USDA seal. Assurances are provided to consumers on the basis of a "participatory guarantee system." Participatory guarantee systems utilize peer review to ensure compliance with agreed-on standards. Farmers who join the system have their farms inspected by other participating farmers, who in turn participate in the inspection of others.[10]

According to Khosla, many small organic farmers were driven away from the federal system because of the cost and paperwork associated with the

NOP. Billiam van Roestenberg of Liberty View Farm in New Paltz confirms that this is an impediment. He uses natural growing methods, but has never sought organic certification. "I can't afford it. It's too expensive. It's way too complicated.... It really doesn't work for small farms"[11]

NOP backers insist that costs and record keeping are not that different from organic certification systems prior to the NOP. They argue that organic farmers (indeed, *all* farmers) should be maintaining detailed records, at least for the purposes of tracking their business, if not for developing goals for improvement as called for in the farm plan requirement that has always been part of most organic certification systems. What's more, the USDA provides aid to organic farmers to offset the monetary cost of certification, although these programs have not always received adequate funding.

Riddle is the founder and first chair of the International Organic Inspectors Association, an organization started in 1991 that was central to systematizing and professionalizing organic inspection. Now he serves as the organic outreach coordinator at the University of Minnesota. He takes issue with the claim that organic certification is too onerous or costly. Riddle contends that cost-share programs make certification affordable for just about anyone. "What they object to, and this gets to the heart of the objection: it's the paperwork, it's the hassle, it's having someone else come to your farm and look over your shoulder who is not necessarily your friend or buddy, who is objective. People don't want to open up their operations to scrutiny."[12]

Blobaum was a big consumer advocate for a federal system. He supports small farmers and feels that certification may not be necessary when consumers have personal knowledge about their farmers. Yet he insists that careful record keeping and rigorous independent oversight are necessary to assure consumers that they are getting what they pay for. He sees shortcomings in certification systems that don't adhere to the strict principles found in the NOP.

They don't have all the steps; all the record keeping, all the paperwork, all the inspections, and the whole decision-making system for certification.... [The NOP has] a separation-of-functions principle. The inspectors cannot be involved in the decision making. The decision makers can't be involved in the inspection, and neither one of them can be involved in running the organization or hiring and firing.... There's this functional requirement that is what makes it truly an arm's-length, third-party system.[13]

Rangan, who oversees ecolabels at the Consumers Union, stresses that the very features that some may want to avoid are those that make for a strong certification system. "The whole reasoning behind organic is to give consumers that confidence that paperwork *is* being maintained for all the different ingredients, and your farm plan is in place, and that it's being done by someone independent." She expresses concern that the inspection of farmers by farmers "takes away the independence of it, and that's one of our criteria for rating ecolabels: Is the label independent?"[14]

NOP advocates like Riddle, Blobaum, and Rangan do not believe that the farmers who opt for the CNG or other alternatives to organic certification are corrupt; they just think that the system must be designed in a way that will provide consumers with the assurance they need. Riddle suggests that resistance by at least some small farmers is rooted in something other than practical barriers. While it is not stated explicitly, the CNG taps into the libertarian sentiment found among many small farmers who look with skepticism at anything that involves the federal government.

But Khosla argues that the CNG system has several real advantages for advancing the practice of organic agriculture and that it avoids many of the dysfunctions that come with a bureaucratized public program. While certifiers working under the USDA system are legally limited in the advice they can offer while inspecting farms, CNG operates similarly to how early organic farmers worked cooperatively with one another. The system encourages the sharing of knowledge and the development of new techniques that improve performance or enhance sustainability. The fact that inspectors are fellow local farmers fosters the development of knowledge about conditions specific to the area in which they are farming—something that is lost in the standardized procedure manuals used by USDA certifiers. The direct farmer interaction required by the system also builds the type of local agriculture community that was eroded when the autonomous collective project carried out by organic farmers' associations was replaced by the USDA-supervised system.

Still, some organic proponents remain guarded about such an alternative system. They worry that although peer review by a community of farmers sounds good, it paves the way for corruption whereby a tight-knit group could secretly agree to look the other way when bending the rules would be advantageous. In short, such a system lacks the fully independent oversight that some consumer groups consider to be important. Corruption may be less likely in a system designed for small farms serving a local market, but when close personal relations foster an environment of trust, then certification of any sort becomes less essential, rendering an inspection system moot.

CNG was not the only alternative mechanism to spring up following the development of the federal organic system. The Food Alliance was formed in Portland, Oregon, in 1997, growing out of a project of Oregon State University, Washington State University, and the Washington State Department of Agriculture. The nonprofit organization developed a certification system and standards for environmental sustainability and social responsibility. It expanded to include over three hundred certified farms and ranches across the United States as well as Canada, but in 2013, it ceased operations due to a lack of grant funding available to sustain the program.[15]

NOFA New York developed a somewhat-different approach. It created what it called the "Farmer's Pledge," a trust-based informal means by which farmers can commit to sustainable growing practices without formal certification.[16] In this case, farmers sign an affidavit attesting to their methods, including not only basic production practices such as the rejection of synthetic fertilizers and pesticides but also social values such as selling their products locally, working with the community, and respecting workers' rights. The adoption of the pledge commits farmers to allowing customers access to inspect the farm, although there is no formal inspection or certification mechanism. According to the NOFA New York Web site, the pledge "arises from the expressed need of growers who have a fundamental disagreement with the usurpation and control of the word 'organic' by the USDA, and those farmers who want to pledge to an additional philosophical statement about their growing practices."[17]

Henderson was involved in developing this alternative.

We wrote it when we realized that probably half the organic farmers in NOFA had never certified, and weren't going to certify under the National Organic Program, and that the National Organic Program, by taking total control of the label organic, was depriving all these farmers of the freedom to call themselves organic. So we wanted to provide them with something that they could do as a group to indicate to their customers that really they were organic, they just couldn't use that wording. So in writing the "Farmer's Pledge" we went way beyond what's in the National Organic Program to the many aspects of organic agriculture that we consider to be central and that were left out of the federal standards, which are just production standards, and the marketing label, but they don't cover the social aspects of organic—humane treatment of animals, humane treatment of people who work on your farm, respectful treatment of interns on your farm, how you deal with your neighbors—all those other things that are part of organic agriculture, and have been for thirty years. So the "Farmer's Pledge" was trying to recapture a broader program.[18]

According to the NOFA New York Web site, over a hundred farmers in the state have opted for the "Farmer's Pledge," but the pledge has been adopted by farmers elsewhere as well.

Those embedded in the organic policy arena are sympathetic to small organic farmers who might be burdened by organic certification costs and record-keeping requirements, yet are wary of alternatives that might compete against or otherwise undermine the organic label. Most don't believe that CNG "has the legs" to catch on and become so widespread as to threaten the NOP, but they do worry about confusion in the marketplace and are even more troubled by those with a more disingenuous agenda.

The activist organic community reacted strongly to a proposal put forth by Scientific Certification Systems (now called SCS Global Services), a California-based company that provides environmental certification, auditing, and testing services.[19] In 2007, the company launched an initiative to develop a "certified sustainably grown" label. Organic proponents were skeptical about such a nebulous claim; there is little consensus regarding what would constitute fully sustainable production. They also took issue with the lack of third-party oversight in the proposed system. And they were further troubled by the inclusion of representatives from major biotech and agribusiness corporations in the program's planning. Natalie Reitman-White, sustainability coordinator for the Organically Grown Company, told a reporter, "The fear of the new label is that it could roll back all the success that we've achieved, if it allows practices prohibited by the National Organic Program."[20]

The development of a certified sustainably grown label is still in the works, and the threat that organic defenders perceive from questionable alternative labeling schemes is ongoing. In a 2012 editorial in the *Packer*, the leading trade publication for the fruit and vegetable industry, the national editor argued that the industry should push for a USDA sustainably grown certification system, noting that such a program would allow "growers and the entire supply chain … to add value to their produce without adding as much cost as strictly organic growing methods."[21] Organic activists see such efforts as a direct threat. According to the NOAP, "The growing number of 'eco-labels' ... pose major challenges for the organic community and government regulators."[22]

Merrigan sees little threat from those outside the organic community, though she does think that alternative schemes and attacks on the NOP could destroy the system from within.

The biggest threat to the organic industry is the organic industry itself saying, "I'm more organic than organic, I'm authentic, I'm beyond organic, the federal law sucks, the rule's imperfect, and I'm disengaging, and picking up stake and calling it something different." That's the biggest threat to organic.... There are a lot of ways you could destroy this market very quickly. Monsanto could try to destroy it. They won't succeed, because it's Monsanto. The American Cattlemen can try to destroy it. They're not going to succeed, because it's not theirs to destroy. But the organic guys can destroy it themselves pretty damn fast.[23]

Therefore, the danger does not just come from bogus labeling schemes that might confuse or deceive consumers. Similar to the way in which conventional food producers initially viewed organic products as a threat, NOP defenders take issue with the suggestion that there is anything wrong with the existing organic system, as implied by an effort to create an alternative. There is concern among these organic advocates that the industry is still too young and vulnerable to withstand competition from alternative certification programs, and that consumers will become cynical about organic as critics seek to bolster their own credentials by denouncing it as a corporate-government sham.

Despite being wary of competition or bogus alternatives that could confuse consumers and undermine the organic label, many organic proponents are not wholly against additional private labels that could supplement, as opposed to displace or compete with, USDA organic. Rangan finds value in this approach.

Organic is something that is along the continuum of sustainable food production. It's not the end all, be all, and there's a lot of other labels you can look for to help add value to organic that can accomplish some of those other things. To jam it all under one label ... is too difficult. [USDA organic] is already too complicated of a label ... so I do believe in cultivating additional labels at this point.... I'm not saying that should always be the case, but when I think about consumer expectations, I try to line it up with what the law outlines in the first place. There are too many aspects of food production to get that in one shot.[24]

Fair trade is an issue that has received a lot of attention from activists in the organic community, and many support the private certification system designed to ensure respect for workers' rights abroad.[25] Some organic activists are even personally involved in efforts to create a certification system designed to address workers' rights domestically and the other social justice issues that have troubled the organic community.

One such attempt is the Agricultural Justice Project. This project was initiated in 1999 in response to the fact that the official NOP did not include criteria for social justice. Since then, a working group headed by representatives of several organizations including RAFI-USA, Comité de Apoyo a los Trabajadores Agrícolas/Farmworker Support Committee, NOFA, Florida Organic Growers/Quality Certification Services, and Fundación RENACE has developed its own standards and certification system. It held a series of meetings around the world to solicit input in an effort to develop such standards.[26] Although a set of "food justice certified" standards have been created and a pilot program has been tested in the Midwest, only a handful of farms and businesses are currently participating.

The Domestic Fair Trade Association is a similar parallel effort that involves some of the same players. Sligh sits on the board along with representatives from the Organic Consumers Association, Organic Valley, and others. It is modeled after the international fair trade movement, but participants recognize that in the agricultural sector, small farmers and farmworkers suffer similar injustices at home. The Domestic Fair Trade Association seeks to foster "a fair, equitable, and sustainable agricultural system that supports family-scale farms ... [and] farmer-led initiatives such as farmer co-ops," and to "ensure that prices received by farmers from the sale of products cover the cost of production, including living wages and benefits for the farmer, working family members, and non-family farmworkers."[27] Rather than create a formal certification system, the association evaluates fair trade programs and offers its endorsement to those that meet its standards.

Interestingly, while many policy-oriented advocates of the NOP want additional standards, especially those addressing social justice issues, most have either given up on ever seeing them incorporated into the organic system or they actually oppose involving the federal government in such certification efforts at this point. Cummins originally wanted labor standards built into the organic system, but now supports independent certification for labor practices. "You don't want [federal officials] involved. These people are not even willing to extend basic labor rights to farmworkers in the US, so we don't want to put them in charge of enforcing fair trade."[28] This is not unlike the arguments originally made by the opponents of federal involvement in the organic system.

Although Hoodes is a steadfast NOP defender, she recognizes that federal involvement means some loss of control over the system and opens the door for co-optation. "We shouldn't do another federal program.... [W]e saw what happened. The more successful it is, the more there is pressure from big business to get in and get their piece of the pie.... I don't see advocating for another federal program, but I would not get rid of this [the NOP] federal program."[29]

The development of more private certification schemes to address issues not included in the NOP presents a number of challenges, even beyond those that originally led organic advocates to seek federal oversight. Today, consumers are inundated with labels claiming to represent certification of various sorts. A directory of labels addressing environmental issues alone lists 455 labels in use in 197 countries.[30] At times it is difficult to distinguish authentic certification programs from standard advertising hype that represents no assessment or recognition by any kind of third party. Some industry groups have created their own certification systems with highly questionable standards and enforcement mechanisms.[31] These systems are often indistinguishable from independent third-party ones. But it is even difficult for many to make sense of the numerous labels that do represent authentic certification schemes.

The Consumers Union now hosts a Web site designed to help shoppers sort through the cacophony of labels and seals each representing different attributes designed to convince consumers of a product's heath qualities, or social or ecological superiority.[32] Salmon-Safe, free range, cruelty free, environmentally friendly, recyclable—these are just a few of the common claims or labeling programs that shoppers confront. Some of them represent meaningful independent third-party certification, and others are essentially meaningless or unverifiable.

"There are a lot of crappy labels out there," says Rangan. "And bad labels are not an accident," she adds. Despite the issue of overwhelming consumers and the problems presented by disingenuous profit-seeking actors, Rangan maintains that if properly designed, private certification systems can be effective. She wants the government to play a role in regulating claims, but is not committed to exclusively publicly administered programs.

There's benefits to having it be regulatory; there's benefits to having the organic program be public, but there are also benefits to private label certification schemes. In many cases you can get to higher standards faster.... [T]o the degree the market wants to innovate, private labels can provide a really fast way to get there. There is a real advantage to having these private certification programs. But ... we want them all to meet [appropriate standards].[33]

Some within the organic movement have adopted alternative private certification systems and supplemental labels. Most, like Khosla, are seeking to address perceived problems with the federal organic system. Some are critical of those they perceive as competing with the USDA organic system, but many see a role for private certification systems that deal with issues not addressed within the NOP. Interestingly, it appears that no one wants new federally administered certification programs or to give the federal government a major role in any new certification efforts. Yet there is a reason that many supported the development of the NOP, and the types of problems encountered within the private organic certification system have not gone away. Consumer confusion along with the risk of corruption and co-optation exist whenever the force of law is not there to oversee marketing claims.[34] In the face of the many problems associated with certification and labeling, be it the federally administrated NOP or private systems, some have opted to abandon this approach altogether. I now turn to the organic activists who have sought other means to advance their values.

### Building Sustainable Community with Dan Guenther

Dan Guenther embodies the third trajectory of the organic movement. Like Hoodes and Khosla, he sees significant problems with the industrial food system. He is part of a branch of the organic movement that prioritizes the relocalization of agriculture.[35] The environmental and health benefits associated with organic methods are important to him, although his efforts have as much to do with connection to community and the land as they do with food per se. Nonetheless, food and agriculture are central vehicles for re-creating those bonds. Most Americans have lost any sense of connection to farming or the source of their food due to transportation improvements, urbanization, consolidation in the food industry, and developments in agricultural technology and food science.[36] Guenther wants to reconnect people with farmers and their food in ways that would surely improve health and environmental protection, but would also foster a renewed sense of community. As mentioned in the opening chapter, he sought to achieve this by starting a CSA project in New Paltz.

CSA programs have been around in the United States since the mid-1980s.[37] The concept was imported from Europe and Japan, where communities seeking access to fresh, healthy food partnered with farmers who were threatened by unstable markets.[38] Sometimes called "subscription farms," these programs allow consumers to prepurchase a share of the farm's harvest. The share, most commonly consisting of a variety of vegetables, is usually picked up weekly or delivered to the subscribers throughout the growing season. Some farms also offer fruit or flower shares under various additional purchasing options, and some specialize in specific products such as meat.

As with farmers' markets, the direct sales allow the farmer to capture a larger share of the purchase price, but CSAs have the added benefit of distributing risk due to bad weather or other problems on the farm. The farmers get their funds up front from subscribers, so there is less need to take out loans. And should there be a bad season, all the members absorb the cost through smaller shares. Thus, CSA farmers are not under the constant threat of market fluctuations or poor weather destroying the value of their assets and plunging them into bankruptcy.

The CSA farms that Guenther started (three in all, in addition to assisting others to establish such programs) were more than just farms. Each of them have a clear educational mission. The first, the Phillies Bridge Farm Project, runs educational programs geared toward an elementary-school-age audience. "Farmer Dan" wanted children to understand that food does not just appear wrapped in plastic or come in a Styrofoam box. He didn't stop with local schoolchildren, though. "After four years I left Phillies Bridge thinking that maybe my mission in life was to start these farms."[39] Guenther went on to create a CSA farm at Vassar College across the Hudson River from New Paltz in the city of Poughkeepsie. He was especially pleased to be able to develop such an operation in an urban area given the potential to help address food justice issues and to reach out to people with little access to farms or farming. From Poughkeepsie, Guenther returned to New Paltz, where he began the Brook Farm Project, a CSA and educational program designed to reach the working- and middle-class students of the State University of New York.

Guenther always used organic methods on his farms, and supported the creation of the federal program to prevent the corruption of the concept at the hands of corporate agribusiness. But he had little use for the NOP himself, and never sought USDA organic certification or certification of any sort. Federal standards and certification are simply not relevant to the approach Guenther takes to fostering agricultural and social transformation. He observes,

The main reason not to be certified is that with CSA, it is unnecessary. With CSA, it is most important that you know the philosophy of the farmer. I can get all sorts of certification and all sorts of documents, and I can still go out there and put poisons on the field and no one is really going to know it. So it's better to know that the person who is growing your food has a certain philosophy than it is to see a certain document.

Guenther's CSA members rarely asked whether or not he was USDA certified. "Organic is not as important to a lot of people anymore."[40]

For Guenther and many like him, building stronger communities and connections to the land are most important; the rest will follow. People who eat locally also eat seasonally. They get to know what the land can produce at different times of the year, and begin to internalize those rhythms. They develop a new appreciation for their surroundings, and with it, a greater awareness of the need for environmental protection. Perhaps more significant are the bonds of community formed around local food systems.

The notion that a local system of food production and distribution can help to strengthen community is not new, nor is local advocacy limited to agriculture. There are campaigns to advance localism in other sectors such as the media, retailing, and energy systems. As such, to some degree local agriculture can be situated within a broader effort by those responding to the social and environmental threats posed by globalization and corporate domination.[41] Whether relocalizing the economy can actually help to bring about sustainability and social justice is open to question, yet proponents like Guenther begin from the position that we need to foster closer social bonds if we are to address many of the problems we currently face.

Observers have long recognized the decline in community connections in recent decades, and have noted the threats that this poses to democracy and community well-being.[42] This is often linked to changes in social organization, and most significantly, a decrease in membership in community associations.[43] People no longer belong to social clubs, political parties, or religious organizations to the extent they did in the past—affiliations through which they developed a strong sense of community. But there is also a substantial literature on the relationship between economic organization and social bonds.[44] Research has shown that in communities where the economy is characterized by small, well-integrated firms, there is greater civic engagement.[45] Some have identified the importance of small, local retail establishments such as cafés, barbershops, and grocery stores, where community members can come together, interact, and discuss matters of community significance. It is through these interactions that citizens form bonds of solidarity and trust.[46]

The idea that this type of economic organization yields social benefits is based on the concept of "embeddedness": the way in which market exchanges are mediated by other social ties.[47] In contemporary society, much economic exchange occurs in an anonymous and detached manner. We purchase mass-produced goods made in distant places by nameless workers while global corporations reap much of the benefit. Given scale and turnover, one might regularly shop at a big-box store and never encounter the same people twice, employees or customers. Communication is typically limited to the most minimal instrumental interactions. It is more common that with smaller, independently owned enterprises, customers may be familiar with employees or even the owner of such establishments. They might encounter fellow shoppers who are also neighbors or friends. This familiarity can influence the character of the economic activity, changing it from one in which buyers and sellers are trying to maximize personal gain, to one where mutual benefit, social support, and community well-being are given greater attention.

Proponents of local food systems see this as an ideal way to reestablish community bonds. This is sometimes referred to as "civic agriculture" based on the belief that having a small-scale, locally oriented farming and food system builds community as well as increases civic engagement.[48] And there is evidence to suggest that this works.[49] Some studies have shown that there are associations between peoples' participation in local food systems and increased involvement in local affairs, such as volunteering and voting.[50]

Sometimes local food advocacy is seen as distinct from organic politics, but it is ultimately rooted in the same tradition. Many of the family farm organizations that sought to save small-scale farms during the 1980s' farm crisis supported organic as a means to provide small farmers with a market of value-added goods that would allow them to compete alongside cheap, conventionally grown food. But the local food movement really took off in the United States after the NOP was implemented. In many instances, local food advocates were responding directly to the perceived shortcomings of the federal system, which included no provisions for farm size or ownership structure.[51]

The term "locavore" was coined in 2005 by some early enthusiasts of trying to orient one's entire diet around local, seasonably available foods.[52] The "slow food" movement is a related development that started in Europe during the late 1980s.[53] While similar in its advocacy of local eating, the slow food movement was primarily a response to globalization and the growing prominence of international fast-food chains in Europe.[54] Cultural pride, as much as environmental stewardship or health, motivated the desire to protect and preserve indigenous food traditions. Yet its appeal overlaps in many ways with the values long held by organic advocates. Slow Food USA was not founded until 2000. It now has over two hundred chapters.[55] The formal objectives of the movement reflect traditional organic ideals.

Guenther is correct when he claims that many people aren't that concerned about organic or other certification schemes and that they are instead seeking connections that can be made through a local food system. A Hartman Group study found that many core organic consumers are not satisfied with USDA-certified organic, noting that

consumers in the Core have moved beyond organic. They still see organic as a positive, but it no longer satisfies their social, environmental and personal health standards. These consumers are seeking foods that are traceable through personal relationships with farmers in order to know that they truly align with all of their standards. These consumers are less concerned with certifications, which they see as being insufficient to encompass all of the attributes they are seeking.[56]

Like Guenther, these conscientious consumers still believe that organic certification is useful, especially as a gateway for those who are less informed about food issues, but first and foremost, the activists associated with this approach want to relocalize the food system.

For a long time the notion of buying local was inherent to buying organic. This was in part because in the early years, there was no significant distribution of organic goods over any distance. Even when goods were not purchased directly from a local farmer, however, organic consumers felt a sense of connection based on the meaning and values associated with the organic label. Kirschenmann, who now works on organic issues at Iowa State University, recalls the day when the organic label conveyed that sense of social connection.

Originally, people who bought organic food ... felt they had some kind of connection to the farmers.... [P]eople had some sense about where the food came from. Now, you can go into a Whole Foods and you pick up a frozen chicken TV dinner, and it doesn't make any difference which company is manufacturing it or processing it. You don't know how the chickens were raised. You don't know who the farmers were that produced it. It's just another product with a label, except it has organic on it. And if you look close at the label, it says a few less processing additives, etc., but it's losing its differentiation as it increasingly becomes just another commodity.[57]

In essence, organic used to serve as its own brand, inherently conveying a local connotation. Whether consumers knew the particular farmer or not, they identified with them in the abstract due to their growing practices. According to Kirschenmann and the survey responses of conscientious consumers, much of that has been lost.

Activists in the local food movement recognize that organic no longer provides that sense of connection, and are seeking to reestablish that tie by creating literal links between farmers and consumers. They are doing so largely in ways that sidestep policy and the state altogether. As Guenther reports, he is not one for lobbying or writing letters.[58] Much of the local food movement is manifest not in organizations or coordinated struggle that would be considered political in the traditional sense. Activists in the local food branch of the movement do things like Guenther, such as starting farms, educating the public, and seeking ways to build connections between consumers and local farmers.

The success that activists have had in spreading this idea is evident in the explosive growth in CSA programs and farmers' markets around the country. The number of farmers' markets in the United States has more than quadrupled in the last twenty years, going from 1,755 in 1994 to 7,864 in 2012. There has been an even greater rise in the number of CSAs since the base was set at zero for this arrangement that did not exist in the United States just a few decades ago. The USDA reported that there were over 12,000 CSA operations in 2007.[59] This type of direct marketing is what has enabled many small farmers to survive and even thrive, thus helping to address one of the primary concerns of the organic movement.[60] The agricultural census showed an increase in the number of small farms between 1997 and 2007.[61] Even more promising, these farms are disproportionately run by women, young people, and immigrants—groups that have been underrepresented in these roles within agriculture.

Yet as with certified organic, the local food effort has not yielded as much benefit for midsize farms. The greatest growth has been in the number of small farms. In the small farm category, as defined by the USDA, those making under $250,000 in sales, the growth comes from those making less than $10,000, while those making between $10,000 and $250,000 in sales saw slight decreases. In terms of acreage, farms smaller than 50 acres or larger than 2,000 increased in number, while the number of those in between declined. Small farms dominate the direct marketing of both local and organic goods, while large farms take the lion's share of indirect sales through conventional supermarkets and natural foods stores.[62] Many midsize farms are still struggling with how to survive in the increasingly bifurcated market. They're too big to sell their goods directly to consumers, but too small to compete against the large players.

Some are seeking ways to make local and organic work for all family farms, not just those small operations dependent on direct marketing. Kirschenmann is among those concerned about the "agriculture of the middle," those midsize family farms that are still disappearing from the US agricultural landscape. He and his colleagues at Iowa State University see hope in creating "values-based food supply chains" in which consumers feel connected to farmers not directly, as is the case with small operations, but instead through brands linked to farmers and a whole set of strategic partners within the supply chain—all committed to certain just and sustainable practices.[63]

Organic Valley, the dairy cooperative composed of family farmers discussed in the last chapter, is an example of such an enterprise. Through such arrangements, midsize family farmers may be able to capitalize on the indirect sales market now dominated by big agribusiness players. This is challenging given the array of actors that must cooperate throughout the supply chain. It is also dependent on these alliances being able to distinguish themselves and attract consumers committed to the principles to which they adhere. This will be difficult because consumers are now subject to so many labels and marketing claims. But Kirschenmann and his colleagues have identified other operators that have successfully deployed the values-based strategy, such as Red Tomato, a nonprofit entity that distributes fair trade fruits and vegetables for thirty-five family farmers in the northeastern United States, and Country Natural Beef, a hundred-member rancher cooperative in the Northwest.[64] The values-based food supply chain concept relies on identifying brands with certain ethical principles such as organic, ecological sustainability, or fair trade. This is akin to what David Hess advocates when he calls for a "global localism," in which local purchasing is preferred, but where other principles are incorporated as consumers buy goods made by small or midsize independent producers in other regions or parts of the world.[65] Yet many of the efforts carried out by advocates of local food on the selling end rely on localness alone as a marketing strategy.

Communities in Support of Sustaining Agriculture (CISA), a program in the Pioneer Valley of Massachusetts, traces its roots back to the early 1990s. It incorporated as a nonprofit organization in 1999, when it launched what it calls "the longest running buy local campaign in the country."[66] As the name implies, though, the campaign is about sustaining agriculture not sustain*able* agriculture. The core of the effort is simply a marketing campaign that taps into consumer interest in purchasing locally produced goods. Affiliates are not required to use organic methods or adopt any particular practices other than producing within the specified region.

The Rondout Valley Growers Association, composed of farmers and small producers in the area around New Paltz, operates similarly. It is primarily a business association with a collective marketing campaign dependent on the "buy local" appeal. The association also offers its affiliates expert symposia on how to diversify income streams through such means as agrotourism and niche marketing. Some farmer members use organic methods, but not all. The primary intent is to garner support for local enterprises. This aspect of local food advocacy is troubling to some proponents of certified organic.

Hoodes is concerned that some local food advocates have forgotten the reason why organic certification was developed in the first place.

I have gone out of my way for 30 years to buy local, so I am a true believer in local. But do not mistake the agricultural practices or the health of the food for how close it's grown to you. The basis of "organic" is a set of codified standards that is transparent; with "local" we don't know. People may forget to ask how [the product] is grown.[67]

As the Hartman study indicates, some hard-core organic consumers looking to go "beyond organic" will likely seek confirmation that local farmers are using sustainable practices, but there is evidence that buy local efforts coupled with critiques of Big Organic are driving many consumers to prioritize local at the expense of organic, and not all of them are aware of the practices being used by their local farmers. A 2008 survey of natural food store shoppers found that 35 percent favor local over organic while only 22 percent said that they would prioritize organic if forced to choose.[68] In another study, a similar preference ordering was discovered among CSA members and farmers' market shoppers.[69]

DiMatteo sees problems in the way in which local is being promoted and is, in some ways, coming to displace organic. She is from the area in which CISA operates.

When CISA first got launched out where I live in Massachusetts, every farmer who had some reason to not like organic jumped all over that. They said, "See? Organic's not good! We're good because we are your local farmer. It doesn't matter that we're still putting chemicals all over the fields." All of a sudden, all of the environmental things that we were trying to do that are, for me, the core of organic, were tossed out the window.... All the people who wouldn't buy organic because it was elitist or hippie or whatever the reasons, they all jumped on the bandwagon and said, "Now I can buy from my local farmer again without feeling guilty. Because local's beautiful." Well local *is* beautiful, I agree, but where's the incentive for these guys to do better than what they were doing before? There isn't any.[70]

Mendelson is concerned that some local food advocates are losing sight of the importance of certification and oversight provided in the organic system.

If I'm at the farmers' market and I know the person, and I've been to their farm and I can trust what they're doing, that's OK. I can understand that. That's great. We want to build those relationships. But not everyone is like that. I've been to farmers' markets where people are buying wholesale produce at some wholesale outlet, and then selling it on a stand with the idea that they produced it, when nothing could be further from the truth.[71]

This raises the same issues that arose when organic was gaining popularity in the 1980s and fraudulent operators grew in number. Consumer interest draws the attention of those willing to capitalize on it via corrupt means. Simply having a stall at a farmers' market or placing a locally grown sign over produce in a grocery store offers consumers little guarantee that the claim is valid or even what it means.

Local is a particularly ambiguous term in regard to distance, and for goods other than fresh produce, one might also question whether and what ingredients in the food product were grown locally, or whether just some aspect of the production was carried out locally. The potential for "localwashing"—that is, fraudulent or questionable claims to localness similar to environmental greenwashing—is great.[72] One college food service provider, for instance, included Pepsi on its list of local products because the soda was bottled nearby. If one considers supporting the local economy as the sole basis for preferring local goods, then perhaps locally bottled Pepsi should qualify. But buying a corn syrup concoction that was mixed with carbonated water locally is not what most consumers have in mind when they seek to express their support for sustainable agriculture. It is this ambiguity and lack of formal oversight that troubles those who believe strongly in the importance of certification systems. The informality and vagaries of the local food cause can be strengths in that powerful political actors cannot commandeer policy. Yet the lack of clear definitions, policy, or even formal organizations that can articulate specific goals and strategy renders the local food movement vulnerable to other forms of co-optation.

Despite being willing to forgo organic certification at times, many local food consumers are motivated by values similar to those that inspire many organic proponents.[73] As with organic, environmental considerations factor into local buying preferences. The concept of "food miles," how far a product must travel from farm to plate, has gained a great deal of currency in recent years based on the fact that it requires energy to transport food great distances. Thus, eating local is thought to yield environmental advantages in addition to social and economic benefits.

The most common statistic cited in this regard is that products typically travel fifteen hundred miles in the United States before reaching the final consumer. This was based on a study conducted at Iowa State University's Leopold Center for Sustainable Agriculture in which researchers tracked the shipping distance for fruits and vegetables grown throughout the United States to Chicago.[74] Although some have questioned the methodology for the calculation, a more significant challenge is presented by those skeptical about the importance of food miles, at least in terms of energy usage and the associated environmental impacts.[75]

Transportation is just one factor in the total energy used in food production, and some research shows energy savings can be had, even when shipping foods a great distance, depending on the growing conditions and farming practices used in different locales. One study asserted that shipping items, such as apples or lamb, from New Zealand to the United Kingdom was more energy efficient than UK-produced goods due to the differences in production and handling methods.[76] A USDA study demonstrated that transportation energy is a relatively small percentage of the overall energy input in food production; the greatest use of energy occurs in the kitchen when people are cooking and doing other preparations before eating at home.[77]

Besides overemphasizing food miles, some analysts criticize the local food movement on several other grounds.[78] Some charge that this branch of the movement has come to "fetishize" the concept of local, assuming that local is always superior, and failing to develop a broader sense of justice and the social good. This risks perpetuating class and racial inequalities as relatively privileged communities close themselves off, securing their own access to desired foods while doing nothing to address injustices in the larger food and agriculture system. Hence, not only are environmental efficiencies potentially lost as farmers attempt to meet local needs by growing crops that may not be ideal for the particular climate or soil; an exclusive focus on the local risks ignoring inequalities that pervade the food and agriculture system.

Many proponents of certified organic also advocate for local buying, but have their own concerns about the overemphasis on local. They fear that gains made through the organic system may be lost. According to Riddle,

[Local and organic] are totally consistent, and any kind of choice that's presented is a false dichotomy. I want local food to be grown organically. I don't want a GMO-confinement hog system polluting the environment next door to me.... To me it needs to be local *and* organic. Those are fully compatible values.... If you focus on local, you're ignoring so many of the ecological benefits that organic brings.[79]

Organic farmers in Vermont have found a way to maintain the association between local and organic. Despite being early skeptics of creating a federal system, Vermont's organic leaders recognize that the NOP is now the only game in town. The organic farming association in the state has rejected the creation of a farmer's pledge or other alternatives to certification. According to Wonnacott, "The issue is consumer confusion.... [I]f you have [an alternative] program, then it puts a shadow on all of the certified growers.... We didn't see any reason to support that kind of defection."

She adds that many dairy farmers in Vermont ship milk out of state, thus out of necessity, they have to participate in the NOP. "Even though we weren't 100 percent behind this federal program, we felt that for us to make a big stink and [develop alternatives], it would really, in the end, confuse consumers and work against the farmers we care about. It's not the farmers fault that there is a national program."[80]

Vermont farmers have been able to take advantage of a provision in the federal organic law that allows for the use of association-specific labels. They are still subject to the same organic standards. They are still USDA certified and able to use the term organic. But rather than use the standard green and white USDA seal, in Vermont most organic products carry a seal that prominently features the state name. In this way, Vermont growers are able to inform customers that they are officially organic with all the assurances that entails, while at the same time tapping into the eat local sensibility by identifying goods produced in the state. "We decided that we really need to make a strong connection between local and organic, that in Vermont you can eat locally *and* organically, that they are really one and the same."[81]

Vermont is unique in using the system to build the connection between local and organic. Most local food advocacy lacks any kind of official structure or association, and this raises interesting questions about social change strategy. For most consumer participants in this effort, all that is involved in being part of the local food movement is purchasing specified goods. Some refer to this and related buying habits as "political consumerism," the act of directing individual spending toward products that in some way embody or support a cause or value.[82] In some sense, this is fundamental to the entire organic movement. But the policy arm of the organic movement relies on more than that. Organized collective action directed toward the state is carried out by several organic advocacy organizations. Those organizations have staff and board members—people who strategize together about how to influence laws and shape policies, not simply how to steer consumer purchasing. Activists actually belong to identifiable groups, even if that membership only means subscribing to a Listserv. They are called on to engage in coordinated collective action, sign petitions, or write letters directed at lawmakers, government officials, or corporate CEOs. Even alternative private labels like CNG require coordinated action on the part of participating farmers and those who manage the system.

More than the other branches of the contemporary organic movement, the local food wing is lacking in the organizational structures that commonly characterize efforts to advance social change. Instead, for most people who identify with this cause, support consists of simply buying local, and for the activists within this branch, activism involves educational efforts that encourage other people to buy local.

Those who would be considered leaders of this diffuse wing of the movement, such as author Michael Pollan, are not necessarily in official positions of any kind. Instead, they serve as high-profile commentators or advocates for a general cultural shift in our relationship to food. Some have raised concerns about a movement led in such a way.[83] But even Pollan is skeptical about whether these efforts actually represent a real movement. He mentions being challenged by President Obama, who although expressing some support in principle, has questioned whether food advocacy is coherent and organized enough to pressure for real change. In an article leading up to the failed 2012 California referendum that would have required labeling on foods containing genetically modified ingredients, Pollan notes a number of scattered policy struggles, but concludes that the food movement is primarily engaged in "soft politics."

[The food] movement has so far had more success in building an alternative food chain than it has in winning substantive changes from Big Food or Washington. In the last couple of decades, a new economy of farmers' markets, community-supported agriculture, … and sustainable farming has changed the way millions of Americans eat and think about food. From this perspective, the food movement is an economic and a social movement, and as such has made important gains. People by the millions have begun, as the slogan goes, to vote with their forks in favor of more sustainably and humanely produced food, and against agribusiness.[84]

Although the local food branch of the movement has little organizational structure or policy focus, there has been the occasional policy response to these sentiments.[85] The USDA's Know Your Farmer, Know Your Food program, initiated by Merrigan when she was President Obama's deputy secretary at the USDA, coordinates grant funding along with other support for the development of local food hubs, farmers' markets, and local marketing programs. There is now more in the way of technical support for small farmers from extension services. Many states have also enacted policies such as local procurement requirements in response to the growing interest in local food.[86]

In many ways, however, the change happening in this regard is still a market-oriented shift for those who choose or can afford to participate in it. As with organic, critics charge that the local food strategy fails to advance an agricultural system that would be truly sustainable and socially just.[87] The strategy can yield some environmental benefits as food miles are reduced. It can help to improve the health of those who consume fresher food. And it can preserve an agricultural landscape for those who live among the farms saved from bankruptcy. The local food strategy can even help to bring justice to some small farmers by providing them with a livable income. Yet critics maintain that the local food strategy is likely to remain the providence of the highly educated and elite. In the meantime, food justice efforts designed to supply poor communities with access to healthy food remain largely separate from this activity.[88] While some see greater potential in this type of local food provision, many of these programs concentrate on self-help strategies that fail to challenge the policies that reinforce an unjust and unsustainable global food system.[89]

Mobilizing around policy changes, like required GMO labeling, or reforms in nutrition and school lunch programs represent broader institutional changes that yield benefits for everyone, and this, from Pollan's perspective, is what would truly constitute a movement. Pollan attributes the lack of policy reform in part to the fact that at least some elements of what he calls a general food movement are relatively young (setting aside the fact that some organic associations are over forty years old). He observes that the food movement does not yet have its Sierra Club or National Rifle Association with real political muscle. He also takes solace in the fact that places like farmers' markets foster the kind of social interaction that can increase civic engagement and lead to more concerted political action— a finding confirmed by recent research.[90] Whether and what kind of institutional changes will emerge out of all this is yet to be seen.

## Conclusion

Taking stock of the present situation, the organic movement, broadly conceived, now has a well-established policy branch consisting of a number of professional advocacy organizations and coalitions with a real presence in Washington, DC. It has a host of groups developing labels and certification schemes addressing everything from fair trade to fish friendliness.

And it has a diffused but formidable wing composed of individual as well as some group efforts to promote local agriculture and food production—a development that has significant consumer support. There are some ties between these groups and those that make up the food justice movement, and strengthening such connections would be mutually beneficial, but given both movements' distinct origins and loose connections, they are appropriately considered separate.

The organic movement has come a long way since it was composed of a handful of back-to-the-land hippies with visions of transforming the food and agriculture system one heirloom tomato at a time. The current state of affairs is the product of decades of struggle on the part of thousands of activists—activists whose efforts have shaped and been shaped by government action, big agribusiness, and the broader community of consumers. In some ways the movement's strategy has evolved, although in other regards, organic activists are using the same essential approach that some were utilizing almost a century ago. In the final analysis the questions to consider are, Has it worked and will it work?

# 8 The Road Not Taken and the Road Ahead

We didn't really have anything clearly in mind.... It's kind of like we were on this whitewater raft with one broken paddle. We managed to come out in the end totally soaked and with some bruises, but we weren't dead.

—Mark Lipson, Organic Farming Research Foundation

The organic movement demonstrates the way in which strategies can shape and are shaped by the circumstances under which they develop. The present state of the organic movement can be understood by examining the ideas and beliefs of the actors involved at various moments, and the way they reacted to the political, economic, and social conditions under which they found themselves. Each strategic turn can be understood in its context, but the overall trajectory was not part of anyone's plan. This is also true for the movement's opponents, some of who, in the end, embraced some of the very ideas and practices they fought so stridently against in prior years.

There is no question that this movement has achieved a great deal. Millions of acres of farmland have been converted to organic production as a result of organic activists' work. Tons of poisonous chemicals have been kept off our food and out of our water and air. Thousands of small farmers have been able to maintain their livelihoods, and thousands more have chosen a life producing healthy food in a sustainable way. More people have access to fresh, nutritious foods grown right in their communities. Indeed, the entire way in which we think about food has been transformed as a result of the dedication and hard work of organic proponents who persevered, even when they were ridiculed and marginalized from much of society. Those who struggled for all those years, from the Rodales through the back-to-the-land hippies right up to those who ultimately created organic associations and formalized the national program, all deserve credit for bringing the importance of what we eat and how we produce it to the public consciousness. Even those conscious eaters not directly associated with organic proper have been influenced by the work of those who toiled for decades to move agricultural sustainability and food issues to the fore.

The preceding chapters provide an analysis of how the movement's actions, understood within their social, economic, and political context, led us to the current state of affairs. Although the organic movement should be credited with many of the positive changes that have occurred in recent years regarding how we view farms and food, some critics express concern about where this movement has taken us. As examined throughout this book, how we evaluate the present circumstances is subject to much debate.

Some see a diverse, thriving movement bringing us closer to the just and sustainable food and agriculture system long sought by organic proponents. The National Organic Program has legitimized the approach advocated by farmers once mocked by establishment figures in industry, government, and research. In addition, the local food movement, food justice advocates, and those developing new standards and certification systems are pushing us beyond organic by bringing additional considerations to the public consciousness, further strengthening a system of more just and sustainable agriculture, and expanding ways in which consumers can drive the cause forward. Many see hope in these developments, and much of this activity can be traced to the work of organic proponents.

Others take a dim view of the present situation. They see a fractured and defeated movement—one that put all its eggs in the organic basket only to see the federal government and big agribusiness co-opt as well as undermine the system painstaking built over several decades. Organic activists have been left wringing their hands about where they went wrong, and casting about to identify new ways to salvage and rebuild their vision. This is not to suggest that those disappointed with the NOP have given up on reforming the food and agriculture system; many of them have made their way into the local food movement or sought alternative mechanisms for fostering sustainable practices. But they place little value on the NOP or official organic, and see much of the effort that the movement expended as wasted.

Opinions about the present situation and how to move forward are inextricably tied to assessments of whether it was strategically wise to involve the federal government in the organic system. Engaging in this type of politics was a major departure for a movement long based on personal and institutional reforms sought outside the state. Looking back at the earliest days of the organic movement in the United States, advocacy was carried out through writing and lectures geared toward the general public. It was an educational effort designed to alter the individual behavior of farmers and consumers. Although early organic adherents occasionally weighed in on policy matters, these pioneers did not intend to establish a national organic agriculture system through government action.

Counterculture radicals who adopted the organic cause in the 1960s and 1970s brought with them many other values and beliefs, but their prefigurative approach to social change was consistent with that of their organic predecessors. Personal transformation and building alternative institutions was the way forward, not getting drawn into policy debates and legislative politics in a system viewed as fundamentally corrupt. The goal was to build a new society from the ground up. Communal farms and food co-ops represented the foundation of a new social order—one that did not have to be legislated but instead could be created by living it. Organic farming practices were part of the new society that was being born.

This engagement in prefigurative politics as a social change strategy made sense given the deep skepticism regarding dominant institutions at that historical juncture. Although part of a different ideological tradition, the antigovernment sentiments were also consistent with agrarian ideals common among independent farmers in the United States, a small percentage of who were already utilizing organic methods. This separateness and alienation from dominant institutions was reinforced by the isolation of organic adherents from establishment figures of all sorts—agricultural scientists, government officials, conventional farmers, and agribusiness firms, many of who were highly critical of organic proponents, viewing them as irrationally opposed to social and technological progress.[1]

Communal farms floundered and food co-ops struggled, but even as the grand social vision of the 1960s' radicals faded, many organic farmers remained committed to transforming the food system, not by reforming conventional agriculture or engaging with dominant institutions, but rather by demonstrating a better way. These farmers developed their own systems, designed to represent the agricultural ideal. This included creating organic associations and eventually standards that would define their

distinct approach. Biodynamic adherents and Rodale had already introduced the idea of having standards and certification systems, so this was
really a continuation of the path laid out decades earlier. True to their
bottom-up grassroots approach to change, these decentralized independent
groups had no national organizations or policy agenda.

It was the threat of fraud that eventually led some organic leaders to
conclude that state involvement of some kind was necessary. Consumer
groups were already pressing for more systematic national oversight, making federal action virtually inevitable. As these groups ventured into the
federal policy arena, organic farmers and their representatives went along,
although in some cases reluctantly.

The inward focus on private certification at the heart of movement strategy left organic proponents ill prepared to deal with conventional politics.
They lacked lobbyists and offices in Washington, DC, or in state capitals.
Organic adherents had no fund-raising apparatus, and lacked expertise in
developing legislation or navigating government bureaucracies. Although
this foray into the policy world was a significant departure in that movement leaders were for the first time engaging with the federal government,
in a broader sense, the legislation that emerged was really the legal manifestation of the strategy long deployed by the movement. Such legislation
sought to address problems in the agricultural sector by creating a separate,
parallel food and agriculture system that organic proponents hoped would
grow and eventually displace conventional agriculture practices.

The specific provisions of the OFPA and regulations derived from it were
hotly debated among the diverse constituents that made up the organic
community. Particular measures enacted through that process would have
decisive implications for the direction of the organic system. If some key
decisions had gone another way, the entire trajectory of the organic market may have been fundamentally altered and the organic world would
be quite different today. Despite the fact that the particular rules adopted
under the OFPA and NOP would determine the fate of the organic market,
though, most debate regarding the pros and cons of the current organic
system center on the more general issue of whether it was the right move
to involve the state at all.

Those pleased with the present state of affairs credit the NOP for having
saved organic from going the way of natural. Organic could have become a
similarly meaningless marketing term used to imply only vague distinctions

among products. Instead, proponents argue, the OFPA gives organic real legal meaning and the NOP provides an institutional mechanism for preserving organic integrity. Federal involvement rescued the system, which is now growing to proportions unthinkable prior to the NOP.

Those unhappy with the status quo tend to blame the federal government and NOP for any shortcomings. Critics are especially dismayed by the incursion of large conventional firms into the organic market—a development that correlates with the federal "takeover" of organic. Many believe that the kind of agriculture now practiced by Big Organic does not reflect the movement's original principles. Agribusiness players have conventionalized organic, stripping it of many environmental and health benefits, and undermining the alternative agricultural vision of the early organic proponents.

## The Real Question

Organic proponents struggle with the question of whether the movement has made more headway with the state supervision of the organic system or whether we would be better off had it been left in private hands. This has been the central debate among activists and observers alike. But this sidesteps a larger question that goes to the core of organic movement strategy. It brings us back to the conflicting perspectives of ecological modernization theorists as opposed to those who subscribe to the treadmill of production theory.

The question is not simply whether the movement could achieve more with state versus private control of the organic certification program. A more basic issue is whether broad social changes can be achieved via this type of market mechanism at all, regardless of how the system is administered. Ecological modernization theorists champion the role of the market, and even that of private firms, in bringing about reform. From this perspective, one could retain the belief that the organic system is working. It may be a small segment of the market now, but organic is still growing. As more enlightened corporate leaders recognize the financial benefits and moral imperative of changing our ecologically destructive ways, and as consumers become more informed, more progress can be expected.

Treadmill theorists see little hope in achieving social reform via the market. The innate drive for profitability and expansion characteristic of capitalism will undermine the pursuit of any other value. From a treadmill vantage point, the state's power must be used to curtail the pathologies inherent in market systems. The only way to achieve that is for social movements to wrest state control from the economic powers that exercise undue influence in our corrupt political order, and then use the power of government to institute policies that fundamentally alter the market's operation.

This would be a different role for the state than it now plays in managing the organic system, and instituting it would require a different approach on the part of organic movement actors. From a treadmill perspective, the issue that caused so much controversy within the organic community— the question of whether or not to create a national organic system under the federal government's authority, or retain the private certification system—is not central. State involvement itself is not the crux of the matter, since the essential social change strategy is the same in any case; both factions within the organic community embraced a strategy that relies on growing the organic market with the intent of displacing the conventional food and agriculture system. The federal government was eventually called on to manage and provide oversight, but its role was one of overseeing a system, the basic contours of which were already established. The government was needed to rationalize the system, foster consistency, and supply enforcement, but it was only strengthening the market-based approach already long in use by the organic movement.

The federal government was not enlisted to play a traditional regulatory role whereby every producer would be required to adhere to a particular standard, as is the case with most policy. Organic standards are not to be applied to all growers and food processors; it is simply a system available to those who *voluntarily* choose to engage in that segment of the market. Nothing in organic law or policy does anything to infringe on the ability of conventional producers to continue as they always have outside the organic system. The fundamental organic strategy remains focused on the expansion of the organic portion of the food and agriculture market.

Thus far the organic sector only represents a tiny percentage of that market. Ecological modernization theorists would point out that this segment continues to grow, however slowly. Treadmill theorists would counter that in the meantime, conventional agriculture, unfettered by organic policy, has developed in dangerous new directions. The use of GMOs in conventional production has spread dramatically in the years since the organic program went into effect. Diseases as well as pests highly resistant to the pesticides and herbicides designed for use in conjunction with GMO crops pose a potentially serious threat to all agricultural production, the environment, and human health.[2] The organic market, bolstered by the organic movement and now a host of agribusiness players, is still growing, providing hope that the dream of an all-organic food and agriculture system is still in the making. But can we say that this strategy for developing a sustainable agriculture system is working when the conventional agriculture sector that supplies 96 percent of the food produced is, in many ways, getting *worse*? Is it viable to change the food and agriculture system as a whole by concentrating effort on nudging up the percentage of organically produced food, while the rest of the industry goes in the opposite direction? Can we expect food justice from an effort that relies on a market-based mechanism that can make access to healthy, sustainable food even more difficult?

Guthman contends that this is the movement's underlying failure. The agrarian dreams of organic proponents cannot come true given the market-based nature of organic movement strategy. Much of her analysis focuses on the way in which organic standards will ultimately be undermined using such an approach. She argues that land speculation in California and the associated costs of real estate have long forced producers to seek crops and growing methods through which added value can be extracted from their goods. The popularity of organic food among some consumers offers the latest opportunity for California growers. Due to their special designation, organic goods provide the needed price premium. But the unrelenting drive to decrease production costs has led producers to intensify organic growing using every allowable industrial farm practice, while perpetuating exploitative labor practices, and leaving traditional organic environmental and social ideals by the wayside. As such, Guthman asserts, that market forces tend to hollow out the essence of organic practices and lead to the conventionalization of organic agriculture.[3]

At the same time, ideologically committed organic producers remain dependent on the price premium that organic goods provide. This is what allows them to survive using the relatively more costly growing methods that do not externalize ecological and health costs, as is the case with destructive and harmful conventional practices. Conventionalizing Big Organic growers add to the supply and force down the premium, thereby undermining the smaller, ideologically committed farmers seeking to maintain high standards. These movement farmers may scramble to identify new crops or marketing techniques that allow them to eek out a living while still producing in accordance with their ideals, yet market forces supply continuous pressure. The implications are not limited to California or its particular agricultural history. The basic failure of the organic movement to achieve its transformative potential can be tied to its essential orientation toward the market.[4]

The fact that organic growers are dependent on the price premiums that organic goods command imposes an inherent limitation to the broader changes that organic proponents are seeking to achieve. According to Guthman, "The capacity to uphold this [legally specified organic] definition turns on entry barriers, which, by definition, are highly antithetical to widespread transformation in agrofood systems." Organic premiums can draw in more farmers, but as those premiums shrink, the incentive to enter that market likewise diminishes. This threatens to perpetually relegate organic to a niche market serving a relatively small segment of affluent consumers, leading some critics to rebrand organic goods as "yuppie chow."[5]

Guthman's analysis focuses on the structural mechanisms that hinder the spread of sustainable organic production, similar to the way in which treadmill theorists see inherent limits to systemic change within a capitalist order. But recognizing that structural barriers are never fully insurmountable, Guthman contends that the movement is also to blame for its limited success in achieving reform. She states that "the direction organic farming has taken in California is not only the legacy of the state's own style of agrarian capitalism. It is also a product of the movement's own choosing."[6]

As can be seen through an analysis of the movement's history, however, the choices made by movement actors are understandable in their historical context. At each turn the decisions made by organic proponents, sometimes defensive and sometimes offensive, made sense given the conditions they were facing and decisions that had been made prior. In some ways, the ongoing retention of the market-oriented certification approach to reform can be seen as a product of inertia or what some scholars refer to as "path dependence."[7]

Decisions early on in the development of a movement, made under particular conditions, become embedded in the participants' ideology and strategy. Those involved become invested in a given approach, making alternatives seem even less appealing. All this shapes strategy into the future, even when different approaches more suitable to new conditions are imaginable. Thus, to the extent that movement participants were making choices, the options they had were limited by circumstances and the direction laid out by those who came before them.

The market-based approach to reform adopted by the movement appeared viable in the social circumstances under which the movement developed, although there was little historical evidence to support that assessment. Organic proponents in many ways were charting a new path. Even if the movement had been centrally planned and coordinated, it would have been difficult to plot a more viable course for advancing the cause given the historical conditions at the time. Organic activists were, in a sense, pioneers of a social change strategy based on market mechanisms.

## Political Consumerism and Social Change via the Market

Despite the prevalence of certification systems today, the approach adopted by organic proponents had little historical precedent.[8] In this light, broad social transformation via a market-based strategy may have appeared more feasible. Many today still consider this to be a viable path toward social change, as evidenced by the plethora of certification systems that have followed the course chartered by the organic movement.[9] Many organic leaders still expect that organic practices will come to overtake conventional agriculture through this strategy.

Scholar Michele Micheletti examines the history of political consumerism, which she defines as the "conscious use of the market as an arena for policy by individuals and groups."[10] While other means of utilizing market forces for political purposes date back to colonial times in the United States, certification schemes were almost nonexistent prior to the 1980s. More often change agents utilized the withholding of spending or labor as a stick to pressure economic targets. Boycotts and strikes were the more common tools for exerting pressure via economic means, and these tactics have played a central role in many historic struggles throughout US history.[11] The country was founded in the midst of a boycott of British tea. Bus boycotts were key to the struggle to achieve racial equality during the civil rights movement.[12] Strikes and work slowdowns are a routine means of exerting economic pressure to achieve social or political goals from the production side. Such efforts in withholding labor have been used to significant effect, especially during the great labor struggles of the early and mid-twentieth century.[13] Divestment, or the withholding of investment, is another means by which activists have sought to exert economic pressure. Institutional divestment proved to be an effective economic strategy when activists around the world were attempting to pressure the South African government to end racial apartheid.

In this sense, the market has been tactically utilized in a number of social change campaigns, at times successfully. Notwithstanding the common use of market-oriented tactics to secure social change goals, private certification systems were almost unheard of until recently. In contrast to tactics that overtly seek to punish those engaged in undesirable practices, these systems *encourage* the purchase of goods made using desired processes, and only indirectly undermine those who do not comply and therefore lose that share of the consumer market. Playing on the term boycott, tactics that support patronizing, compliant firms are sometimes referred to as "buycotts."[14] Certification schemes essentially systematize buycotts.

The National Consumers League's "white label campaign" is a rare, early example of a private certification system, founded at the dawn of the twentieth century. The campaign was an effort to challenge sweatshop employers by identifying garments produced under better working conditions and to encourage consumers to select goods labeled as such. It lasted for several years and achieved some success, but ultimately legislation was necessary to address the deplorable health and safety conditions in the garment industry.[15] Unions have also long used a labeling scheme of sorts in advocating that consumers "look for the union label" when purchasing goods. It is not clear whether that campaign had any effect, although if current US unionization rates are any indicator, few consumers are looking for the union label.

Other forms of private regulation can be found in professional associations such as the American Medical Association and American Bar Association, which play central roles in setting standards and monitoring conduct among these professionals. Certified professionals who fail to live up to the standards set by these associations can lose their right to practice. Underwriters Laboratories (UL) represents another early case of a type of private regulatory system. Founded in 1894 with the goal of advancing consumer product safety, UL relies on its public reputation to draw businesses into adhering to the safety standards that it establishes. UL then utilizes testing and certification mechanisms to ensure compliance, and allows products to carry the UL seal if they meet the test.[16]

In spite of these examples, private regulation was never the centerpiece of social movement strategy, and organic activists were essentially charting new territory in seeking to advance their cause in this way. It was also the first instance in which the producers of a product created the standards system. Given that, one might be tempted to characterize organic certification as primarily a marketing scheme.[17] Rodale was certainly an entrepreneur, although like several organic industry titans today, he was obviously committed to spreading the organic gospel regardless of its financial reward. Notwithstanding the economic interests of some individuals, an examination of the history of organic clearly demonstrates all the characteristics of a social movement, and certification has long been crucial to organic movement strategy.[18]

Like many certification systems created since then, organic activists developed this private market-based approach at least in part because of their alienation from the state. Government officials were hostile to organic from the beginning. Movement founders like Rodale were treated as charlatans. Sixties' activists who later took up the cause were disinclined toward policy solutions because of their hostility toward what they viewed as a corrupt state and the general disillusionment with established institutions that followed the turmoil of that period. Opportunities for agricultural reform through the state, especially of the scope sought by organic proponents, appeared remote. Formalizing certification systems was at first a defensive measure to prevent co-optation and fraud, but it was really the continuation of the nonstatist approach.

Other movements have since adopted this tactic more consciously.[19] Some critics charge that this is a reflection of the pervasive influence of neoliberal ideology and the abandonment of hope, even by progressive activists, for achieving the collective good through democratic participation with the state.[20] Indeed, ecological modernization theorists champion the role that the free market can play in advancing ecological sustainability, and thus support certification schemes on those grounds.[21] But others argue that these strategies have emerged as a *reaction* to liberalization and the associated limitations on state power.[22] If it appears unfeasible to secure the passage of legislation that would directly restrict or prohibit an undesirable practice by private actors, as is commonly the case in the United States today, then private regulation through certification systems may be the most viable alternative.

Some contend that this is especially true given global economic developments and international trade treaties. Free trade agreements have hampered activists' ability to impose standards on economic actors via traditional government action. One certification organization developed directly out of a thwarted attempt to create state regulation for forest products. When Austrian lawmakers reversed course while considering restrictions on the sale of timber produced in ecologically damaging ways, sustainable forestry advocates established the Forest Stewardship Council.[23] The government backed down when challenged on the grounds that such a restriction would violate international trade rules.[24] The Forest Stewardship Council was designed to pressure timber producers via a private certification system not subject to challenge based on trade agreements.

Michael Conroy, who has dedicated much effort to analyzing market-based movement strategy, notes that the international trade regime coordinated through the World Trade Organization has rules that explicitly prohibit trade restrictions on the basis of "production and process methods"—that is, how a product is made.[25] This encompasses a vast array of practices that activists may have previously challenged on a policy-by-policy basis through their respective governments. Conroy maintains that "in an increasingly privatized world, with restrictions on what the global trading system will allow local and national governments to legislate, these movements may be the *only* alternative to the competitive downgrading of social and environmental practices by firms worldwide."[26]

As discussed in chapter 7, labeling systems are now in use for numerous products based on many different criteria ranging from ecological concerns to human rights. Micheletti argues that this is a positive development, not just because of its necessity in the current neoliberal context, but also because it provides consumers with new ways to engage politically. She writes that "political consumerism gives simple folks a political tool to civilize capitalism and create regulatory mechanisms in areas where the state is unable or does not want to act effectively."[27]

The "everyday activism" of steering the market by purchasing goods consistent with one's values opens new channels for political engagement. Micheletti says this is especially important for groups traditionally marginalized from conventional state politics, such as women and young people—two groups that are better represented in causes like organic that rely on market-based strategies. She also contends that the ability for everyone to participate in political action through their purchases gives political consumer movements more of a bottom-up quality relative to the hierarchical structures characteristic of policy-focused advocacy.[28]

In other respects, however, market-based change strategies have significant limitations.[29] In his book *Brewing Justice*, Daniel Jaffee examines the fair trade movement—one of several causes for which activists adopted the private certification strategy pioneered by the organic movement.[30] Similar to organic standards, fair trade began with little clarity as to what specifically qualified as fair. Originally it was a network of importers bringing handcrafted products directly to retail outlets from producers in less developed countries. Fairly traded agricultural goods were not brought until the 1980s, and it was not until the late 1990s that a fair trade certification system was established.

Today, coffee is by far the largest fairly traded product, but other certified goods such as chocolate, bananas, and flowers can also be found. Unlike the hodgepodge of standards initially created by many different organic farmers' associations, fair trade activists managed to create a fairly centralized certification system with one main organization, Fair Trade USA (formerly Transfair USA), as the primary fair trade certifier.[31]

Yet the fair trade movement has encountered problems similar to those faced by organic proponents. Fair trade activists must confront the same strategic dilemmas when it comes to expanding the market while trying to maintain high standards. Most early fair trade operations were composed of ideologically committed activists whose entire enterprise was based on fairly traded goods. As a result of a combination of consumer pressure and the lure of a profitable niche market, larger enterprises eventually developed side operations in accordance with fair trade standards and sought certification for these products. Similar to the way in which spreaders in the organic movement welcomed the interest of large companies with wide market reach, some fair trade leaders saw the entry of big players as a sign of the success of their efforts. But like organic, this too came with a threat to the integrity of the label and a weakening of standards.[32]

Originally, the rules required that 5 percent of a company's products must meet fair trade standards in order for any of its goods to secure certification.

Jaffee describes the concession that was made in allowing some big firms, such as Starbucks, to obtain certification for a smaller percentage of their products. Some defended this move on the grounds that even a small percentage of Starbucks' enormous business would mean greatly expanded opportunities for fair trade growers. Plus, involving large national firms like Starbucks would draw attention to the fair trade cause and create opportunities for still-greater expansion. Others saw this as co-optation and the undermining of the movement's core principles.[33]

In this case, criticism was leveled at not only the actual weakening of the fair trade standard but also the behind-the-scenes process by which it occurred. While political consumerism provides a sense of grassroots empowerment by allowing individuals to express their values in the marketplace, the certification systems on which they rely are not necessarily democratically structured or open. Grassroots activists and individual fair trade consumers were not privy to the decision to make exceptions to the minimum threshold rule for fair trade certification. Jaffee also points out that producers in less developed countries, those who the fair trade system is designed to serve, have little control over the fair trade certification body. In light of these problems and the potential for corruption within private certification systems, Jaffee calls for the development of oversight bodies that will, as it were, police the police.

Jaffee concludes that private certification mechanisms are still less prone to co-optation than publicly administered systems such as the NOP.[34] But the strengths and weaknesses of these approaches are subject to debate. Despite the legitimate grievances expressed by organic advocates regarding the USDA's unresponsiveness or the undue influence of agribusiness interests in Congress, as a publicly operated system, the NOP still affords a measure of accountability and democratic control that does not exist with organizations that wholly reside in private hands. Proponents of the federally supervised organic system are quick to remind NOP critics of the time when anyone could create a private organic certification organization and hand out organic labels on the basis of whatever standards (and for whatever fee) they wanted. Public administration of the certification system draws activists into the policy arena, where they face powerful adversaries, but attempting to maintain control over a private system presents its own challenges.[35]

## The Niche Problem

The entry of firms like Starbucks into the fair trade system or the foray of General Mills into organic brings us back to the market niche problem. Large operators such as these typically commit only a small percentage of their business to the certified cause. With rare exceptions, there is little ideological commitment to be found among the CEOs and business strategists at these firms.[36] The big players are willing to target that segment of conscious consumers, but as treadmill theorists and other market critics observe, they cannot be counted on to advance the cause when doing so is less profitable than promoting other products that they also sell.

This has not deterred proponents of this type of private regulation. It can be exciting to see the significant jump in the market share of certified goods when big firms move in. In terms of dollars, the value of the certified sector may double or triple once big companies commit resources to this market. The organic sector grew by 20 percent annually for several years following the NOP's creation when conventional firms bought into the system. This gives the impression that the strategy is working, the market for certified goods is expanding, and the practices that activists promote will eventually become the industry-wide standard.

Nevertheless, despite rapid growth rates, the base from which these products are starting is tiny, and the overall market share remains quite small for most products years or decades after the launch of the system. Over a decade following the establishment of the NOP and over forty years since the first certified organic products appeared on store shelves, organic makes up just 4 percent of the overall food and beverage market in the United States.[37]

Growth rates fell into the single digits during the recession that began in 2007, suggesting that economic factors play a decisive role in consumers' willingness to pay a premium for organic goods. When economic conditions warrant it, conventional firms are likely to shift marketing strategies to emphasize value and de-prioritize organic products. At current rates, it would take decades of steady growth to achieve the dream of an all-organic system. But it is more likely that growth in this sector is finite and certified products will remain a niche market indefinitely.

The niche problem is *the* fundamental flaw in the market-based strategy used by organic proponents and others who have adopted the standards and certification approach. The implications of private regulatory schemes for democracy and citizen engagement can be debated, yet the basic question for social change strategists is, Does it work? While political consumerism may allow citizens to express their values in the marketplace, the question remains as to whether this is a viable means of bringing about *broad* social change, or if it is just that: a means for some individual consumers to express values and identity in the same way that many consumer goods serve as forms of personal self-expression.[38]

In his final assessment of the fair trade movement, Jaffee shows that the fair trade system does indeed benefit those producers in less developed parts of the world that are integrated into these networks. Others have found that private certification systems or even the threat of the development of a certification system can yield at least some performance improvements within the targeted industry.[39] But the vast majority of the products on the market are not subject to any private regulatory conditions simply because producers choose not to involve themselves in or target the market of consumers who act on such values. Most goods are *not* part of the fair trade system, just as the vast majority of our crops are *not* grown organically. The organic system has done a great deal of good, but it poses no fundamental challenge to the conventional food and agriculture system. Former organic resources coordinator for the USDA Garth Youngberg and his coauthor, Suzanne DeMuth, wrote that, while the creation of the organic system that culminated in the NOP "was, in many ways, an enormously positive and widely supported accomplishment, it did not greatly alter the overall agricultural political, environmental or structural landscape."[40]

The evidence suggests that certification systems, be they wholly privately administered or managed by the state, are effective at altering the way in which *some* products are produced. They provide a means for *some* consumers to signal a preference and secure goods produced in accordance with their values. Consistent with analyses put forth by treadmill of production theorists, though, they are unlikely to ever work as a means of altering the larger system. Even when proponents are able to defend standards and prevent them from being instrumentally degraded by profit-seeking firms—a significant challenge in itself—production of goods outside the certified niche will carry on in accordance with expansionary tendencies that undermine the achievement of social and environmental goals. As such, a critical perspective on the organic movement and any similar market-based certification scheme is that they change the world, but only a small percentage of it.[41]

A still more cynical outlook suggests that the whole market-based strategy used by the organic movement actually undermined progress toward a more sustainable system of agriculture. Youngberg and DeMuth indicate that on the producer end, in the process of creating a separate organic system, those engaged in conventional agriculture who were interested in making incremental changes toward more sustainable methods of farming may have lost out.[42] The existence of the organic system presented an either-or option for growers; they could adhere to organic principles or not. From a marketing standpoint, there was little incentive to adopt *some* ecologically preferable measures if one could not secure full certification and capture the associated price premiums.

On the consumer side, by providing a small segment of highly motivated consumers with a means to reliably secure the goods that meet their desires, as the NOP does, the incentive for these relatively privileged individuals to demand broader systemic change is diminished. Health conscious shoppers need not be concerned with what toxins may be found in conventional food since they can avoid these risks by buying only USDA-certified organic.[43]

This point is tied to the matter of social justice as it pertains to political expression and participation. If consumer support for organic products is what drives firms to change their practices, and if certified goods inevitably come with a price premium, then many less affluent consumers are unfairly being left out of the social change effort and denied access to desired products. In contrast to being an inclusive forum where ordinary people can engage in political action daily, political consumerism is plagued by the same power imbalances that characterize institutional politics. The common plea for consumers to "vote with their dollars" by shopping for products that reflect their values is an open disenfranchisement of those who lack dollars and a debasing of democratic principles, which include fair representation. As inequitable as the US system of democracy may be in many respects, citizens rich and poor are still allowed one vote per person. The same cannot be said when dollars are equated to votes in the marketplace. And when social change only comes in the form of the availability of certain products, as opposed to universal standards applied to all goods, the fruits of this change are only accessible to those who can afford them.

## The Road Not Taken?

Critics are correct to challenge market-based social change strategies, and organic activists themselves have been deeply self-critical when examining the shortcomings of the organic system. But perhaps this critique is unjustified. While it requires a certain amount of speculation, we must assess the effectiveness of the standards and certification strategy *relative to what may have been achieved using a different approach*. In terms of the fundamental question, Did it work? it is more appropriate to consider the question, Did it work *as well as alternatives may have*?

There are certainly flaws in market-based social change strategies. But could the organic movement have achieved more had it focused on traditional state regulation of the food and agriculture industry as a whole? Was the market-based certification strategy a massive diversion of political energy, or was it an essential phase in the course of the development of a truly transformative food and agriculture movement?

Whether sustainable agriculture advocates could have achieved more by pushing for universal regulation and greater state control over all agricultural production is an open question. Historically, advocates working on other aspects of the food system certainly made significant gains by seeking to reform the system as a whole rather than create alternative markets subject to their own idealized set of rules.[44] Regulations regarding food safety, nutritional labeling, and the ban of certain pesticides and food additives are examples of reforms to the food system that were universally imposed through state mandate.

It is possible that sustainable agriculture advocates could have made similar gains had they not rejected state politics and looked inward to develop a wholesale alternative. Seeking bans on the worst pesticides, tighter regulation of concentrated animal feeding operations, and reforming the agriculture subsidy system are all policy changes that could move conventional production in the direction of a sustainable agriculture system. Activists could have strategically identified measures that when implemented, would have placed large agribusiness at a disadvantage relative to smaller regional or local producers. Instead of having a small niche market of "pure" organic goods alongside an increasingly hazardous conventional system that dominates the vast majority of the market, political energy could have been directed toward making the whole of the industry more sustainable and healthy for all consumers and farmworkers.

The question arises, however, as to whether the movement could have achieved any success at all using this approach. Looking at the political power configuration within the food and agriculture policy system from the 1970s onward, this is highly doubtful. Sustainable agriculture advocates were extremely weak and marginal during most of this period. Conventional agribusiness was already firmly entrenched, and those interests dominated federal agriculture policy. Agencies designed to address food and agriculture issues, like the USDA and the Food and Drug Administration, were essentially captured by the industry. It is difficult to imagine a movement of small farmers making gains under the Nixon administration with Earl Butz leading the USDA and openly calling for small operators to "get out" of farming. Small gains may have been possible during the Carter years, but most progressive change during his administration was reversed in the Reagan era. Tentative explorations into sustainable agriculture options at the time were summarily buried once Reagan came into office.

Consumer and environmental organizations paid little attention to agriculture for most of this period, and given this, it would have been a challenge for sustainable agriculture advocates to win allies with greater influence in Washington, DC. Perhaps sustainable agriculture advocates could have had some success in building ties with animal welfare organizations as the industrialization of livestock agriculture was bringing new horrors to bear within that sector. Yet many humane organizations have been reluctant to embrace the sustainable agriculture cause since it still depends on the exploitation of animals. While all this is speculative, assessing the constellation of political interests during the 1970s and 1980s, it is difficult to see how sustainable agriculture advocates could have achieved any significant political victories had they concentrated their energies on federal policy reform.

## Preparing the Ground and the Season to Come

The marginal political status of sustainable agriculture advocates through the 1980s suggests that developing an alternative system outside the state policy apparatus may have been the most advantageous course for the movement to take under the circumstances. Building the organic system had a number of strategic advantages for a fringe movement, which may have positioned it to make real gains later.[45]

First, by creating the organic system independently, advocates demonstrated that a real sustainable alternative to the conventional food and agriculture industry in fact was possible. While critics and conventional agribusiness spokespeople would continuously denounce organic farming as a fantasy, organic farmers and consumers were empirically demonstrating the feasibility of such a system. In addition, despite the dearth of public funding, independent organic research organizations and a handful of university scholars were confirming that organic agriculture *can* work. Studies showing that organic yields were comparable to conventional methods debunked claims that the adoption of such practices would lead to mass starvation. Had organic proponents spent their time lobbying for the investment of federal research dollars instead of engaging in their own research based on practicing organic farmers within a functioning organic system, they probably would have wasted limited time and effort.

The organic strategy proved effective in other ways as well. Public outreach and mobilization efforts based on hypothetical possibilities for an alternative to the industrial food system would not likely have gone far without a tangible product to demonstrate. While critics are correct to be concerned about the creep of neoliberal thinking into all aspects of life, the advancement of movement ideology through the marketing of products may have been beneficial in this instance. Patricia Allen and Martin Kovach argue that every organic consumer is potentially a new organic activist, so cultivating a consumer base at least has the potential to grow the movement.

These analysts also point out the organizational capacity that can develop out of a market.[46] Few movements have ready-made forums through which they can spread their message to the public or even means by which to identify likely supporters. The existence of certified organic goods for sale made every food co-op and health food store a venue for trumpeting the organic cause and recruiting activists. Although commercial activity complicates the exchange of ideas, the sale of organic products nonetheless allowed proponents to reach thousands of potential supporters, and health food stores and co-ops provided the organizational infrastructure that is invaluable for movement building. It should come as no surprise that tens of thousands of the postcards delivered to the USDA during the battle over the Big Three were distributed via these outlets.

Perhaps the most important component of the market-based strategy was the way in which organic, as a wholly alternative system, captured the imagination of sympathizers and brought them to the cause. In her analysis of the sustainable agriculture movement, Merrigan laments the way in which the alternative identity of organic adherents limited their ability to compromise with others in the agriculture policy world.[47] There is indeed evidence from European cases suggesting that the organic market grows more slowly when proponents situate themselves in opposition to conventional actors.[48]

Yet from a movement-building perspective, these distinctions can play a critical role. Identity is a crucial component for social movements of all sorts.[49] A shared identity can help create a sense of solidarity among members, a sense of unity in the face of adversaries. That sense of "we" is key to inspire individuals to act collectively for any cause. The stronger that sense of solidarity is, the greater the potential for collective action. Conversely, a less distinct group identity can make it more difficult to foster unity and a sense of urgency to act collectively. By creating a system starkly distinct from the conventional way of doing things, organic activists fostered an essential sense of identity among movement proponents. This allowed the movement to build a base of supporters that eventually numbered in the hundreds of thousands.

Had the movement taken a traditional state-focused policy reform approach and sought piecemeal change, starting, for example, with the elimination of a certain hazardous pesticide, would the cause have captured the public imagination in the same way as having a wholly alternative organic system? Would shoppers have been drawn to the cause if activist farmers were marketing a product that was otherwise conventionally produced, save for the targeted chemical? This is impossible to say with certainty. Analogous causes have proven effective in some cases. For instance, the effort to eliminate synthetic bovine growth hormone use in dairy cows was largely successful. Although the hormone is still legal in the United States, activists generated enough concern among consumers that most dairy producers will not use it, and many label their products accordingly.[50]

Thus, it is possible that targeting one undesirable practice at a time could have brought gradual reform in the agriculture sector. Given the broader political landscape during the period of the organic movement's development, though, it is difficult to see this piecemeal approach working often. In contrast, the creation of a separate organic system not only provided activists with an identity that drew them together with like-minded individuals but also offered a vision that a real alternative to the conventional food and agriculture system was indeed possible.

## The Road Ahead

The hope that the organic sector will grow and eventually displace conventional agriculture via market mechanisms alone is unfounded. Organic will likely remain a niche market indefinitely. Yet the organic movement, those activists who toiled for years to build the organic system and continue to do so, have made a lasting contribution regardless of the future of organic as such. The organic concept provided the foundation for the development of a real sustainable agriculture movement—one that otherwise may never have gotten off the ground. Its development was far from a diversion from the cause. Rather, the creation of this alternative should be viewed as an important and necessary step toward broader reform.

Issues of food and agriculture today receive more attention than at almost any time in US history, and organic movement activists can be credited with that achievement. The development of CSA programs, the renaissance of farmers' markets, and the creation of the entire local food movement—none of this would have occurred without the organic movement having paved the way. Even some of the urban food justice efforts were responding to the broader farm and food issues brought to the fore by organic activists. On the commercial side, the many businesses that are part of this multi-billion dollar industry owe their success to the organic proponents who worked for decades to bring these issues to public attention.

The fact that the movement built an alternative agricultural order instead of seeking piecemeal policy reforms probably contributed significantly to that success given the conditions under which the movement was operating. Issues of agricultural policy are obscure, and have long been far removed from the knowledge or experience of most citizens.[51]

The commodity agriculture support system, which maintains the entire food and agriculture order in the United States and influences much of the world, is largely hidden from public view. As agriculture was industrialized over the course of the twentieth century, Americans lost any sense of connection to their food, how it is grown, where it comes from, and from what it is made.[52] As the local farming infrastructure was whittled away, food appeared to emerge spontaneously in supermarkets or from behind the counter of fast-food restaurants. Few asked about and fewer knew of its origin or the system that kept it in place. As the food system became more opaque, the prospects for mobilizing against it diminished. It is unlikely that a mass movement for agricultural reform could have been based on calls to alter the incomprehensible subsidy system. But the organic movement presented something that was understandable and captured the public imagination: growing food naturally.

The nuances of crop insurance and the extraction of fructose were not the stuff of mass mobilization. Yet many people had a sense that something was wrong with their food. The fact that some farmers were growing crops organically and that there was a whole system in place to ensure the purity of their products suggested that there was a real alternative to the troublesome conventional food order. Without that example, people may never have been inspired to ask whether there is another way or whether there was anything wrong with industrial agriculture to begin with.

Now those organic activists and the organizations they created are tackling the policy issues necessary to advance the sustainable agriculture cause. They are a major force in the struggle for federal agriculture policy reform. Consider the organizational basis of the sustainable agriculture movement today. Groups such as the OCA, NOC, and NOFA now have affiliates that together number in the hundreds of thousands. This dwarfs the membership of the small farmer and sustainable agriculture organizations that preceded them. Several organic organizations now have a real Washington presence—something that was almost unheard of prior to the mobilization around the organic cause. Perhaps most significantly, these groups mobilize their members to take action on issues that extend well beyond organic proper. Granted, many still have the word organic in their names, and still spend considerable time fighting over the minutiae of national organic policy. But they also mobilize their supporters to speak out on matters ranging from school nutrition programs to banning GMOs.

This is where the organic movement is currently making its most significant contribution. Unlike ever before in history, there is now a sizable and mobilized sustainable agriculture constituency, and organic advocacy organizations are central to that effort. Consumer organizations and environmental groups are more engaged with agriculture issues than at any time previously. These groups, frequently taking cues from organic advocacy groups, now regularly weigh in on matters of agriculture policy. Changes in obscure provisions of the Farm Bill draw national media attention given that this vast, mobilized constituency is aware of how these policies affect it. It is difficult to imagine a movement ever developing around such issues had organic activists not worked for decades raising the public consciousness, building a concrete vision of an alternative way to feed ourselves, and finally, challenging the conventional agriculture system on the policy front where the entire dysfunctional order is upheld.

The other efforts that have grown out of the organic movement also have an important role to play. Alternative certification and labeling schemes face the same limitations as the organic system and any purely market-based approach to social change. At the same time, those developing such systems help to spread awareness about food and agriculture issues. The creation of formal organizations associated with these systems also builds political capacity in the same way that early organic farmers' associations proved to be politically significant.

Local food advocates are another critical part of the movement for sustainable agriculture. They fill a crucial role in helping to raise consciousness about food and farming issues in their communities. Local food supporters can play a greater part if they move beyond individual political consumerism and further develop organizational structure. Farmers' markets and CSAs provide important forums through which potential recruits can be drawn into more active engagement, but individuals need ways in which they can be engaged beyond buying healthy food.

Local food advocacy organizations can aid in facilitating the growth of sustainable agriculture markets, yet there may be ways in which such groups could be more directly politically engaged. Some have suggested that just as local municipalities have departments that deal with such social needs as transportation, water, and waste removal, they could have departments dedicated to addressing issues of food and agriculture, too.[53] This would provide a means through which local food proponents could seek policy mechanisms that would facilitate the development of regional food hubs beyond the educational efforts and purely private, market-based appeals that characterize much of local food advocacy today.

Most important, all these sustainable agriculture proponents should be tied in with the organizations that are fighting on the federal policy front. One need not choose between individual political consumption, community-based action, and involvement with state and federal policy. But sustainable agriculture advocates cannot afford to ignore the policy domain, and every attempt should be made to draw individuals, newly conscious of the significance of food and agriculture, into organizations addressing the central structural mechanisms that perpetuate an unjust, ecologically destructive, and unhealthy food system. Too many sustainable agriculture proponents still cling to the belief that consumers alone can transform society just by "voting with their dollars" (or their forks). Consumer action must be supplemented by greater concerted action to change policies in ways that can strengthen the small farmers that organic advocates seek to support and weaken the ability of industrial-scale agribusiness to practice as it does—practices that will enable the big players to continue to dominate the market without structural reform. Policy-oriented local food groups could provide crucial grassroots support to national organic organizations that have expanded their agendas to include federal policies that can aid in local food system development.

While all the branches of the sustainable agriculture movement must dedicate at least some of their efforts to policy work, in the meantime, those groups focused on issues at the federal level need to strategically identify those policies most likely to advance the cause. Like most advocacy groups, organic organizations tend to lunge from issue to issue, putting out fires as they arise and seizing opportunities when they present themselves.

Perhaps more than in most movements, organic advocates have tried to avoid this crisis mobilization tendency, and to reflect and plan long-term strategy. Under the Rural Advancement Foundation International's leadership, organic advocates have come together regularly over the years to assess the status of the movement and consider ways to advance the cause. But much of this strategizing still revolves around the organic system itself, even as the work of these organizations expands out in broader policy areas that are potentially more fruitful in terms of advancing sustainable agriculture.

Even as the organic system commands much of their time and energy, proponents of sustainable agriculture have made advances on a number of issues. One area in which sustainable agriculture advocates have been making headway is in regard to GMOs. Activists, including many organic advocacy organizations, have long demanded labeling for products containing GMOs. Despite some setbacks on state referenda and in the courts, there are indications that they will eventually succeed, and this would be a significant victory. An outright GMO ban would be optimal, but may not be achievable at this time, and mandatory labeling is the next best thing. Labeling still depends on market forces to yield the desired result, although if successful, it can set the stage for a legal prohibition. As consumers learn about the issue and are alerted through a clear label, they will likely steer away from such products, and producers would likewise abandon their use. Once the use of technologies like GMOs is reduced and their proponents weakened, sustainable agriculture advocates would be better positioned to seek an outright ban.

Meanwhile, since the GMO labeling strategy does rely on consumer action in the marketplace, there are risks. The GMO labeling campaign could backfire if the labels are so pervasive that consumers come to accept such ingredients as normal. Price differentials would also play a role. If foods containing GMOs were considerably cheaper than those without, consumers may still be drawn to these inexpensive products and the labeling campaign might fail in its intended effect. Nonetheless, GMO labels would greatly enhance the market position of small and midsize regional farmers who do not use GMOs. The GMO campaign is an example of how the movement can direct energy at federal policy reforms that could have significant implications for the entire food and agriculture system, not just the organic sector.

Another promising development comes in the growing concern about the effect of the routine use of antibiotics in the livestock industry. Restrictions on antibiotic use could have far-reaching implications for industrial livestock production. Conditions in concentrated animal feeding operations may have to be dramatically altered if animals cannot be given the drugs necessary to prevent the rapid spread of disease that would likely occur under those horrific conditions. This would again hamper industrial production in ways that would make small and midsize local producers more competitive.

Measures like GMO labeling and restrictions on antibiotic use are the kinds of policy reforms that have the potential to shift the entire food and agriculture system in more positive directions. Advocates are wise to strategically target the technologies that enable industrial agriculture, while fighting to redirect federal support toward small farmers using sustainable growing practices. These measures make healthier, more sustainably produced products competitive in the market and more available to all consumers, not just those who can afford to pay a high premium for purely organic goods.

Through this policy-oriented strategy, the food and agriculture system could be gradually transformed. The worst manifestations of the industrial agriculture system could be eliminated while bolstering the position of those who use more sustainable practices, even if they are not purely organic. This contrasts with the strategy that emphasizes transforming the agriculture system by trying to grow the organic sector. That approach relies too much on consumers voluntarily paying a fairly substantial price premium for goods produced under the idealized organic system even as the conventional sector discretely introduces dangerous new technologies that make its products even cheaper.

A strategy based on local purchasing faces similar challenge in that sustainably produced local goods often come with a price premium. This creates a real barrier for those who truly cannot afford to pay more and a psychological barrier for the majority of consumers for whom price is always a consideration. This is in addition to issues of accessibility for those in areas lacking in healthy food outlets. Once again this threatens to relegate sustainably produced goods to a small niche forever while leaving the conventional system to proceed largely unchallenged. Sustainable agriculture advocates are likely to achieve the greatest success when they focus on policy reforms that target the worst practices for elimination or that alter the overall incentive matrix by making the price of industrially produced foods truly reflect the high cost they impose on human health and the environment.

At the same time, policies that support small and midsize sustainable farmers are needed. The federal supports that make industrially produced commodities artificially cheap must be redirected toward those who are producing in a sustainable manner. Organic activists are doing this when they lobby for greater research funding for sustainable practices and when they demand cost-share support for organic certification. But financial support for organic farmers should not have to come in the form of assistance to make compliance with the organic mandates somewhat less expensive. In essence, such aid is a slight reduction in the tax paid by farmers who are doing that which all farmers should be required to do. Small farmers using environmentally superior methods should receive support as a matter of course, in the way that large industrial producers do now. Reforms to crop insurance policies alone would do much to help small farmers growing diverse crops.

Similarly, reform advocates should seek interim measures to penalize industrial producers for using unsustainable practices. Taxes on synthetic fertilizers and pesticides would make the price of their goods better reflect their true cost to the environment and human health. Ideally with the proper taxation, redirected federal support, and appropriate regulation, the price of goods on the store shelves would come to mirror the true costs to society. Ecologically harmful and unhealthy products would be more expensive, and those that are produced using more sound methods would be comparatively affordable. Consumers, still primarily attuned to price considerations, would migrate toward those products that are better for the environment *and* less expensive. They would not be asked to pay more out of personal virtue or private health concerns as they are now through the organic system.

Such a shift would require an expanded role for government in the transition to a sustainable agriculture system. Federal officials would not just be managing a distinct program that allows organic producers to differentiate themselves (at significant cost); they would be enforcing regulations, taxing destructive practices and products, and actively supporting those who are adopting sustainable methods.

Whether sustainable agriculture advocates have the power to achieve real reforms in the food and agriculture system is yet to be seen. The organic movement has done much to bring these issues to public consciousness, and the constituency supporting changes to the system has grown enormously. Although the standards and certification mechanism that was at the heart of movement strategy for decades has significant limitations, the approach demonstrated that a real alternative to industrial agriculture is possible. Moreover, the mobilization around the OFPA and NOP was central to the development of a federal policy orientation and the creation of a national presence for sustainable agriculture advocates. There is promise in the fact that organic associations, which at one time concentrated almost exclusively on building an alternative market, are now engaged in policy matters relevant to the entire food and agriculture system.

The development of the NOP and the certification systems that preceded it was a necessary stage for the sustainable agriculture movement. The organic system will not disappear, but neither is it likely to grow and displace the dominant agricultural order. It will remain a small segment of the market serving as an example of how farming can be done. If the rest of the agriculture system is to be brought up to those standards, however, it will not happen through market forces alone. Organic activists and local food advocates should not abandon the work they have been doing. They should defend organic integrity, start farmers' markets, join CSA projects, and buy sustainably produced goods. But we will not buy our way to a just and sustainable food and agriculture system. These disparate groups, activists, and consumers must also engage in the policy arena. They should become active with the national organic organizations and other sustainable agriculture advocacy groups fighting in Washington for the reforms that will bring healthy food produced using just and ecologically sustainable means to all people.

# Notes

## Chapter 1: Introduction

1. Deborah Rich, "Organic and Sustainable Up for Review ... Again," *Organic Farming Research Foundation Information Bulletin* (Fall 2008): 4–13, http://www.organic-center.org/reportfiles/ib16%20organic%20and%20sustainable%20up%20for%20review.pdf (accessed July 22, 2014).

2. Personal communication, December 11, 2007.

3. The couple eventually sold their farm, and left farming in 2011 after training a young farmer to replace them at their CSA, which is still thriving.

4. John D. McCarthy and Mayer N. Zald, *The Trend of Social Movements in America: Professionalization and Resource Mobilization* (Morristown, NJ: General Learning Press, 1973); Suzanne Staggenborg, "The Consequences of Professionalization and Formalization in the Pro-Choice Movement," *American Sociological Review* 53, no. 4 (August 1988): 585–605.

5. Jill M. Bystydzienski and Steven P. Schacht, eds., *Forging Radical Alliances across Difference: Coalition Politics for the New Millennium* (Lanham, MD: Rowman and Littlefield, 2001); Nella Van Dyke and Holly McCammon, eds., *Strategic Alliances: Coalition Building and Social Movements* (Minneapolis: University of Minnesota Press, 2010).

6. Organic Trade Association, "2011 Organic Industry Survey," http://www.ota.com/pics/documents/2011OrganicIndustrySurvey.pdf (accessed July 23, 2014).

7. Patricia Allen and Martin Kovach, "The Capitalist Composition of Organic: The Potential of Markets in Fulfilling the Promise of Organic Agriculture," *Agriculture and Human Values* 17, no. 3 (September 2000): 221–232; Julie Guthman, "Regulating Meaning, Appropriating Nature: The Codification of California Organic Agriculture," *Antipode* 30, no. 2 (April 1998): 135–154; Brian Obach, "Theoretical Interpretations of the Growth in Organic Agriculture: Agricultural Modernization or an Organic Treadmill?" *Society and Natural Resources* 20, no. 3 (2007): 229–244.

8. David Meyer, Valerie Jenness, and Helen Ingram, *Routing the Opposition: Social Movements, Public Policy, and Democracy* (Minneapolis: University of Minnesota Press, 2005); Brian Obach, "Political Opportunity and Social Movement Coalitions: The Role of Policy Segmentation and Non-Profit Tax Law," in *Strategic Alliances: Coalition Building and Social Movements*, ed. Nella Van Dyke and Holly McCammon, 197–218 (Minneapolis: University of Minnesota Press, 2010).

9. Magnus Boström, "Environmental Organisations in New Forms of Political Participation: Ecological Modernisation and the Making of Voluntary Rules, *Environmental Values* 12, no. 2 (May 2003): 175–193; Arthur P. J. Mol, *The Refinement of Production: Ecological Modernization Theory and the Dutch Chemical Industry* (Utrecht: Jan van Arkel/International Books, 1995); Arthur P. J. Mol, "Ecological Modernization: Industrial Transformations and Environmental Reform," in *International Handbook of Environmental Sociology*, ed. Michael Redclift and Graham Woodgate, 138–149 (Cheltenham, UK: Edward Elgar, 1997); Arthur P. J. Mol and Gert Spaargaren, "Ecological Modernization Theory in Debate: A Review," in *Ecological Modernization around the World*, ed. Arthur P. J. Mol and David A. Sonnenfeld, 17–49 (London: Frank Cass, 2000).

10. For a more in-depth analysis of shifts in environmental policy, see Daniel Mazmanian and Michael Kraft, *Toward Sustainable Communities: Transitions and Transformations in Environmental Policy* (Cambridge, MA: MIT Press, 1999).

11. Karin Hofer, "Labelling of Organic Food Products," in *The Voluntary Approach to Environmental Policy: Joint Environmental Policy-Making in Europe*, ed. Arthur P. J. Mol, Volkmar Lauber, and Duncan Liefferink (Oxford: Oxford University Press, 2000), 156–191.

12. Susanne Padel, "Conversion to Organic Farming: A Typical Example of the Diffusion of an Innovation?" *Sociologia Ruralis* 41, no. 1 (2001): 40–61.

13. Frederick H. Buttel, "The Treadmill of Production: An Appreciation, Assessment, and Agenda for Research," *Organization and Environment* 17, no. 3 (2004): 323–336; John Bellamy Foster, "The Treadmill of Accumulation: Schnaiberg's Environment and Marxian Political Economy," *Organization and Environment* 18, no. 1 (2005): 7–18; Kenneth Gould, David Pellow, and Allan Schnaiberg, "Interrogating the Treadmill of Production: Everything You Wanted to Know about the Treadmill but Were Afraid to Ask," *Organization and Environment* 17, no. 3 (2004): 296–313; Allan Schnaiberg, *The Environment: From Surplus to Scarcity* (New York: Oxford University Press, 1980); Allan Schnaiberg and Kenneth Gould, *Environment and Society* (New York: St. Martin's Press, 1994); David A. Sonnenfeld, "Contradictions in Ecological Modernization: Pulp and Paper Manufacturing in Southeast Asia," in *Ecological Modernization around the World*, ed. Arthur P. J. Mol and David A. Sonnenfeld (London: Frank Cass, 2000), 235–256; Richard York and Eugene A. Rosa, "Key Challenges to Ecological Modernization Theory," *Organization and Environment* 16, no. 3 (2003):

273–288; Richard York, "The Treadmill of (Diversifying) Production," *Organization and Environment* 17, no. 3 (2004): 355–362.

14. Julie Guthman, *Agrarian Dreams: The Paradox of Organic Farming in California* (Berkeley: University of California Press, 2004).

15. Philip Howard, "Consolidation in the North American Organic Food Processing Sector, 1997–2007," *International Journal of Agriculture and Food* 16, no. 1 (2009): 13–30.

16. Alan Hall and Veronika Mogyorody, "Organic Farmers in Ontario: An Examination of the Conventionalization Argument," *Sociologia Ruralis* 41, no. 4 (October 2001): 399–422; Douglas H. Constance, Jin Youn Choi, and Holly Lyke-Ho-Gland, "Conventionalization, Bifurcation, and Quality of Life: Certified and Non-Certified Organic Farmers in Texas," *Southern Rural Sociology* 23, no. 1 (2008): 208–234; Daniel Buck, Christina Getz, and Julie Guthman, "From Farm to Table: The Organic Vegetable Commodity Chain in Northern California," *Sociologia Ruralis* 44, no. 3 (1997): 3–20; Julie Guthman, "The Trouble with 'Organic Lite': A Rejoinder to the 'Conventionalization' Debate," *Sociologia Ruralis* 44, no. 3 (2004): 301–316.

17. Howard, "Consolidation in the North American Organic Food Processing Sector."

18. Jeff Goodwin and James M. Jasper, *The Social Movements Reader: Cases and Concepts*, 2nd ed. (West Sussex, UK: Blackwell Publishing, 2009), 4.

19. US Department of Agriculture, "Organic Production and Organic Food: Information Access Tools," http://www.nal.usda.gov/afsic/pubs/ofp/ofp.shtml (accessed July 24, 2014).

20. Dietlind Stolle, Marc Hooghe, and Michele Micheletti, "Politics in the Supermarket: Political Consumerism as a Form of Political Participation," *International Political Science Review* 26, no. 3 (2005): 245–269.

21. Daniel Jaffee and Philip H. Howard, "Corporate Cooptation of Organic and Fair Trade Standards," *Agriculture and Human Values* 27 (2010): 387–399.

22. Warren James Belasco, *Appetite for Change: How the Counterculture Took on the Food Industry* (Ithaca, NY: Cornell University Press, 2007); Jeffrey Haydu, "Cultural Modeling in Two Eras of U.S. Food Protest: Grahamites (1830s) and Organic Advocates (1960s–70s)," *Social Problems* 58, no. 3 (2011): 461–487. Note that even though Haydu characterizes the organic movement as state focused relative to the Grahamite food movement to which he is contrasting organic, as I will argue, organic activists only reluctantly turned to the state, and even then it was only for the purpose of better enabling consumers to advance social change goals via the market.

23. Robert Gottlieb and Anupama Joshi, *Food Justice* (Cambridge, MA: MIT Press, 2010). See also Alison Hope Alkon and Julian Agyeman, *Cultivating Food Justice: Race,*

*Class, and Sustainability* (Cambridge, MA: MIT Press, 2011); Alison Hope Alkon and Kari Marie Norgaard, "Breaking the Food Chains: An Investigation of Food Justice Activism," *Sociological Inquiry* 79, no. 3 (August 2009): 289–305; David J. Hess, *Localist Movements in a Global Economy: Sustainability, Justice, and Urban Development in the United State*s (Cambridge, MA: MIT Press, 2009).

24. McCarthy and Zald, *The Trend of Social Movements in America*; Staggenborg, "The Consequences of Professionalization and Formalization in the Pro-Choice Movement."

25. Robert Michels, *Political Parties: A Sociological Study of the Oligarchical Tendencies of Modern Democracy* (New York: Evergreen Review Inc., 2008); Frances Fox Piven and Richard Cloward, *Poor People's Movements* (New York: Pantheon, 1977).

26. Laura B. DeLind, "Are Local Food and the Local Food Movement Taking Us Where We Want to Go? Or Are We Hitching Our Wagons to the Wrong Stars?" *Agriculture and Human Values* 28, no. 2 (June 2011): 273–283.

27. Hess, *Localist Movements in a Global Economy*.

## Chapter 2: The Birth of the Organic Movement

1. For a detailed historical account of the organic movement, see Philip Conford, *The Origins of the Organic Movement* (Edinburgh: Floris Books, 2001).

2. Jerome Irving Rodale, "Introduction to Organic Farming," *Organic Farming and Gardening* 1, no. 1 (May 1942): 3.

3. William Cronon, "The Trouble with Wilderness; or Getting Back to the Wrong Nature," *Environmental History* 1, no.1 (1996): 7–28.

4. Gunter Vogt, "The Origins of Organic Farming," in *Organic Farming: An International History*, ed. William Lockeretz (Oxfordshire, UK: CABI, 2007), 9–29.

5. Frederick H. Buttel, "Agricultural Change, Rural Society, and the State in the Late Twentieth Century: Some Theoretical Observations," in *Agricultural Restructuring and Rural Change in Europe*, ed. David Symes and Anton J. Jansen (Wageningen: Agricultural University, 1994), 13–31.6. Vogt, "Origins of Organic Farming."

7. Ibid.

8. William W. Cochrane, *The Development of American Agriculture: A Historical Analysis* (Minneapolis: University of Minnesota Press, 1979); Deborah Fitzgerald, *Every Farm a Factory: The Industrial Ideal in American Agriculture* (New Haven, CT: Yale University Press, 1979); William Friedland, "The New Globalization: The Case of Fresh Produce," in *From Columbus to Conagra: The Globalization of Agriculture and Food*, ed. Alessandro Bonanno, Lawrence Busch, William Friedland, Lourdes Gouveia, and Enzo Mingione, (Lawrence: University of Kansas Press, 1994), 210–231; Linda Lobao

and Katherine Meyer, "The Great Agricultural Transition: Crisis, Change, and Social Consequences of Twentieth Century US Farming," *Annual Review of Sociology* 27 (2001): 103–125; Fred Magdoff, John Bellamy Foster, and Frederick H. Buttel, eds., *Hungry for Profit: The Agribusiness Threat to Farmers, Food, and the Environment* (New York: Monthly Review Press, 2000); Michael Pollan, *The Omnivore's Dilemma: A Natural History of Four Meals* (New York: Penguin, 2006).

9. Justus Von Liebig, *Organic Chemistry in Its Applications to Agriculture and Physiology* (London: Bradbury and Evans, 1840).

10. Conford, *Origins of the Organic Movement*; Albert Howard, *Organic Gardening* 3, no. 2 (July 1943).

11. Cochrane, *Development of American Agriculture*; D. D. Treadwell, D. E. McKinney, and N. G. Creamer, "From Philosophy to Science: A Brief History of Organic Horticulture in the United States," *Hortscience* 38, no. 5 (2003): 1009–1014; Pollan, *Omnivore's Dilemma*.

12. Kevin F. Goss, Richard D Rodefeld, and Frederick H. Buttel, "The Political Economy of Class Structure in US Agriculture: A Theoretical Outline," in *The Rural Sociology of the Advanced Societies: Critical Perspectives*, ed. Frederick H. Buttel and Howard Newby (Montclair, NJ: Allanheld Osmun, 1980), 83–132; Lobao and Meyer, "Great Agricultural Transition."13. Conford, *Origins of the Organic Movement*, 48.

14. Albert Howard, "The Good Earth" *Organic Farming and Gardening* 1, no. 2 (June 1942): 5.

15. Albert Howard, 1943, quoted in D. H. Stinner, "The Science of Organic Farming," in *Organic Farming: An International History*, ed. William Lockeretz (Oxfordshire, UK: CABI, 2007), 40–72.

16. Stinner, "Science of Organic Farming."

17. Albert Howard, "The Cause of Plant Disease," *Organic Gardening and Farming* 2, no. 1 (December 1942): 4

18. Quoted in Conford, *Origins of the Organic Movement*, 54.

19. Hilmar Moore, "Rudolf Steiner: A Biographical Introduction for Farmers," *Biodynamics* 214 (November–December 1997): 29–32.

20. Rudolf Steiner Archive, http://www.rsarchive.org/ (accessed July 28, 2014).

21. Vogt, "Origins of Organic Farming."

22. "Biodynamic Farm Standard," Demeter Association, Inc., http://demeter-usa .org/downloads/Demeter-Farm-Standard.pdf (accessed July 28, 2014).

23. Lord Northbourne, *Look to the Land* (London: J. M. Dent and Sons, 1940).

24. Eve Balfour, *The Living Soil* (London: Faber and Faber, 1943).

25. Conford, *Origins of the Organic Movement*, 90.

26. Eve Balfour, "Towards a Sustainable Agriculture—The Living Soil" (talk at the International Federation of Organic Agriculture Movements conference, Sissach, Switzerland, 1977), http://www.soilandhealth.org/01aglibrary/010116Balfourspe ech.html (accessed July 28, 2014).

27. Albert Howard, "The Good Earth," *Organic Farming and Gardening* 1, no. 2 (June 1942): 5.

28. Conford, *Origins of the Organic Movement*, 45.

29. Albert Howard, "The Wheel of Life," *Organic Gardening* 3, no. 5 (October 1943): 20.

30. Conford, *Origins of the Organic Movement*.

31. Ibid., 143. Green Party adherents commonly refer to their ideology as "neither left nor right, but out in front."

32. David Danbom, "Past Visions of American Agriculture," in *Visions of American Agriculture,* ed. William Lockeretz (Iowa City: Iowa State University Press, 2000), 3–16.

33. William Lockeretz, *Visions of American Agriculture* (Iowa City: Iowa State University Press, 2000).

34. Conford, *Origins of the Organic Movement*, 100.

35. Rodale altered the name of this publication several times, reversing the order of the words farming and gardening, or dropping one term or the other completely.

36. David M. Tucker, *Kitchen Gardening in America: A History* (Iowa City: Iowa State University Press, 1993), 148.

37. "Organic Gardening Clubs of America," *Organic Gardening and Farming* 2 (October 1955): 76–79.

38. "Organic Club Notes," *Organic Gardening and Farming* (March 1957): 18–19.

39. "Organic Market Opens," *Organic Gardening* (December 1951): 46.

40. Richard Gericke, "The Forum: Know Your Organic Food Producers," *Organic Gardening* (September 1951): 4–6.

41. Jerome Irving Rodale, "With the Editor: Do Chemical Fertilizers Kill Earthworms," *Organic Gardening* (February 1948): 12–17.

42. "Organic World," *Organic Gardening and Farming* (June 1958): 2–3; "Organic World," *Organic Gardening and Farming* (August 1958): 3; "Organic World," *Organic Gardening and Farming* (April 1959): 2.

43. Treadwell, McKinney, and Creamer, "From Philosophy to Science."

44. Robert Gottlieb, *Forcing the Spring: The Transformation of the American Environmental Movement* (Washington, DC: Island Press, 2005); Robert Brulle, "Environmental Discourse and Social Movement Organizations: A Historical and Rhetorical Perspective on the Development of US Environmental Organizations," *Sociological Inquiry* 66, no. 1 (January 1996): 58–83.

45. Gottlieb, *Forcing the Spring*.

46. Robert Gordon, "Poisons in the Fields: The United Farm Workers, Pesticides, and Environmental Politics," *Pacific Historical Review* 68, no. 1 (1999): 51–77.

47. Michael Sligh and Thomas Cierpka, "Organic Values," in *Organic Farming: An International History*, ed. William Lockeretz (Oxfordshire, UK: CABI, 2007), 30–39; Treadwell, McKinney, and Creamer, "From Philosophy to Science."

48. Warren James Belasco, *Appetite for Change: How the Counterculture Took on the Food Industry* (Ithaca, NY: Cornell University Press, 2007).

49. Treadwell, McKinney, and Creamer, "From Philosophy to Science."

50. Ronnie Cummins, personal communication, November 25, 2007.

51. Wade Greene, "Guru of the Organic Food Cult," *New York Times*, June 6, 1971, SM30.

52. Belasco, *Appetite for Change*, 72.

53. Ibid., 87.

54. Julie Guthman, "Fast Food/Organic Food: Reflexive Tastes and the Making of 'Yuppie Chow,'" *Social and Cultural Geography* 4, no. 1 (2003): 45–58.

55. Belasco, *Appetite for Change*.

56. Barbara Epstein, *Political Protest and Cultural Revolution: Nonviolent Direct Action in the 1970s and 1980s* (Berkeley: University of California Press, 1991); Enrique Laraña, Hank Johnston, and Joseph R. Gusfield, eds., *New Social Movements: From Ideology to Identity* (Philadelphia: Temple University Press, 1994); Verta A. Taylor and Nancy Whittier, "Collective Identity in Social Movement Communities: Lesbian Feminist Mobilization," in *Frontiers in Social Movement Theory*, ed. Aldon Morris and Carol McClurg Mueller (New Haven, CT: Yale University Press, 1992), 104–130.

57. Laura B. DeLind, "Transforming Organic Agriculture into Industrial Organic Products: Reconsidering National Organic Standards," *Human Organization* 59, no. 2 (2000): 198–208; Julie Guthman, *Agrarian Dreams: The Paradox of Organic Farming in California* (Berkeley: University of California Press, 2004).

58. Kathleen Merrigan, "Negotiating Identity within the Sustainable Agriculture Advocacy Coalition" (PhD diss., Massachusetts Institute of Technology, September 2000).

59. Vogt, "Origins of Organic Farming."

60. Sligh and Cierpka, "Organic Values."

61. Suzanne Staggenborg, "The Consequences of Professionalization and Formalization in the Pro- Choice Movement," *American Sociological Review* 53, no. 4 (1988): 585–605.

## Chapter 3: Certification and the State: The Dilemmas of Growth

1. Margaret D. Pacey, "Nature's Bounty: Merchandizers of 'Health Foods' Are Cashing in on It," *Barron's National Business and Financial Weekly*, May 10, 1971, P5.

2. Ibid.

3. Quoted in Julius Duscha, "Up, Up, Up—Butz Makes Hay Down on the Farm," *New York Times*, April 16, 1972, SM34.

4. Michael Pollan, *The Omnivore's Dilemma: A Natural History of Four Meals* (New York: Penguin, 2006), 52.

5. Michael F. Jacobson, "Feeding the People, Not Food Producers," *New York Times*, August 31, 1972, 33.

6. Garth Youngberg and Suzanne P. DeMuth, "Organic Agriculture in the United States: A 30-Year Retrospective," *Renewable Agriculture and Food Systems* 28, no. 4 (December 2013): 1–35, doi:10.1017/S1742170513000173.

7. Kenneth Beeson, "Spring Gardens: What about the 'Organic Way'?" *New York Times*, April 16, 1972, 33.

8. Quoted in Jean Hewitt, "Organic Food Fanciers Go to Great Lengths for the Real Thing," *New York Times*, September 7, 1970, 23.

9. Quoted in Grace Lichtenstein, "'Organic' Food Study Finds Pesticides," *New York Times*, December 2, 1972, 39.

10. Elizabeth M. Whelan and Frederick John Stare, *Panic in the Pantry: Food Facts, Fads, and Fallacies* (New York: Atheneum, 1977).

11. Warren James Belasco, *Appetite for Change: How the Counterculture Took on the Food Industry* (Ithaca, NY: Cornell University Press, 2007).

12. Hewitt, "Organic Food Fanciers Go to Great Lengths for the Real Thing"; Wade Greene, "Guru of the Organic Food Cult," *New York Times*, June 6, 1971, SM30.

13. Lichtenstein, "'Organic' Food Study Finds Pesticides."

14. E. B. Weiss, "Look Out for Coming Scandal in Surging Organic Foods." *Advertising Age*, December 6, 1971, 44.

15. Federal Trade Commission, *Federal Register* (Washington, DC: US Government Printing Office, 1978), 234.

16. Federal Trade Commission, *Proposed Trade Regulation Rule on Food Advertising, 16 CFR Part 437, Phase I: Staff Report and Recommendations* (Washington, DC: Government Printing Office, 1978), 231.

17. Federal Trade Commission, *Federal Register* 39842 (Washington, DC: Government Printing Office, November 11, 1974).

18. Federal Trade Commission, *Proposed Trade Regulation Rule on Food Advertising*, 231, 234.

19. Lynn Coody, personal communication, July 9, 2007.

20. Virginia Lee Warren, "Organic Foods: Spotting the Real Thing Can Be Tricky," *New York Times*, April 9, 1972, 72.

21. Jeffrey Haydu, "Cultural Modeling in Two Eras of U.S. Food Protest: Grahamites (1830s) and Organic Advocates (1960s–70s)," *Social Problems* 58, no. 3 (2011): 461–487; Youngberg and DeMuth, "Organic Agriculture in the United States."

22. Wini Breines, *Community and Organization in the New Left, 1962–1968: The Great Refusal* (New Brunswick, NJ: Rutgers University Press, 1981).

23. D. D. Treadwell, D. E. McKinney, and N. G. Creamer, "From Philosophy to Science: A Brief History of Organic Horticulture in the United States," *Hortscience* 38, no. 5 (2003): 1009–1014.

24. California Certified Organic Farmers, "Our History," http://www.ccof.org/ccof/history (accessed March 9, 2012).

25. Bernward Geier, "IFOAM and the History of the International Organic Movement," in *Organic Farming: An International History*, ed. William Lockeretz (Oxfordshire, UK: CABI, 2007), 175–186.

26. Youngberg and DeMuth, "Organic Agriculture in the United States."

27. John D. McCarthy and Mayer N. Zald, *The Trend of Social Movements in America: Professionalization and Resource Mobilization* (Morristown, NJ: General Learning Press, 1973); Suzanne Staggenborg, "The Consequences of Professionalization and Formalization in the Pro-Choice Movement," *American Sociological Review* 53, no. 4 (August 1988): 585–605.

28. Youngberg and DeMuth, "Organic Agriculture in the United States."

29. Elizabeth Henderson, personal communication, July 15, 2007.

30. Coody, personal communication.

31. Kathleen Merrigan, "Negotiating Identity within the Sustainable Agriculture Advocacy Coalition" (PhD diss., Massachusetts Institute of Technology, September 2000).

32. Enid Wonnacott, personal communication, January 8, 2008.

33. Coody, personal communication.

34. Treadwell, McKinney, and Creamer, "From Philosophy to Science."

35. Julie Guthman, *Agrarian Dreams: The Paradox of Organic Farming in California* (Berkeley: University of California Press, 2004).

36. Denise Webb, "Eating Well: Food Isn't Organic Just Because the Label Says So," *New York Times*, June 7, 1989, C10.

37. Julie Guthman, "Fast Food/Organic Food: Reflexive Tastes and the Making of 'Yuppie Chow,'" *Social and Cultural Geography* 4, no. 1 (2003): 45–58.

38. T. Robert Fetter and Julie A. Caswell, "Variation in Organic Standards Prior to the National Organic Program," *American Journal of Alternative Agriculture* 17, no. 2 (2002): 55–75.

39. Grace Gershuny, personal communication, January 8, 2007.

40. Jake Lewin, personal communication, November 28, 2007; Gershuny, personal communication, January 8, 2007.

41. Pat Stone, "Organic Agriculture: Turning a Movement into an Industry," *Mother Earth News*, September–October 1989, http://www.motherearthnews.com/homesteading-and-livestock/organic-agriculture-movement-industry-zmaz89sozshe.aspx#axzz2jeKbB1WH (accessed August 1, 2014).

42. Katherine DiMatteo and Grace Gershuny, "The Organic Trade Association," in *Organic Farming: An International History*, ed. William Lockeretz (Oxfordshire, UK: CABI, 2007), 254.

43. Ibid.

44. Gershuny, personal communication, January 8, 2007.

45. Katherine DiMatteo, personal communication, November 29, 2007.

46. Ibid.

47. Youngberg and DeMuth, "Organic Agriculture in the United States."

48. Jim Riddle, personal communication, November 2, 2007.

49. Coody, personal communication, July 9, 2007.

50. Treadwell, McKinney, and Creamer, "From Philosophy to Science."

51. Guthman, *Agrarian Dreams*.

52. University of California at Santa Cruz, University Library, "Timeline: Cultivating a Movement, An Oral History Series on Organic Farming and Sustainable Agriculture on California's Central Coast," http://library.ucsc.edu/reg-hist/cultiv/ timeline (accessed August 1, 2014); Bob Scowcroft, "The Organic Conversation Begins Anew (Again)," *GreenMoney Journal*, http://archives.greenmoneyjournal.com/ article.mpl?newsletterid=39&articleid=505 (accessed August 30, 2014).

53. Scowcroft, "The Organic Conversation Begins Anew (Again)."

54. Webb, "Eating Well."

55. California Certified Organic Farmers, "Our History"; Stone, "Organic Agriculture."

56. Wonnacott, personal communication.

57. Belasco, *Appetite for Change*; Robert Gottlieb, *Forcing the Spring: The Transformation of the American Environmental Movement* (Washington, DC: Island Press, 2005).

58. Lynn R. Goldman et al., "Pesticide Food Poisoning from Contaminated Watermelons in California, 1985," *Archives of Environmental Health: An International Journal* 45, no. 4 (1990): 229–236.

59. Laura Shapiro and Linda Wright, "Suddenly, It's a Panic for Organic," *Newsweek*, March 27, 1989.

60. Stone, "Organic Agriculture."

61. Coody, personal communication.

62. Youngberg and DeMuth, "Organic Agriculture in the United States."

63. USDA Study Team on Organic Farming, "Report and Recommendations on Organic Farming" (Washington, DC: US Department of Agriculture, July 1980), http://www.nal.usda.gov/afsic/pubs/USDAOrgFarmRpt.pdf (accessed August 4, 2104).

64. Deborah Rich, "Organic and Sustainable Up for Review … Again," *Organic Farming Research Foundation Information Bulletin* (Fall 2008): 4–13, http://www .organic-center.org/reportfiles/ib16%20organic%20and%20sustainable%20up%20 for%20review.pdf (accessed July 22, 2014).

65. Joseph Heckman, "A History of Organic Farming: Transitions from Sir Albert Howard's War in the Soil to USDA National Organic Program," *Renewable Agriculture and Food Systems* 21, no. 3 (September 2006): 143–150, doi.org/10.1079/RAF2005126.

66. Rich, "Organic and Sustainable Up for Review … Again."

67. Youngberg and DeMuth, "Organic Agriculture in the United States."

68. In a revealing indication of the culture within the agriculture policy community, Merrigan, senior staff member of the US Senate Committee on Agriculture, Nutrition, and Forestry at the time, noted that the conventional agribusiness lobbyists and their legislative staff supporters would mock the "girly"-sounding LISA program, thereby in part providing the impetus for the name change. Kathleen Merrigan, personal communication, August 1, 2008.

69. Youngberg and DeMuth, "Organic Agriculture in the United States"; Rich, "Organic and Sustainable Up for Review … Again."

70. Merrigan, "Negotiating Identity within the Sustainable Agriculture Advocacy Coalition."

71. Sandra S. Batie, *Soil Erosion: Crisis in America's Croplands?* (Washington, DC: Conservation Foundation, 1983); Committee on the Role of Alternative Farming Methods in Modern Production Agriculture, National Research Council, *Alternative Agriculture* (Washington, DC: National Academies Press, 1989).

72. Keith Schneider, "Science Academy Recommends Resumption of Natural Farming," *New York Times*, September 8, 1989, A1.

73. Committee on the Role of Alternative Farming Methods, *Alternative Agriculture*.

74. Bob Scowcroft, personal communication, August 7, 2008.

75. California Certified Organic Farmers, "Our History."

76. Mark Lipson, personal communication, August 7, 2008.

77. Merrigan, personal communication.

78. US Department of Agriculture, Food Safety and Inspection Service, "If a Label Bears a Halal or Kosher Statement, Does FSIS Have to Monitor the Production of the Product?," http://askfsis.custhelp.com/app/answers/detail/a_id/375/related/1/session/L2F2LzEvdGltZS8xMzc3ODc3NjY5L3NpZC9Ea2lHXzN6bA%3D%3D (accessed August 4, 2014).

79. Merrigan, personal communication.

80. Stone, "Organic Agriculture."

81. Webb, "Eating Well."

82. Henderson, personal communication.

83. Scowcroft, personal communication, August 7, 2008.

84. Coody, personal communication, July 9, 2007.

85. Coody, personal communication; Lipson, personal communication.

86. Merrigan, "Negotiating Identity within the Sustainable Agriculture Advocacy Coalition."

87. Henderson, personal communication.

88. Wonnacott, personal communication.

89. Coody, personal communication.

90. Amy Little, personal communication, June 24, 2013.

91. Wonnacott, personal communication.

92. Henderson, personal communication.

93. Fred Kirschenmann, personal communication, November 16, 2007.

94. Coody, personal communication.

95. Over the years of working on this legislation, Merrigan would continuously encourage organic activists to create more a centralized and coordinated structure— something that most organic groups were reluctant to do. For her analysis of the weaknesses of the broader sustainable agriculture movement, see Merrigan, "Negotiating Identity within the Sustainable Agriculture Advocacy Coalition."

96. Organic Farmers Association Council, "Original Invitation to Join OFAC," December 6, 1990, http://rogerblobaum.com/656 (accessed August 4, 2014).

97. Merrigan, "Negotiating Identity within the Sustainable Agriculture Advocacy Coalition."

98. Jay Feldman, personal communication, July 19, 2007.

99. John D. McCarthy and Mayer N. Zald, "Resource Mobilization and Social Movements: A Partial Theory," *American Journal of Sociology* 82, no. 6 (May 1977): 1212–1241.

## Chapter 4: The Organic Coalition: United and Divided

1. Paul Cienfuegos, "The Organic Foods Movement: Led by Heinz Corporation or We the People?" CommonDreams.org, May 31, 2004, http://www.commondreams .org/views04/0531-11.htm (accessed August 6, 2014); Eliot Coleman, "The Benefits of Growing Organic Food," *Mother Earth News*, December 2001–January 2002, http://www.motherearthnews.com/real-food/benefits-of-growing-organic-food-zmaz01djzgoe.aspx?PageId=1#axzz2hAIokGcY (accessed August 6, 2014).

2. Hilary Chop, "Whose Label?" *Alternatives Journal* 29, no. 4 (2003): 19–20; Vijay Cuddeford, "When Organics Go Mainstream," *Alternatives Journal* 29, no. 4 (2003): 14–19.

3. Julie Guthman, *Agrarian Dreams: The Paradox of Organic Farming in California* (Berkeley: University of California Press, 2004); Kathleen Merrigan, "Negotiating Identity within the Sustainable Agriculture Advocacy Coalition" (PhD diss., Massachusetts Institute of Technology, September 2000); Garth Youngberg and Suzanne P. DeMuth, "Organic Agriculture in the United States: A 30-Year Retrospective," *Renewable Agriculture and Food Systems* 28, no. 4 (May 2013): 1–35, doi:10.1017/S1742170513000173.

4. Merrigan, "Negotiating Identity within the Sustainable Agriculture Advocacy Coalition."

5. Daniel Jaffee and Philip H. Howard, "Corporate Cooptation of Organic and Fair Trade Standards," *Agriculture and Human Values* 27 (2010): 387–399.

6. David S. Meyer and Catherine Corrigall-Brown, "Coalitions and Political Context: U.S. Movements against Wars in Iraq," *Mobilization* 10, no. 3 (2005): 327–344.

7. Amy Little, personal communication, June 24, 2013.

8. Samuel B. Bacharach and Edward J. Lawler, *Power and Politics in Organizations* (San Francisco: Jossey-Bass, 1980); William Gamson, *The Strategy of Social Protest*, 2nd ed. (Belmont, CA: Wadsworth, 1990); Barbara Hinckley, *Coalitions and Politics* (New York: Harcourt Brace Jovanovich, 1981); Kevin W. Hula, *Lobbying Together: Interest Group Coalitions in Legislative Politics* (Washington, DC: Georgetown University Press, 1999); Edward J. Lawler and George A. Youngs Jr., "Coalition Formation: An Integrative Model," *Sociometry* 38, no.1 (March 1975): 1–17; Nelson W. Polsby, "Coalition and Faction in American Politics: An Institutional View," in *Emerging Coalitions in American Politics*, ed. Seymour Martin Lipset (San Francisco: Institute for Contemporary Studies, 1978), 103–126; William H. Riker, *The Theory of Political Coalitions* (New Haven, CT: Yale University Press, 1962); John R. Wright and Arthur S. Goldberg, "Risk and Uncertainty as Factors in the Durability of Political Coalitions," *American Political Science Review* 79, no. 3 (September 1985): 704–718; John D. McCarthy and Mayer N. Zald, *The Trend of Social Movements in America: Professionalization and Resource Mobilization* (Morristown, NJ: General Learning Press, 1973); John D. McCarthy and Mayer N. Zald, "Resource Mobilization and Social Movements: A Partial Theory," *American Journal of Sociology* 82, no. 6 (May 1977): 1212–1241; Nella Van Dyke, "Crossing Movement Boundaries: Factors That Facilitate Coalition Protest by American College Students, 1930–1990," *Social Problems* 50 (2003): 226–250.

9. Paul A. Sabatier and Hank C. Jenkins-Smith, "Evaluating the Advocacy Coalition Framework," *Journal of Public Policy* 14, no. 2 (April 1994): 175–203; Bill Keller, "Coalitions and Associations Transform Strategy, Methods of Lobbying in Washington," *Congressional Quarterly Weekly Report* 40 (1982): 119–123; Burdett Loomis, "Coalitions of Interests: Building Bridges in the Balkanized State," in *Interest Group Politics*, ed. Allen J. Cigler and Burdett Loomis, 2nd ed. (Washington, DC: CQ Press,

1986), 258–274; Robert Salisbury, "The Paradox of Interest Groups in Washington: More Groups, Less Clout," in *The New American Political System,* ed. Antony King (Washington, DC: AEI Press, 1990), 203–230; Robert Salisbury, John Heinz, Edward Laumann, and Robert Nelson, "Who Works with Whom? Interest Group Alliances and Opposition," *American Political Science Review* 81 (1987): 1217–1234.

10. Nella Van Dyke and Holly M. McCammon, eds., *Strategic Alliances: Coalition Building and Social Movements* (Minneapolis: University of Minnesota Press, 2010).

11. Sabatier and Jenkins-Smith, "Evaluating the Advocacy Coalition Framework."

12. Marie Hojnacki, "Interest Groups' Decisions to Join Alliances or Work Alone," *American Journal of Political Science* 41 (January 1997): 61–87; Brian Obach, *Labor and the Environmental Movement: The Quest for Common Ground* (Cambridge, MA: MIT Press, 2004); Jill M. Bystydzienski and Steven P. Schacht, eds., *Forging Radical Alliances across Difference: Coalition Politics for the New Millennium* (Lanham, MD: Rowman and Littlefield, 2001); Klaus Eder, *The New Politics of Class: Social Movements and Cultural Dynamics in Advanced Societies* (Thousand Oaks, CA: SAGE, 1993); Fred Rose, *Coalitions across the Class Divide: Lessons from the Labor, Peace, and Environmental Movements* (Ithaca, NY: Cornell University Press, 2000); Suzanne Staggenborg, "Coalition Work in the Pro-Choice Movement: Organizational and Environmental Opportunities and Obstacles," *Social Problems* 33, no. 5 (June 1986): 374–390.

13. Merrigan, "Negotiating Identity within the Sustainable Agriculture Advocacy Coalition."

14. US Department of Agriculture, Agricultural Marketing Service, National Organic Program, "National Organic Standards Board," http://www.ams.usda.gov/AMSv1.0/ NOSB (accessed August 7, 2014).

15. US Department of Agriculture, National Organic Program, Organic Foods Production Act of 1990, 7 U.S.C. 6501, http://www.ams.usda.gov/AMSv1.0/getfile?dDo cName=STELPRDC5060370&acct=nopgeninfo (accessed August 7, 2014).

16. US Department of Agriculture, "National Organic Standards Board."

17. Cornucopia Institute, "The Organic Watergate—White Paper; Connecting the Dots: Corporate Influence at the USDA's National Organic Program" (Cornucopia, WI: Cornucopia Institute, May 2012), http://www.cornucopia.org/USDA/ OrganicWatergateWhitePaper.pdf (accessed August 7, 2014).

18. Guthman, *Agrarian Dreams,* 3.

19. Carolyn Dimitri and Lydia Oberholtzer, "Market-Led versus Government-Facilitated Growth: Development of the U.S. and EU Organic Agricultural Sectors," US Department of Agriculture, Economic Research Service, WRS-05–05, August 2005, http://www.ers.usda.gov/publications/wrs-international-agriculture-and-trade-outlook/wrs0505.aspx#.UopmG8QqhAQ (accessed August 7, 2014); Philip Conford,

*The Origins of the Organic Movement* (Edinburgh: Floris Books, 2001); Warren James Belasco, *Appetite for Change: How the Counterculture Took on the Food Industry* (Ithaca, NY: Cornell University Press, 2007).

20. Katherine DiMatteo, personal communication, November 29, 2007.

21. Urvashi Rangan, personal communication, July 18, 2007.

22. Kathleen Merrigan, personal communication, August 1, 2008.

23. Katherine DiMatteo, personal communication, November 29, 2007.

24. Bob Scowcroft, "From the Director," *Organic Farming Research Foundation Information Bulletin*, Fall 2008, 4.

25. Jay Feldman, personal communication, June 27, 2007.

26. Guthman, *Agrarian Dreams*.

27. Michael Sligh, personal communication, July 2, 2007.

28. Liana Hoodes, personal communication, July 6, 2006.

29. Sligh, personal communication, July 2, 2007.

30. Damien C. Adams and Matthew J. Salois, "Local versus Organic: A Turn in Consumer Preferences and Willingness-to-Pay," *Renewable Agriculture and Food Systems* 25, no. 4 (December 2010): 331–341.

31. Roger Blobaum, personal communication, June 27, 2007.

32. Fred Kirschenmann, personal communication, November 16, 2007.

33. Robert Gottlieb, *Forcing the Spring: The Transformation of the American Environmental Movement* (Washington, DC: Island Press, 2005).

34. Grace Gershuny, personal communication, January 8, 2007.

35. Hartman Group, "The Many Faces of Organic 2008," http://www.hartman-group.com/publications/reports/the-many-faces-of-organic-2008 (accessed August 11, 2014); Stewart Lockie, Kristen Lyons, Geoffrey Lawrence, and Kerry Mummery, "Eating 'Green': Motivations behind Organic Food Consumption in Australia," *Sociologia Ruralis* 42, no. 1 (2002): 23–41; Maria Magnusson, Anne Arvola, Ulla-Kaisa Koivisto Hursti, Lars Aberg, and Per-Olow Sjoden, "Choice of Organic Foods Is Related to Perceived Consequences for Human Health and to Environmentally Friendly Behaviour," *Appetite* 40, no. 2 (2003): 109–118; Anna Saba and Federico Messina, "Attitudes towards Organic Foods and Risk/Benefit Perception Associated with Pesticides," *Food Quality and Preference* 14, no. 8 (2003): 637–646.

36. Blobaum, personal communication.

37. Dimitri and Oberholtzer, "Market-Led versus Government-Facilitated Growth"; Hartman Group "The Many Faces of Organic 2008."

38. Lynn Coody, personal communication, July 9, 2007.

39. In 1988, during this debate, the Organic Trade Association was still called the OFPANA.

40. Katherine DiMatteo and Grace Gershuny, "The Organic Trade Association," in *Organic Farming: An International History*, ed. William Lockeretz (Oxfordshire, UK: CABI, 2007), 256–257.

41. Mark Lipson, personal communication, August 7, 2008.

42. Kirschenmann, personal communication, November 16, 2007.

43. Patricia Allen and Martin Kovach, "The Capitalist Composition of Organic: The Potential of Markets in Fulfilling the Promise of Organic Agriculture," *Agriculture and Human Values* 17, no. 3 (September 2000): 221–232; Julie Guthman, "Regulating Meaning, Appropriating Nature: The Codification of California Organic Agriculture," *Antipode* 30, no. 2 (April 1998): 135–154.

44. Kirschenmann, personal communication.

45. Joan Dye Gussow, personal communication, July 18, 2007.

46. Coody, personal communication.

47. "Organic Foods Production Act of 1990."

48. National Organic Standards Board, Joint Committee Meetings, May 1–2, 4–6, 1992, http://www.ams.usda.gov/AMSv1.0/getfile?dDocName=STELPRDC5057489 (accessed December 7, 2013).

49. Gussow noted the implicit sexism in her appointment to the board: "Even though I was a college professor, I was a woman and not a scientist, so I was considered a consumer representative." Gussow, personal communication.

50. Joan Dye Gussow, "Can an Organic Twinkie Be Certified?" http://joansgarden.org/Twinkie.pdf (accessed August 11, 2014).

51. DiMatteo, personal communication.

52. Jack Kittredge, "Sligh: 'Stay the Course,'" *Natural Farmer*, Spring 2006, 38, http://www.nofa.org/tnf/2006spring/Sligh%20-%20Stay%20the%20Course.pdf (accessed August 17, 2014).

53. Samuel Fromartz, *Organic, Inc.: Natural Foods and How They Grew* (Orlando, FL: Harcourt, 2007).

54. Rangan, personal communication.

55. Kirschenmann, personal communication.

56. Elizabeth Henderson, personal communication, July 15, 2007.

57. Jake Lewin, personal communication, November 28, 2007.

58. Hoodes, personal communication.

59. Dimitri and Oberholtzer, "Market-Led versus Government-Facilitated Growth."

60. Parasiticides are agents used to fight parasites.

61. Hoodes, personal communication.

62. In addition to the USDA organic label, certified organic goods identify the certifying organization. This is typically found in small print on the back of the package, though, in contrast to the large green and white USDA organic label that most consumers use to identify an organic product. The particular certifier has no special meaning to most consumers since they are all supposedly adhering to the same federal standard.

63. Some certifying organizations accredited by the USDA to do organic certification also provide other types of certification services, but these are private systems distinct from organic.

64. DiMatteo, personal communication.

65. Braydon G. King, "Reputational Dynamics of Private Regulation," *Socioeconomic Review* 12 (2014): 200–206; Marc Schneiberg and Tim Bartley, "Organizations, Regulation, and Economic Behavior: Regulatory Dynamics and Forms from the Nineteenth to Twenty-First Century," *Annual Review of Law and Social Science* 4 (2008): 31–61.

66. Frederick H. Buttel, "Environmentalization: Origins, Processes, and Implications for Rural Social Change," *Rural Sociology* 57, no. 1 (Spring 1992): 1–27.

67. Feldman, personal communication.

68. Merrigan, personal communication.

69. NOSB farmer rep, personal communication, 2008.

70. Merrigan, "Negotiating Identity within the Sustainable Agriculture Advocacy Coalition," 124.

71. Todd Gitlin, "The Left, Lost in the Politics of Identity," *Harper's Magazine*, September 1993, 16

72. Hoodes, personal communication.

73. Sabatier and Jenkins-Smith, "Evaluating the Advocacy Coalition Framework."

74. Hoodes, personal communication.

75. National Organic Coalition, "About Us," http://www.nationalorganiccoalition.org/about (accessed December 7, 2013).

76. "USDA Abandons Three Contentious Issues in Proposed Organic Standards," *American Journal of Alternative Agriculture* 13, no. 2 (1998): 1.

77. Gershuny, personal communication.

78. Lon S. Hatamiya to Michael V. Dunn, "Informational Memorandum for the Department Secretary," May 1, 1997, US Department of Agriculture, Agricultural Marketing Services, http://www.motherjones.com/politics/1998/05/organic-engineering-memo-6 (August 21, 2014).

79. Merrigan, "Negotiating Identity within the Sustainable Agriculture Advocacy Coalition."

80. Nella Van Dyke and Holly McCammon, eds., *Strategic Alliances: Coalition Building and Social Movements* (Minneapolis: University of Minnesota Press, 2010); Kevin W. Hula, *Lobbying Together: Interest Group Coalitions in Legislative Politics* (Washington, DC: Georgetown University Press, 1999).

81. Organic Consumers Association, "About the OCA: Who We Are and What We're Doing," http://www.organicconsumers.org/aboutus.cfm (accessed August 21, 2014).

82. Ronnie Cummins, personal communication, November 25, 2007.

83. Ibid.

84. Ibid.

85. William Fantle, personal communication, November 30, 2007.

86. Enid Wonnacott, personal communication, January 8, 2008.

87. Interview with organization member, 2008.

88. Robert Michels, *Political Parties: A Sociological Study of the Oligarchical Tendencies of Modern Democracy* (New York: Evergreen Review, Inc., 2008).

89. David S. Meyer and Catherine Corrigall-Brown, "Coalitions and Political Context: U.S. Movements against Wars in Iraq," *Mobilization* 10, no. 3 (2005): 327–344.

90. Herbert Haines, "Black Radicalization and the Funding of Civil Rights, 1957–1970," *Social Problems* 32, no. 1 (October 1984): 31–43.

91. Philip Howard, "Consolidation in the North American Organic Food Processing Sector, 1997–2007," *International Journal of Agriculture and Food* 16, no. 1 (2009): 13–30.

## Chapter 5: Are We Better Off? Movement Achievements and the Threat from Big Organic

1. Hartman Group, "Beyond Organic and Natural 2010: Resolving Confusion in Marketing Foods and Beverages," Hartman Group Syndicated Survey, February 2010.

2. Abigail Haddad, "The Problem with Organic Food," *American*, June 8, 2008, http://www.american.com/archive/2008/june-06-08/the-problem-with-organic-food (accessed August 27, 2014); Robert Paarlberg, "Attention Whole Foods Shoppers," *Foreign Policy*, May–June 2010, http://www.foreignpolicy.com/articles/2010/04/26/attention_whole_foods_shoppers (accessed August 27, 2014).

3. Dennis Avery, "The Hidden Dangers in Organic Food," Hudson Institute, November 1, 1998, http://www.hudson.org/index.cfm?fuseaction=publication_details&id=1196 (accessed August 27, 2014); Dennis Avery, "The Silent Killer in Organic Foods," Hudson Institute, April 8, 1999, http://www.hudson.org/index.cfm?fuseaction=publication_details&id=299 (accessed August 27, 2014); Alex Avery and Dennis Avery "Organic Food Campaign Goes Sharply Negative," Hudson Institute, November 16, 2002, http://www.hudson.org/index.cfm?fuseaction=publication_details&id=2075 (accessed August 27, 2014).

4. Roger Cohen, "The Organic Fable," *New York Times*, September 6, 2012, http://www.nytimes.com/2012/09/07/opinion/roger-cohen-the-organic-fable.html?_r=1&emc=tnt&tntemail1=y (accessed August 27, 2014); Joshua Gilder, "Science Reporting on Organic Food Is Out to Lunch," *U.S. News and World Report*, March 2, 2012, http://www.usnews.com/opinion/blogs/joshua-gilder/2012/03/02/science-reporting-on-organic-food-is-out-to-lunch (accessed August 27, 2014); Joanna Pearlstein, "Surprise! Conventional Agriculture Can Be Easier on the Planet," *Wired*, May 19, 2008, http://www.wired.com/science/planetearth/magazine/16-06/ff_heresies_03organics (accessed August 27, 2014); James McWilliams, "Organic Crops Alone Can't Feed the World," *Slate*, March 10, 2011, http://www.slate.com/articles/health_and_science/green_room/2011/03/organic_crops_alone_cant_feed_the_world.html (accessed August 27, 2014); Kim Severson, "More Choice, and More Confusion, in Quest for Healthy Eating," *New York Times*, September 8, 2012, http://www.nytimes.com/2012/09/09/us/would-be-healthy-eaters-face-confusion-of-choices.html (accessed August 27, 2014).

5. Organic Center, http://www.organic-center.org (accessed August 27, 2014).

6. Catherine Greene and Amy Kremen, *US Organic Farming in 2000–2001: Adoption of Certified Systems*, Agriculture Information Bulletin no. 780 (Washington, DC: US Department of Agriculture, Economic Research Services, Resource Economics Division, 2003); Mark Lipson, "Searching for the 'O-Word': Analyzing the USDA Current Research Information System for Pertinence to Organic Farming" (Santa Cruz, CA: Organic Farming Research Foundation, 1997); James F. Parr, "USDA Research on

Organic Farming: Better Late Than Never," *American Journal of Alternative Agriculture* 18, 3 (September 2003): 171–172.

7. National Sustainable Agriculture Coalition, "Organic Agriculture Research and Extension Initiative," http://sustainableagriculture.net/publications/grassrootsguide/ sustainable-organic-research/organic-research-extension-initiative/ (accessed August 27, 2014).

8. Oregon Tilth, "Farm Bill Victories for Organic Farmers," http://tilth.org/news/ farm-bill-victories-for-organic-farmers (accessed August 27, 2014).

9. Center for Agroecology and Sustainable Food Systems, "Sustainable Agriculture at UC Santa Cruz," http://casfs.ucsc.edu/about/history/sustainable-agriculture-at-uc-santa-cruz (accessed August 27, 2014).

10. Charles Benbrook, Xin Zhao, Jaime Yáñez, Neal Davies, and Preston Andrews, *New Evidence Confirms the Nutritional Superiority of Plant-Based Organic Foods* (Washington, DC: Organic Center, March 2008).

11. Catherine Greene, Carolyn Dimitri, Biing-Hwan Lin, William McBride, Lydia Oberholtzer, and Travis Smith, *Emerging Issues in the U.S. Organic Industry*, Economic Information Bulletin no. 55 (Washington, DC: US Department of Agriculture, Economic Research Service, June 2009), http://www.ers.usda.gov/publications/eib-economic-information-bulletin/eib55.aspx#.UpVLvbW1F8E (accessed August 27, 2014).

12. Emily E. Marriott and Michelle M. Wander, "Total and Labile Soil Organic Matter in Organic and Conventional Farming Systems," *Soil Science Society of America* 70 (May 2006): 950–959.

13. Mark Shwartz, "New Study Confirms the Ecological Virtues of Organic Farming," *Stanford Report*, March 10, 2006, http://news.stanford.edu/news/2006/ march15/organics-030806.html (accessed August 27, 2014).

14. Sydney Lupkin, "Pesticides in Tap Water Linked to Food Allergies," *ABC News*, December 3, 2012, http://abcnews.go.com/blogs/health/2012/12/03/pesticides-in-tap-water-linked-to-food-allergies (accessed August 27, 2014).

15. Environmental Protection Agency, "What Is a CAFO?" http://www.epa.gov/ region7/water/cafo (accessed August 27, 2014); Daniel Imhoff, *CAFO: The Tragedy of Industrial Animal Factories* (San Rafael, CA: Earth Aware Editions, 2010).

16. Pete Smith, Daniel Martino, Zucong Cai, Daniel Gwary, Henry Janzen, Pushpam Kumar, Bruce McCarl, Stephen Ogle, Frank O'Mara, Charles Rice, Bob Scholes, and Oleg Sirotenko, "Agriculture," in *Climate Change 2007: Mitigation; Contribution of Working Group III to the Fourth Assessment Report of the Intergovernmental Panel on Climate Change*, ed. Bert Metz, Ogunlade Davidson, Peter Bosch, Rutu Dave, and Leo Meyer (Cambridge: Cambridge University Press, 2007), 498–540.

17. David Pimentel, "Impacts of Organic Farming on the Efficiency of Energy Use in Agriculture" (Washington, DC: Organic Center, August 2006), http://organic-center .org/reportfiles/EnergyExecSummary.pdf (accessed August 27, 2014).

18. Carolyn Dimitri, Anne Effland, and Neilson Conklin, *Transformation of U.S. Agriculture and Farm Policy*, Economic Information Bulletin no. 3 (Washington, DC: US Department of Agriculture, Economic Research Service, June 2005).

19. Gerold Rahmann, "Biodiversity and Organic Farming: What Do We Know?" *Agriculture and Forestry Research* 3 (2011): 189–208.

20. Ivonne Audirac, ed., *Rural Sustainable Development in America* (New York: John Wiley and Sons, 1997).

21. Ian Heap, "International Survey of Herbicide Resistant Weeds," WeedScience .org, http://www.weedscience.org/summary/home.aspx (accessed August 27, 2014).

22. Miguel Altieri, "Ecological Impacts of Industrial Agriculture and the Possibilities for Truly Sustainable Farming," in *Hungry for Profit: The Agribusiness Threat to Farmers, Food, and the Environment*, ed. Fred Magdoff, John Bellamy Foster, and Frederick H. Buttel (New York: Monthly Review Press, 2000), 77–92.

23. Union of Concerned Scientists, "Genetic Engineering," http://www.ucsusa.org/ food_and_agriculture/our-failing-food-system/genetic-engineering/risks-of-genetic-engineering.html (accessed August 27, 2014).

24. Aaron J. Gassmann, Jennifer L. Petzold-Maxwell, Ryan S. Keweshan, and Mike W. Dunbar, "Field-Evolved Resistance to Bt Maize by Western Corn Rootworm," *PLoS ONE* 6, no. 7 (July 29, 2011), doi:10.1371/journal.pone.0022629.

25. Jeffrey M. Smith, *Genetic Roulette: The Documented Health Risks of Genetically Engineered Foods* (Fairfield, IA: Yes! Books, 2007).

26. Vandana Shiva, *Stolen Harvest: The Hijacking of the Global Food Supply* (Cambridge, MA: South End Press, 2000); Anthony Trewavas, "Urban Myths of Organic Farming," *Nature* 410 (March 22, 2001): 409–410.

27. Catherine Badgley, Jeremy Moghtader, Eileen Quintero, Emily Zakem, M. Jahi Chappell, Katia Avilés-Vázquez, Andrea Samulon, and Ivette Perfecto, "Organic Agriculture and the Global Food Supply," *Renewable Agriculture and Food Systems* 22, no. 2 (June 2007): 86–108; Joachim Raupp, Carola Pekrun, Meike Oltmanns, and Ulrich Köpkes, eds., *Long Term Field Experiments in Organic Farming* (Berlin: International Society of Organic Agriculture Research, 2006); John P. Reganold, Jerry D. Glover, Preston K. Andrews, and Herbert R. Hinman, "Sustainability of Three Apple Production Systems," *Nature* 410 (April 2001), doi:10.1038/35073574.

28. Crystal Smith-Spangler, Margaret L. Brandeau, Grace E. Hunter, J. Clay Bavinger, Maren Pearson, Paul J. Eschbach, Vandana Sundaram, Hau Liu, Patricia Schirmer, Christopher Stave, Ingram Olkin, and Dena M. Bravata, "Are Organic Foods Safer or

Healthier Than Conventional Alternatives? A Systematic Review," *Annals of Internal Medicine* 157, no. 5 (September 2012): 348–366. See also Alan D. Dangour, Karen Lock, Arabella Hayter, Andrea Aikenhead, Elizabeth Allen, and Ricardo Uauy, "Nutrition-Related Health Effects of Organic Foods: A Systematic Review," *American Journal of Clinical Nutrition* 92, no. 1 (2010): 203–210.

29. Alan D. Dangour, Sakhi Dodhia, Arabella Hayter, Elizabeth Allen, Karen Lock, and Ricardo Uauy, "Nutritional Quality of Organic Foods: A Systematic Review," *American Journal of Clinical Nutrition* 90, no. 3 (2009): 680–685.

30. Tom Philpott, "5 Ways the Stanford Study Sells Organics Short," *Mother Jones*, September 5, 2012, http://www.motherjones.com/tom-philpott/2012/09/five-ways-stanford-study-underestimates-organic-food (accessed August 28, 2014).

31. Mark Bittman, "That Flawed Stanford Study," *New York Times*, October 2, 2012, http://opinionator.blogs.nytimes.com/2012/10/02/that-flawed-stanford-study/ (accessed August 28, 2014).

32. Katrin Woese, Dirk Lange, Christian Boess, and Klaus Werner Bogl, "A Comparison of Organically and Conventionally Grown Foods: Results of a Review of the Relevant Literature," *Journal of the Science of Food and Agriculture* 74, no. 3 (July 1997): 281–293; Benbrook et al., *New Evidence Confirms the Nutritional Superiority of Plant-Based Organic Foods*.

33. Kirsten Brandt, Carlo Leifert, Roy Sanderson, and Chris Seal, "Agroecosystem Management and Nutritional Quality of Plant Foods: The Case of Organic Fruits and Vegetables," *Critical Reviews in Plant Sciences* 30, no. 1–2 (2011): 177–197.

34. John P. Reganold, Preston K. Andrews, Jennifer R. Reeve, Lynne Carpenter-Boggs, Christopher W. Schadt, J. Richard Alldredge, Carolyn F. Ross, Neal M. Davies, and Jizhong Zhou, "Fruit and Soil Quality of Organic and Conventional Strawberry Agroecosystems," *PLoS ONE* 5, no. 9 (September 2010), doi: 10.1371/journal.pone.0012346; Benbrook et al., *New Evidence Confirms the Nutritional Superiority of Plant-Based Organic Foods*.

35. USDA National Organic Program, USDA Science and Technology Programs, "2010–2011 Pilot Study: Pesticide Residue Testing of Organic Produce" (Washington, DC: Agriculture Marketing Service, November 2012), http://www.ams.usda.gov/AMSv1.0/getfile?dDocName=STELPRDC5101234 (accessed August 28, 2014).

36. Chensheng Lu, Dianne Knutson, Jennifer Fisker-Andersen, and Richard A. Fenske, "Biological Monitoring Survey of Organophosphorus Pesticide Exposure among Preschool Children in the Seattle Metropolitan Area," *Environmental Health Perspectives* 109, no. 3 (March 2001): 299–303.

37. Marian Burros, "Obamas to Plant Vegetable Garden at White House," *New York Times*, March 19, 2009, A1.

38. UN Food and Agriculture Organization, "Priority Areas for Interdisciplinary Action: Medium Term Plan, 2002–2007," August 2000, http://www.fao.org/docrep/X7572E/X7572e02.htm (accessed August 28, 2014).

39. Fred Magdoff, John Bellamy Foster, and Frederick H. Buttel, eds., *Hungry for Profit: The Agribusiness Threat to Farmers, Food, and the Environment* (New York: Monthly Review Press, 2000); Raj Patel, *Stuffed and Starved: The Hidden Battle for the World Food System* (Brooklyn: Melville House Publishing, 2007); Shiva, *Stolen Harvest.*

40. Rachel Hine and Jules Pretty, "Organic Agriculture and Food Security in Africa," (New York: United Nations, 2008).

41. Badgley et al., "Organic Agriculture and the Global Food Supply."

42. Carolyn Dimitri and Lydia Oberholtzer, "Market-Led versus Government-Facilitated Growth: Development of the U.S. and EU Organic Agricultural Sectors," US Department of Agriculture, Economic Research Service, WRS 05–05, August 2005, http://www.ers.usda.gov/publications/wrs-international-agriculture-and-trade-outlook/wrs0505.aspx#.UopmG8QqhAQ (accessed August 7, 2014); Johannes Michelsen, "Recent Development and Political Acceptance of Organic Farming in Europe," *Sociologia Ruralis* 41, no. 1 (January 2001): 3–20.

43. While organic purchasing is widespread among the consuming public, the extent to which consumers are committed to organic varies substantially. The Hartman Group has conducted research on the organic market for nearly twenty years. It separates organic consumers into three categories: core organic consumers who primarily consume organic foods, mid-level consumers who regularly buy certain organic products, and periphery organic consumers who only occasionally purchase organic goods. The majority of organic consumers fall in the mid-level category (62 percent), while core organic consumers make up 24 percent of the organic-buying population and 14 percent reside in the organic periphery. Hartman Group, "Beyond Organic and Natural 2010."

44. Organic Trade Association, "2011 Organic Industry Survey Overview," http://www.ota.com/pics/documents/2011OrganicIndustrySurvey.pdf (accessed July 23, 2014).

45. Organic Trade Association, "Industry Statistics and Projected Growth," June 2011, http://www.ota.com/organic/mt/business.html (accessed December 1, 2013).

46. US Department of Agriculture, Economic Research Services, "Organic Production: Table 2–U.S. Certified Organic Farmland Acreage, Livestock Numbers, and Farm Operations, 1992–2011," http://www.ers.usda.gov/data-products/organic-production.aspx#25762 (accessed August 28, 2014).

47. Greene et al., *Emerging Issues in the U.S. Organic Industry.*

48. Some research indicates that the most crucial support that government can offer to advance organic agriculture is the establishment of production standards and certification systems. Subsidies for farmers to convert to organic have relatively less effect. Kennet S. C. Lynggaard, "The Farmer within an Institutional Environment: Comparing Danish and Belgian Organic Farming," *Sociologia Ruralis* 41, no.1 (January 2001): 85–111. For more on the role of European governments in organic development, see Karin Hofer, "Labelling of Organic Food Products," in *The Voluntary Approach to Environmental Policy: Joint Environmental Policy-Making in Europe*, ed. Arthur P. J. Mol, Volkmar Lauber, and Duncan Liefferink (Oxford: Oxford University Press, 2000), 156–191; Lynggaard, "The Farmer within an Institutional Environment"; Johannes Michelsen, "Organic Farming in a Regulatory Perspective," *Sociologia Ruralis* 41, no. 1 (2001): 62–84; Matthew Reed, "Fight the Future! How the Contemporary Campaigns of the UK Organic Movement Have Arisen from Their Composting Past," *Sociologia Ruralis* 41, no. 1 (2001): 131–145.

49. Philip Howard, "Consolidation in the North American Organic Food Processing Sector, 1997–2007," *International Journal of Agriculture and Food* 16, no. 1 (2009): 13–30.

50. Michael Sligh, personal communication, July 2, 2007.

51. Michael Sligh and Carolyn Christman, "Who Owns Organic? The Global Status, Prospects, and Challenges of a Changing Organic Market" (Pittsboro, NC: Rural Advancement Foundation International—USA, 2003); Michael Pollan, *The Omnivore's Dilemma: A Natural History of Four Meals* (New York: Penguin, 2006).

52. Vijay Cuddeford, "When Organics Go Mainstream," *Alternatives Journal* 29, no. 4 2003): 14–19; Daniel Jaffee and Philip H. Howard, "Corporate Cooptation of Organic and Fair Trade Standards," *Agriculture and Human Values* 27 (2010): 387–399; Kristen Lyons, "Corporate Environmentalism and Organic Agriculture in Australia: The Case of Uncle Toby's," *Rural Sociology* 64, no. 2 (1999): 251–266.

53. Daniel Buck, Christina Getz, and Julie Guthman, "From Farm to Table: The Organic Vegetable Commodity Chain in Northern California," *Sociologia Ruralis* 44, no. 3 (1997): 3–20; Douglas H. Constance, Jin Youn Choi, and Holly Lyke-Ho-Gland, "Conventionalization, Bifurcation, and Quality of Life: Certified and Non-Certified Organic Farmers in Texas," *Southern Rural Sociology* 23, no. 1 (2008): 208–234; Daniel Imhoff, "Organic Incorporated," *Whole Earth* 92, no. 4 (1998); Karen Klonsky, "Forces Impacting the Production of Organic Foods," *Agriculture and Human Values* 17 (2000): 233–243; Laura B. Delind, "Transforming Organic Agriculture into Industrial Organic Products: Reconsidering National Organic Standards," *Human Organization* 59, no. 2 (2000): 198–208; Stewart Lockie and Darren Halpin, "The 'Conventionalization' Thesis Reconsidered: Structural and Ideological Transformation of Australian Organic Agriculture," *Sociologia Ruralis* 45, no. 4 (2005): 284–307.

54. Organic Trade Association, "2011 Organic Industry Survey Overview."

55. Klonsky, "Forces Impacting the Production of Organic Foods."

56. Hartman Group, "Beyond Organic and Natural 2010."

57. Howard, "Consolidation in the North American Organic Food Processing Sector."

58. Ibid., 16.

59. Ibid., 18.

60. Sligh and Christman, "Who Owns Organic?"

61. Howard, "Consolidation in the North American Organic Food Processing Sector."

62. Associated Press, "Target Rolling Out Organic, Natural Grocery Brand," June 7, 2013, http://www.cbsnews.com/8301-505145_162-57588227/target-rolling-out-organic-natural-grocery-brand (accessed August 29, 2014).

63. George Southworth, "Natural/Organic Industry Outlook," *Cooperative Grocer*, September–October 2001, http://www.cooperativegrocer.coop/articles/index.php?id=333 (accessed August 29, 2014); Stephen Hannaford, *Market Domination! The Impact of Industry Consolidation on Competition, Innovation, and Consumer Choice* (Westport, CT: Praeger Publishers, 2007).

64. Klonsky, "Forces Impacting the Production of Organic Foods."

65. Liana Hoodes, Michael Sligh, Harriet Behar, Roger Blobaum, Lisa J. Bunin, Lynn Coody, Elizabeth Henderson, Faye Jones, Mark Lipson, and Jim Riddle, *National Organic Action Plan: From the Margins to the Mainstream–Advancing Organic Agriculture in the U.S.* (Pittsboro, NC: Rural Advancement Foundation International–USA, January 2010), http://www.rafiusa.org/docs/noap.pdf (accessed August 29, 2014).

66. Ibid.

67. William Fantle, personal communication, November 30, 2007.

68. Hoodes et al., *National Organic Action Plan*.

69. Jim Riddle, personal communication, November 2, 2007.

70. Howard, "Consolidation in the North American Organic Food Processing Sector."

71. Bob Scowcroft, "The Organic Conversation Begins Anew (Again)," *GreenMoney Journal*, http://archives.greenmoneyjournal.com/article.mpl?newsletterid=39&articleid=505 (accessed August 30, 2014).

72. James Traub, "Into the Mouths of Babes," *New York Times*, July 24, 1988. SM18.

73. Cornucopia Institute, "The Organic Watergate–White Paper; Connecting the Dots: Corporate Influence at the USDA's National Organic Program" (Cornucopia, WI: Cornucopia Institute, 2012), 3,

74. Organic Consumers Association, "USDA Attempts to Pack Organic Standards Board with Corporate Agribusiness Reps: Organic Consumers Fight Hijacked Seats on NOSB," December 7, 2006, http://www.organicconsumers.org/articles/article_3526.cfm (accessed August 30, 2014).

75. Liana Hoodes, letter to the NOP and NOSB, January 29, 2013, regarding the NOSB discussion during public meetings.

76. Hoodes et al., *National Organic Action Plan*.

77. Sligh, personal communication, July 2, 2007.

78. Andrew Martin, "How to Add Oomph to Organic," *New York Times*, August 19, 2007, http://www.nytimes.com/2007/08/19/business/yourmoney/19feed.html?_r=0 (accessed August 30, 2014).

79. Jack Kittredge, "Sligh: 'Stay the Course,'" *Natural Farmer*, Spring 2006, 38, http://www.nofa.org/tnf/2006spring/Sligh%20-%20Stay%20the%20Course.pdf (accessed August 17, 2014).

80. Quoted in ibid.

81. Liz Henderson, personal communication, July 15, 2007.

82. Merrigan abruptly resigned from her post in 2013, leaving many in the organic community speculating about the sudden move.

83. Hoodes et al., *National Organic Action Plan*.

84. Andrew Mollison, "Georgia Rep. Nathan Deal behind Attack on Organics," *Atlanta Journal and Constitution*, February 15, 2003.

85. Organic Trade Association, "Organic Foods Production Act Backgrounder," http://www.ota.com/pp/legislation/backgrounder.html (accessed August 30, 2014).

86. Stephanie Strom, "Has 'Organic' Been Oversized?" *New York Times*, July 7, 2012, http://www.nytimes.com/2012/07/08/business/organic-food-purists-worry-about-big-companies-influence.html?pagewanted=all&_r=0 (accessed August 30, 2014).

87. Ibid.

88. Title 7 U.S. Code Section 6510(a)(1).

89. Samuel Fromartz, *Organic, Inc.: Natural Foods and How They Grew* (Orlando, FL: Harcourt, 2007).

90. Organic Trade Association, "Lawsuit Chronology," http://www.ota.com/LawsuitChronology.html (accessed August 30, 2014).

91. Fromartz, *Organic Inc.*

92. Suzanne Shelton, "Executive Interview: Q&A with Katherine DiMatteo, Executive Director, OTA," *Engredea News and Analysis*, December 22, 2005, http://newhope360.com/supply-news-amp-analysis/executive-interview-qa-katherine-dimatteo-executive-director-ota (accessed August 30, 2014).

93. Emily Brown Rosen, "The Devil Is in the Details: or Why Organic Standards Matter," *Natural Farmer*, Spring 2006, http://www.nofa.org/tnf/2006spring/The%20 Devil%20Is%20In%20the%20Details.pdf (accessed August 30, 2014).

94. Ibid.

95. William Neuman, "New Pasture Rules Issued for Organic Dairy Producers," *New York Times*, February 12, 2010, http://www.nytimes.com/2010/02/13/business/13organic.html (accessed August 30, 2014).

96. Michael Jarvis and Billy Cox, "USDA Issues Final Rule on Organic Access to Pasture," *USDA Agricultural Marketing Service*, February 12, 2010, http://www.ams.usda .gov/AMSv1.0/ams.fetchTemplateData.do?template=TemplateU&navID=&page= Newsroom&resultType=Details&dDocName=STELPRDC5082658&dID=126904&wf =false&description=USDA+Issues+Final+Rule+on+Organic+Access+to+Pasture+&top Nav=Newsroom&leftNav= (accessed August 30, 2014).

97. Liana Hoodes, personal communication, 2013.

98. Wayne F. Wilcox, "Fire Blight," New York State Integrated Pest Management Program, Cornell University, http://www.nysipm.cornell.edu/factsheets/treefruit/diseases/fb/fb.asp (accessed August 30, 2014).

99. Julie Guthman, *Agrarian Dreams: The Paradox of Organic Farming in California* (Berkeley: University of California Press, 2004).

100. David J. Hess, "Organic Food and Agriculture in the U.S.: Object Conflicts in a Health-Environmental Social Movement," *Science as Culture* 13, no. 4 (December 2004): 493–513.

101. Diane Brady, "The Organic Myth," *Business Week*, October 16, 2006, 50–56.

102. Quoted in Melanie D. G. Kaplan, "Stonyfield Farm CEO: How an Organic Yogurt Business Can Scale," *Smart Planet*, May 17, 2010, http://www.smartplanet .com/blog/pure-genius/stonyfield-farm-ceo-how-an-organic-yogurt-business-can-scale/3638 (accessed August 31, 2014).

103. Quoted in Brady, "The Organic Myth."

104. Buck, Getz, and Guthman, "From Farm to Table"; Tracey Clunies-Ross, "Organic Food: Swimming against the Tide?" in *Political, Social, and Economic Perspectives on the International Food System*, ed. Terry Marsden and Jo Little (Aldershot, UK: Avebury Press, 1990), 200–214; Tracey Clunies-Ross and Graham Cox, "Challenging the

Productivist Paradigm: Organic Farming and the Politics of Agricultural Change," in *Regulating Agriculture*, ed. Philip Lowe, Terry Marsden, and Sarah Whatmore (London: David Fulton Publishers, 1994), 53–74; Julie Guthman, "Regulating Meaning, Appropriating Nature: The Codification of California Organic Agriculture," *Antipode* 30, no. 2 (April 1998): 135–154; Brad Coombes and Hugh Campbell, "Dependent Reproduction of Alternative Modes of Agriculture: Organic Farming in New Zealand," *Sociologia Ruralis* 38, no. 2 (1998): 127–145; David Goodman, "Organic and Conventional Agriculture: Materializing Discourse and Agro-Ecological Managerialism," *Agriculture and Human Values* 17, no. 3 (September 2000): 215–219; Timothy Vos, "Visions of the Middle Landscape: Organic Farming and the Politics of Nature," *Agriculture and Human Values* 17 (2000): 245–256; Hugh Campbell and Ruth Liepins, "Naming Organics: Understanding Organic Standards in New Zealand as a Discursive Field," *Sociologia Ruralis* 41, no. 1 (2001): 21–39; Stewart Lockie, Kristen Lyons, and Geoffrey Lawrence, "Constructing 'Green' Foods: Corporate Capital, Risk, and Organic Farming in Australia and New Zealand," *Agriculture and Human Values* 17, no. 4 (December 2000): 315–322.

105. US Department of Agriculture, Agriculture Marketing Service, "National Count of Farmers Markets, Directory Listing Graph: 1994–2013," http://www.ams.usda .gov/AMSv1.0/ams.fetchTemplateData.do?template=TemplateS&navID=Wholesale andFarmersMarkets&leftNav=WholesaleandFarmersMarkets&page=WFMFarmers MarketGrowth&description=Farmers%20Market%20Growth&acct=frmrdirmkt (accessed August 31, 2014); US Department of Agriculture, "Know Your Farmer, Know Your Food: Our Mission," http://www.usda.gov/wps/portal/usda/usdahome ?navid=KYF_MISSION (accessed August 31, 2014).

## Chapter 6: Searching for Social Justice

1. David J. Connell, John Smothers, and Alun Joseph, "Farmers' Markets and the 'Good Food' Value Chair: A Preliminary Study," *Local Environment: The International Journal of Justice and Sustainability* 13, no. 3 (2008): 169–186.

2. Alison Hope Alkon and Julian Agyeman, eds., *Cultivating Food Justice: Race, Class, and Sustainability* (Cambridge, MA: MIT Press, 2011); Laura B. DeLind, "Transforming Organic Agriculture into Industrial Organic Products: Reconsidering National Organic Standards," *Human Organization* 59, no. 2 (2000): 198–208.

3. Fred Magdoff, John Bellamy Foster, and Frederick H. Buttel, eds., *Hungry for Profit: The Agribusiness Threat to Farmers, Food, and the Environment* (New York: Monthly Review Press, 2000); Linda Lobao and Katherine Meyer, "The Great Agricultural Transition: Crisis, Change, and Social Consequences of Twentieth Century US Farming," *Annual Review of Sociology* 27 (2001): 103–125.

4. Environmental Protection Agency, "Agriculture: Demographics," http://www .epa.gov/agriculture/ag101/demographics.html (accessed August 31, 2014).

5. Jorge Fernandez-Cornejo, *Off-Farm Income, Technology Adoption, and Farm Economic Performance*, Economic Research Report no. ERR-36 (Washington, DC: US Department of Agriculture, Economic Research Service, February 2007), http://www.ers.usda.gov/publications/err-economic-research-report/err36.aspx#.UrThs8r8IyE (accessed August 31, 2014).

6. Magdoff, Foster, and Buttel, *Hungry for Profit*; Patrick Canning, *A Revised and Expanded Food Dollar Series: A Better Understanding of Our Food Costs*, Economic Research Report no. ERR-114 (Washington, DC: US Department of Agriculture, Economic Research Service, February 2011), http://www.ers.usda.gov/publications/err-economic-research-report/err114.aspx#.UrToM7W1F8E (accessed August 31, 2014).

7. Julie Guthman, *Agrarian Dreams: The Paradox of Organic Farming in California* (Berkeley: University of California Press, 2004).

8. Samuel Fromartz, *Organic, Inc.: Natural Foods and How They Grew* (Orlando, FL: Harcourt, 2007); Julie Guthman, "Fast Food/Organic Food: Reflexive Tastes and the Making of 'Yuppie Chow,'" *Social and Cultural Geography* 4, no. 1 (2003): 45–58.

9. Katherine DiMatteo, personal communication, November 29, 2007.

10. Guthman, "Fast Food/Organic Food."

11. Ibid.

12. Fromartz, *Organic, Inc.*, 137.

13. Will Fantle, personal communication, November 30, 2007.

14. United Natural Foods, Inc., "UNFI History," https://www.unfi.com/Company/Pages/UNFIHistory.aspx (accessed September 1, 2014).

15. Lynn Coody, personal communication, July 9, 2007.

16. Walmart, "Global Responsibility: Sustainable Agriculture," http://corporate.walmart.com/global-responsibility/environment-sustainability/sustainable-agriculture (accessed September 1, 2014).

17. Organic Valley, "Transparency: Organic Valley Dairy Herd Size," http://www.organicvalley.coop/about-us/transparency/herd-chart (accessed September 1, 2014).

18. Organic Valley, "Transparency: Additional Farm Standards," http://www.organicvalley.coop/about-us/transparency/additional-farm-standards/ (accessed September 1, 2014).

19. "USDA Backs Down on Curtailing Group Certification for Coops and Grower Groups," *Sustainable Food News*, May 2, 2007, http://www.organicconsumers.org/articles/article_5061.cfm (accessed September 1, 2014).

20. Warren James Belasco, *Appetite for Change: How the Counterculture Took on the Food Industry* (Ithaca, NY: Cornell University Press, 2007).

21. Catherine Greene, Carolyn Dimitri, Biing-Hwan Lin, William McBride, Lydia Oberholtzer, and Travis Smith, *Emerging Issues in the U.S. Organic Industry*, Economic Information Bulletin no. 55 (Washington, DC: US Department of Agriculture, Economic Research Service, June 2009), http://www.ers.usda.gov/publications/eib-economic-information-bulletin/eib55.aspx#.UpVLvbW1F8E (accessed August 27, 2014).

22. Organic Trade Association, "2009 U.S. Families' Organic Attitudes and Beliefs Study," June 2009, http://www.ota.com/pics/documents/2009OTA-KiwiExecutive Summary.pdf (accessed September 1, 2014).

23. Roger Cohen, "The Organic Fable," *New York Times*, September 6, 2012, http://www.nytimes.com/2012/09/07/opinion/roger-cohen-the-organic-fable.html?_r=0 (accessed August 27, 2014); Abigail Haddad, "The Problem with Organic Food," *American*, June 8, 2008, http://www.american.com/archive/2008/june-06-08/the-problem-with-organic-food (accessed August 27, 2014); Robert Paarlberg, "Attention Whole Foods Shoppers," *Foreign Policy*, May–June 2010, http://www.foreignpolicy.com/articles/2010/04/26/attention_whole_foods_shoppers (accessed August 27, 2014).

24. Alkon and Agyeman, *Cultivating Food Justice*.

25. Nathan McClintock, "From Industrial Garden to Food Desert: Demarcated Devaluation in the Flatlands of Oakland, California," in *Cultivating Food Justice: Race, Class, and Sustainability*, ed. Alison Hope Alkon and Julian Agyeman (Cambridge, MA: MIT Press, 2011), 89–120.

26. US Department of Agriculture, Agricultural Marketing Service, "Creating Access to Healthy, Affordable Food: Food Deserts," http://apps.ams.usda.gov/fooddeserts/foodDeserts.aspx (accessed September 1, 2014).

27. Robert Gottlieb and Anupama Joshi, *Food Justice* (Cambridge, MA: MIT Press, 2010). See also Alkon and Agyeman, *Cultivating Food Justice*; Alison Hope Alkon and Kari Marie Norgaard, "Breaking the Food Chains: An Investigation of Food Justice Activism," *Sociological Inquiry* 79, no. 3 (August 2009): 289–305.

28. David J. Hess, *Localist Movements in a Global Economy: Sustainability, Justice, and Urban Development in the United States* (Cambridge, MA: MIT Press, 2009).

29. For a history of urban gardening, see Laura J. Lawson, *City Bountiful: A Century of Community Gardening in America* (Berkeley: University of California Press, 2005).

30. Nuestras Raices, http://www.nuestras-raices.org/home.html (accessed December 21, 2013).

31. Hess, *Localist Movements in a Global Economy*.

32. Growing Power, "Grow," http://www.growingpower.org/growing.htm (accessed September 1, 2014).

33. Julian Agyeman and Jesse McEntee, "Moving the Field of Food Justice Forward through the Lens of Urban Political Ecology," *Geography Compass* 8, no. 3 (2014): 211–220.

34. Alkon and Agyeman, *Cultivating Food Justice*; Robert Gottlieb, *Forcing the Spring: The Transformation of the American Environmental Movement* (Washington DC: Island Press, 2005).

35. US Department of Agriculture, Food and Nutrition Service, "Women, Infants, and Children: Farmers' Market Nutrition Program (FMNP)," http://www.fns.usda.gov/fmnp (accessed September 1, 2014).

36. Some CSA farms have even more expansive social justice commitments. The W. Rogowski Farm, run by MacArthur Fellow ("Genius Grant" winner) Cheryl Rogowski in Orange County, New York, works with Just Food in New York City to give low-income consumers access to fresh produce. Rogowski provides English-language lessons for native Spanish speakers and has helped several immigrant farmers start their own operations. See W. Rogowski Farm, http://www.rogowskifarm.com/index.html (accessed September 1, 2014).

37. Gottlieb and Joshi, *Food Justice*.

38. Margaret Gray, *Labor and the Locavore: The Making of a Comprehensive Food Ethic* (Berkeley: University of California Press, 2013).

39. Ibid.

40. Linda Majka and Theo Majka, "Organizing US Farmworkers: A Continuous Struggle," in *Hungry for Profit: The Agribusiness Threat to Farmers, Food, and the Environment*, ed. Fred Magdoff, John Bellamy Foster, and Frederick H. Buttel (New York: Monthly Review Press, 2000), 161–174.

41. Robert Gottlieb, *Environmentalism Unbound* (Cambridge, MA: MIT Press, 2001).

42. Linda Majka and Theo Majka, "Organizing US Farmworkers: A Continuous Struggle," in *Hungry for Profit: The Agribusiness Threat to Farmers, Food and the Environment,* edited by Fred Magdoff, John Bellamy Foster, and Frederick H. Buttel (New York: Monthly Review Press, 2000), 161–174.

43. Guthman, *Agrarian Dreams*.

44. Katherine DiMatteo and Grace Gershuny, "The Organic Trade Association," in *Organic Farming: An International History*, ed. William Lockeretz (Oxfordshire, UK: CABI, 2007), 253–263.

45. International Federation of Organic Agriculture Movements, *The IFOAM Norms for Organic Production and Processing* (Bonn: Die Deutsche Bibliothek, 2012).

46. Aimee Shreck, Christy Getz, and Gail Feenstra, "Farmworkers in Organic Agriculture: Toward a Broader Notion of Sustainability," *Newsletter of the University of*

*California Sustainable Agriculture Research and Education Program* 17, no. 1 (Winter–Spring 2005): 1–3.

47. Guthman, *Agrarian Dreams*.

48. Quoted in Jason Mark, "Workers on Organic Farms Are Treated as Poorly as Their Conventional Counterparts," *Grist*, August 2, 2006, http://grist.org/article/mark (accessed September 1, 2014).

49. Swanton Berry Farm, http://www.swantonberryfarm.com (accessed September 1, 2014).

50. Quoted in Andrea Blum, "Organic Farming's Labor Problem," *Common Ground*, February 2006, http://www.columbia.org/pdf_files/cainstituteforruralstudies.pdf (accessed September 1, 2014).

51. Shreck, Getz, and Feenstra, "Farmworkers in Organic Agriculture," 3.

52. Liana Hoodes, Michael Sligh, Harriet Behar, Roger Blobaum, Lisa J. Bunin, Lynn Coody, Elizabeth Henderson, Faye Jones, Mark Lipson, and Jim Riddle, *National Organic Action Plan: From the Margins to the Mainstream- Advancing Organic Agriculture in the U.S.* (Pittsboro, NC: Rural Advancement Foundation International–USA, January 2010), http://www.rafiusa.org/docs/noap.pdf (accessed August 29, 2014).

## Chapter 7: Strategic Innovation: The Three Trajectories of the Organic Movement

1. For other examples as well as more theoretical analysis of the interactive dynamics between policy and movement development, see Doug McAdam, "Conceptual Origins, Current Problems, Future Directions," in *Comparative Perspectives on Social Movements*, ed. Doug McAdam, John D. McCarthy, and Mayer Zald (Cambridge: Cambridge University Press, 1996), 23–40; John D. McCarthy, David Britt, and Mark Wolfson, "The Institutional Channeling of Social Movements by the State in the United States," *Research in Social Movements, Conflicts, and Change* 13 (1991): 45–76; David S. Meyer, Valerie Jenness, and Helen Ingram, *Routing the Opposition: Social Movements, Public Policy, and Democracy* (Minneapolis: University of Minnesota Press, 2005); Sidney Tarrow, "States and Opportunities," in *Comparative Perspectives on Social Movements*, ed. Doug McAdam, John D. McCarthy, and Mayer Zald, 41–61 (Cambridge: Cambridge University Press, 1996).

2. David Porter and Chester L. Mirsky, *Megamall on the Hudson: Planning, Walmart, and Grassroots Resistance* (Victoria, BC: Trafford Publishing, 2003).

3. Liana Hoodes, Michael Sligh, Harriet Behar, Roger Blobaum, Lisa J. Bunin, Lynn Coody, Elizabeth Henderson, Faye Jones, Mark Lipson, and Jim Riddle, *National Organic Action Plan: From the Margins to the Mainstream—Advancing Organic*

*Agriculture in the U.S.* (Pittsboro, NC: Rural Advancement Foundation International—USA, January 2010), http://www.rafiusa.org/docs/noap.pdf (accessed August 29, 2014), 5.

4. Ronnie Cummins, personal communication, November 25, 2007.

5. Suzanne Staggenborg, "Coalition Work in the Pro-Choice Movement: Organizational and Environmental Opportunities and Obstacles," *Social Problems* 33, no. 5 (June 1986): 374–390.

6. Joe Mendelson, personal communication, July 30, 2007.

7. Eliot Coleman, "The Benefits of Growing Organic Food," *Mother Earth News*, December 2001–January 2002, http://www.motherearthnews.com/real-food/benefits-of-growing-organic-food-zmaz01djzgoe.aspx?PageId=1#axzz2hAIokGcY (accessed September 2, 2014).

8. Hoodes et al., *National Organic Action Plan.*

9. Certified Naturally Grown, "Brief History of Certified Naturally Grown," http://www.naturallygrown.org/about-cng/brief-history-of-certified-naturally-grown (accessed September 2, 2014).

10. Measures are included to prevent corruption such as a prohibition on reciprocal inspection between two member farmers and the requirement that all reports be publicly available.

11. Quoted in Emily Stewart, "Organic in the Valley: Few Producers, Growing Demand," *Poughkeepsie Journal*, June 2, 2013.

12. Jim Riddle, personal communication, November 2, 2007.

13. Roger Blobaum, personal communication, June 27, 2007.

14. Urvashi Rangan, personal communication, July 18, 2007.

15. Food Alliance, "About Food Alliance," http://foodalliance.org/about (accessed September 2, 2014).

16. Mark Dunau and Elizabeth Henderson, "Growing the Farmer's Pledge," *Natural Farmer* (Spring 2006): 44.

17. "The Farmer's Pledge," *Natural Farmer* (Fall 2003): 35.

18. Elizabeth Henderson, personal communication, July 15, 2007.

19. SCS Global Services, "About SCS: Company," http://www.scsglobalservices.com/company (accessed September 2, 2014).

20. Quoted in Joel Preston Smith, "A Sustainable Agriculture Label: Coming to Foods near You?" *PCC Natural Markets*, September 2008, http://www.pccnaturalmarkets.com/sc/0809/sc0809-sus-ag-label.html (accessed September 2, 2014).

21. Tom Karst, "Time to Push for USDA Certification of 'Sustainably Grown,'" *Packer*, September 12, 2012, http://www.thepacker.com/opinion/fresh-talk-blog/ Time-to-push-for-USDA-certification-of--sustainably-grown-169457096.html (accessed September 2, 2014).

22. Hoodes et al., *National Organic Action Plan*.

23. Kathleen Merrigan, personal communication, August 1, 2008.

24. Rangan, personal communication.

25. Michael E. Conroy, "Can Advocacy-Led Certification Systems Transform Global Corporate Practices? Evidence and Some Theory" (working paper, Political Economy Research Institute, University of Massachusetts at Amherst, September 2001); Daniel Jaffee, "Weak Coffee: Certification and Co-optation in the Fair Trade Movement," *Social Problems* 59, no. 1 (2012): 94–116.

26. Agricultural Justice Project, "History: The Development of Food Justice Certi-fied," http://agriculturaljusticeproject.org/?page_id=18 (accessed December 21, 2013).

27. Domestic Fair Trade Association, "What We Do: Evaluations," http://www .thedfta.org/what-we-do/evaluations (accessed December 21, 2013).

28. Cummins, personal communication.

29. Liana Hoodes, personal communication, July 6, 2006.

30. Ecolabel Index, http://www.ecolabelindex.com (accessed June 25, 2014).

31. Brayden G. King, "Can Private Politics Effectively Replace Government Regula-tion?" (paper presented at Filling the Governance Gap: Aligning Enterprise and Advocacy, Kellogg School of Management/Aspen Institute Business and Society Leadership Summit, Evanston, IL, February 27–28, 2014).

32. Consumers Reports, GreenerChoices, http://www.greenerchoices.org/home.cfm (accessed December 21, 2013).

33. Urvashi Rangan, "Putting the Label on the Table" (keynote address at the Har-vard Food Law Society Forum on Food Labeling, Cambridge, MA, March 2013).

34. King, "Can Private Politics Effectively Replace Government Regulation?"; Marc Schneiberg and Tim Bartley, "Organizations, Regulation, and Economic Behavior: Regulatory Dynamics and Forms from the Nineteenth to Twenty-First Century," *Annual Review of Law and Social Science* 4 (2008): 31–61.

35. Mary K. Hendrickson and William D. Heffernan, "Opening Spaces through Relo-calization: Locating Potential Resistance in the Weaknesses of the Global Food System," *Sociologia Ruralis* 42, no. 4 (October 2002): 347–369; Jack Kloppenburg Jr., John Hendrickson, and G. W. Stevenson, "Coming into the Foodshed," *Agriculture and Human Values* 13, no. 3 (1996): 33–42.

36. Michael Meyerfield Bell, *Farming for Us All: Practical Agriculture and the Cultivation of Sustainability* (State College: Pennsylvania State University Press, 2004); Kate Clancy, "Reconnecting Farmers and Citizens in the Food System," in *Visions of American Agriculture*, ed. William Lockeretz (Ames: Iowa State University Press, 1997), 47–58.

37. Cynthia Cone and Andrea Myhre, "Community-Supported Agriculture: A Sustainable Alternative to Industrial Agriculture?" *Human Organization* 59, no. 2 (Summer 2000): 187–197; Sean Jacques and Lyn Collins, "Community Supported Agriculture: An Alternative to Agribusiness," *Geography Review* 16, no. 5 (2003): 30–33; Steven M. Schnell, "Food with a Farmer's Face: Community-Supported Agriculture in the United States," *Geographical Review* 97, no. 4 (October 2007): 550–564.

38. Suzanne P. DeMuth, *Community Supported Agriculture (CSA): An Annotated Bibliography and Resource Guide* (Beltsville, MD: Alternative Farming Systems Information Center, September 1993), http://www.nal.usda.gov/afsic/pubs/csa/at93-02.shtml (accessed September 2, 2014).

39. Dan Guenther, personal communication, December 11, 2007.

40. Ibid.

41. David J. Hess, *Localist Movements in a Global Economy: Sustainability, Justice, and Urban Development in the United States* (Cambridge, MA: MIT Press, 2009).

42. National Conference on Citizenship, "America's Civic Health Index: Broken Engagement" (Washington, DC: National Conference on Citizenship, September 18, 2006); National Conference on Citizenship. "Civic Health Assessment" (Washington, DC: National Conference on Citizenship, 2010); Miller McPherson, Lynn Smith-Lovin, and Matthew E. Brashears, "Social Isolation in America: Changes in Core Discussion Networks over Two Decades," *American Sociological Review* 71 (2006): 353–375.

43. Robert D. Putnam, *Bowling Alone* (New York: Simon and Schuster, 2000); Robert D. Putnam, *Making Democracy Work: Civic Traditions in Modern Italy* (Princeton, NJ: Princeton University Press, 1993).

44. Ray Oldenburg, *The Great Good Place* (New York: Paragon House, 1991); Theda Skocpol, *Diminished Democracy: From Membership to Management in American Civic Life* (Norman: University of Oklahoma Press, 2003); Alexis de Tocqueville, *Democracy in America* (London: Penguin Classics, 2003).

45. Michael J. Piore and Charles F. Sabel, *The Second Industrial Divide: Possibilities for Prosperity* (New York: Basic Books, 1984).

46. Charles M. Tolbert II, "Minding Our Own Business: Local Retail Establishments and the Future of Southern Civic Community," *Social Forces* 83, no. 4 (2005): 1309–1328.

47. Fred Block, *Postindustrial Possibilities: A Critique of Economic Discourse* (Berkeley: University of California Press, 1990); C. Clare Hinrichs, "Embeddedness and Local Food Systems: Notes on Two Types of Direct Agriculture Market, *Journal of Rural Studies* 16 (2000): 295–303; Jonathan Murdoch, Terry Marsden, and Jo Banks, "Quality, Nature, and Embeddedness: Some Theoretical Considerations in the Context of the Food Sector," *Economic Geography* 76, no. 2 (2000): 107–125.

48. Walter Goldschmidt, "Large-Scale Farming and the Rural Social Structure," *Rural Sociology* 43, no. 3 (1978): 362–366; Thomas A. Lyson, *Civic Agriculture: Reconnecting Farm, Food, and Community* (Medford, MA: Tufts University Press, 2012); Thomas A. Lyson, Robert J. Torres, and Rick Welsh, "Scale of Agricultural Production, Civic Engagement, and Community Welfare," *Social Forces* 80, no. 1 (September 2001): 311–327; Hinrichs, "Embeddedness and Local Food Systems"; Brandi Janssen, "Local Food, Local Engagement: Community-Supported Agriculture in Eastern Iowa," *Culture and Agriculture* 32, no. 1 (2010): 4–16.

49. For a more critical assessment of CSAs, see Antoinette Pole and Margaret Gray, "Farming Alone? What's Up with the 'C' in Community Supported Agriculture?" *Agriculture and Human Values* 30, no. 1 (March 2013): 85–100. They find that CSA members are often operating on the same instrumental principles as conventional shoppers and frequently experience little sense of community with fellow members. See also E. Melanie DuPuis, David Goodman, and Jill Harrison, "Just Values or Just Value? Remaking the Local in Agro-Food Studies," in *Between the Local and the Global: Confronting Complexity in the Contemporary Agri-Food Sector*, ed. Terry. Marsden and Jonathan Murdoch (Bingley, UK: Emerald, 2006), 241–268; Laura B. DeLind, "Close Encounters with a CSA: The Reflections of a Bruised and Somewhat Wiser Anthropologist," *Agriculture and Human Values* 16 (1999): 3–9; Robert Feagan and Amanda Henderson, "Devon Acres CSA: Local Struggles in a Global Food System," *Agriculture and Human Values* 26 (2009): 203–217.

50. , "Embeddedness and Local Food Systems"; Brian Obach and Kathleen Tobin, "Civic Agriculture and Community Engagement," *Agriculture and Human Values* 31, no. 2 (2014): 307–322; J. A. Sheets, "Food—By the People, for the People: The Study of the Relationship between Local Food Networks and Civic Engagement" (paper presented at the joint annual meetings of the Agriculture, Food, and Human Values Society, Association for the Study of Food and Society, and Society for Anthropology of Food and Nutrition, Missoula, MT, June 2011).

51. Damien C. Adams and Matthew J. Salois, "Local versus Organic: A Turn in Consumer Preferences and Willingness-to-Pay," *Renewable Agriculture and Food Systems* 25 (2010): 331–341; Helen LaTrobe and Tim Acott, "Localizing the Global Food System," *International Journal for Sustainable Development and World Ecology* 7, no. 4 (2000): 309–320; Jim Minick, "Beyond Organic," *Counterpunch*, November 27–29, 2004, http://www.counterpunch.org/2004/11/27/beyond-organic (accessed September 3, 2014).

52. Jessica Prentice, "The Birth of Locavore," Oxford University Press blog, November 20, 2007, http://blog.oup.com/2007/11/prentice (accessed September 3, 2014).

53. Carlo Petrini, *Slow Food: The Case for Taste* (New York: Columbia University Press, 2003).

54. E. Melanie DuPuis and David Goodman, "Should We Go Home to Eat? Toward a Reflexive Politics of Localism," *Journal of Rural Studies* 21, no. 3 (2005): 359–371.

55. Slow Food. "About Us." http://slowfood.com/international/1/about-us?-session=query_session:47A9264213ff82D20BOm4DD0A54A (accessed August 19, 2012).

56. Hartman Group, *The Many Faces of Organic 2008*, http://www.hartman-group.com/publications/reports/the-many-faces-of-organic-2008 (accessed August 11, 2014).

57. Fred Kirschenmann, personal communication, November 16, 2007.

58. Despite his modesty, in actuality, Guenther has been a prolific letter writer in the local newspaper and cofounded a local environmental group that regularly advocates on climate policy. He took up this type of activism since his farming days, though.

59. US Department of Agriculture, National Agricultural Statistics Service, *Census of Agriculture—State Data, 2007* (Washington, DC: US Department of Agriculture, February 2009), 605–606, table 44. It should be noted that the wording of the CSA question on the agriculture census may have led to the conclusion that there are more CSA farms than there are in actuality. Nonetheless, the number has grown substantially in recent years.

60. Alida Cantor and Ron Strochlic, *Breaking Down Market Barriers for Small and Mid-Sized Organic Growers* (Davis: California Institute for Rural Studies, November 2009), http://www.ams.usda.gov/AMSv1.0/getfile?dDocName=STELPRDC5081306 (accessed September 3, 2014).

61. US Department of Agriculture, National Agricultural Statistics Service, *Census of Agriculture—State Data, 2007*, 7–8, table 1.

62. Sarah A. Low and Stephen Vogel, *Direct and Intermediated Marketing of Local Foods in the United States*, Economic Research Report no. 128 (Washington, DC: US Department of Agriculture, Economic Research Service, November 2011).

63. Agriculture of the Middle, "Characterizing Ag of the Middle and Values-Based Supply Chains," http://www.agofthemiddle.org/archives/2012/01/characterizing.html (accessed September 3, 2014).

64. Agriculture of the Middle, "Case Studies Profile Mid-Scale Food Enterprises," http://www.agofthemiddle.org/archives/2009/11/value_chain_cas.html (accessed September 3, 2014).

65. David J. Hess, "Declarations of Independents: On Local Knowledge and Localist Knowledge," *Anthropological Quarterly* 83, no. 1 (2010): 153–176.

66. Community Involved in Sustaining Agriculture, "Our Programs," http://www.buylocalfood.org (accessed September 3, 2014).67. Quoted in Tracy Frisch, "Up Close: Liana Hoodes," *Valley Table* 58 (June–August 2012), http://www.valleytable.com/article.php?article=008+Up+Close%2FLiana+Hoodes (accessed September 3, 2014).

68. Catherine Greene, Carolyn Dimitri, Biing-Hwan Lin, William McBride, Lydia Oberholtzer, and Travis Smith, *Emerging Issues in the U.S. Organic Industry*, Economic Information Bulletin no. 55 (Washington, DC: US Department of Agriculture, Economic Research Service, June 2009), http://www.ers.usda.gov/publications/eib-economic-information-bulletin/eib55.aspx#.UpVLvbW1F8E (accessed August 27, 2014).

69. Brian Obach and Kathleen Tobin, *Agriculture Supporting Community in the Mid-Hudson Region*, Discussion Brief no. 5 (New Paltz: Center for Research, Regional Education, and Outreach, State University of New York, Spring 2011).

70. Katherine DiMatteo, personal communication, November 29, 2007.

71. Mendelson, personal communication.

72. Hess, *Localist Movements in a Global Economy*.

73. This is not the case for all local food consumers, just as some organic consumers are seeking only personal benefit such as health advantages through their purchase of organic goods. For example, Antoinette Pole and Margaret Gray found that many CSA members were motivated only by their desire to secure fresh, organic produce. They were not seeking a sense of community or to advance other social values. Pole and Gray, "Farming Alone?"

74. Rich Pirog, Timothy Van Pelt, Kamyar Enshayan, and Ellen Cook, *Food, Fuel, and Freeways: An Iowa Perspective on How Far Food Travels, Fuel Usage, and Greenhouse Gas Emissions* (Ames: Leopold Center for Sustainable Agriculture, Iowa State University, June 2001).

75. Jane Black, "What's in a Number? How the Press Got the Idea That Food Travels 1,500 Miles from Farm to Plate," *Slate*, September 17, 2008, http://www.slate.com/articles/life/food/2008/09/whats_in_a_number.html (accessed September 4, 2014); Branden Born and Mark Purcell, "Avoiding the Local Trap Scale and Food Systems in Planning Research," *Journal of Planning Education and Research* 26, no. 2 (2006): 195–207; Hess, *Localist Movements in a Global Economy*.

76. Caroline Saunders, Andrew Barber, and Greg Taylor, *Food Miles—Comparative Energy/Emissions Performance of New Zealand's Agriculture Industry*, Research Report no. 285 (Lincoln, NZ: Agribusiness and Economics Research Unit, Lincoln University, July 2006).

77. Patrick Canning, Ainsley Charles, Sonya Huang, and Karen R. Polenske, *Energy Use in the U.S. Food System*, Economic Research Report no. 94 (Washington, DC: US Department of Agriculture, Economic Research Service, March 2010).

78. DeLind, "Close Encounters with a CSA"; DeLind, Laura B., and Jim Bingen, "Place and Civic Culture: Re-Thinking the Context for Local Agriculture," *Journal of Agricultural and Environmental Ethics* 21 (2008): 127–151; Laura B. DeLind, Are Local Food and the Local Food Movement Taking Us Where We Want to Go? Or Are We Hitching Our wagons to the Wrong Stars?" *Agriculture and Human Values* 28, no. 2 (2011): 273–283; DuPuis and Goodman, "Should We Go Home to Eat?"; DuPuis, Goodman, and Harrison, "Just Values or Just Value?"; Feagan and Henderson, "Devon Acres CSA"; C. Clare Hinrichs and Patricia Allen, "Selective Patronage and Social Justice: Local Food Consumer Campaigns in a Historical Context," *Journal of Agriculture and Environmental Ethics* 21, no. 4 (2008): 329–352; Michael Winter, "Embeddedness, the New Food Economy, and Defensive Localism," *Journal of Rural Studies* 19 (2003): 23–32.

79. Riddle, personal communication.

80. Enid Wonnacott, personal communication, January 8, 2008.

81. Ibid.

82. Michele Micheletti, *Political Virtue and Shopping: Individuals, Consumerism, and Collective Action* (New York: Palgrave Macmillan, 2003).

83. DeLind, "Are Local Food and the Local Food Movement Taking Us Where We Want to Go?"

84. Michael Pollan, "Vote for the Dinner Party: Is This the Year That the Food Movement Finally Enters Politics?" *New York Times*, October 10, 2012, http://www.nytimes.com/2012/10/14/magazine/why-californias-proposition-37-should-matter-to-anyone-who-cares-about-food.html?pagewanted=all&_r=0 (accessed September 4, 2014).

85. Hess, *Localist Movements in a Global Economy*.

86. Greene et al., *Emerging Issues in the U.S. Organic Industry*; Hess, "Declarations of Independents."

87. DuPuis and Goodman, "Should We Go Home to Eat?"; DuPuis, Goodman, and Harrison, "Just Values or Just Value?"

88. Alison Hope Alkon, "Paradise or Pavement: The Social Constructions of the Environment in Two Urban Farmers' Markets, and Their Implications for Environmental Justice and Sustainability," *Local Environment* 13, no. 3 (April 2008): 271–289.

89. Hess, *Localist Movements in a Global Economy*.

90. Obach and Tobin, "Civic Agriculture and Community Engagement."

## Chapter 8: The Road Not Taken and the Road Ahead

1. Warren James Belasco, *Appetite for Change: How the Counterculture Took on the Food Industry* (Ithaca, NY: Cornell University Press, 2007); Kathleen Merrigan, "Negotiating Identity within the Sustainable Agriculture Advocacy Coalition" (PhD diss., Massachusetts Institute of Technology, September 2000).

2. Doug Gurian-Sherman and Margaret Mellon, "The Rise of Superweeds and What to Do about It," Union of Concerned Scientists Policy Brief, 2013, http://www.ucsusa.org/assets/documents/food_and_agriculture/rise-of-superweeds.pdf (accessed September 4, 2014).

3. Julie Guthman, *Agrarian Dreams: The Paradox of Organic Farming in California* (Berkeley: University of California Press, 2004).

4. Julian Agyeman and Jesse McEntee, "Moving the Field of Food Justice Forward through the Lens of Urban Political Ecology," *Geography Compass* 8, no. 3 (2014): 211–220.

5. Julie Guthman, "Fast Food/Organic Food: Reflexive Tastes and the Making of 'Yuppie Chow,'" *Social and Cultural Geography* 4, no. 1 (2003): 45–58.

6. Guthman, *Agrarian Dreams*, 111.

7. Tim Bartley, "Certifying Forests and Factories: States, Social Movements, and the Rise of Private Regulation in the Apparel and Forest Products Fields," *Politics and Society* 32, no. 3 (2003): 433–464.

8. Michael E. Conroy, "Can Advocacy-Led Certification Systems Transform Global Corporate Practices? Evidence and Some Theory" (working paper, Political Economy Research Institute, University of Massachusetts at Amherst, September 2001); Jim Salzman, "Product and Raw Material Eco-Labeling: The Limits of a Transatlantic Approach" (working paper no. 117, Berkeley Roundtable on the International Economy, University of California at Berkeley, May 1998); Kerstin Tews, Per-Olof Busch, and Helge Jorgens, "The Diffusion of New Environmental Policy Instruments," *European Journal of Political Research* 42, no. 4 (June 2003): 569–600.

9. Michael E. Conroy, *Branded! How the "Certification Revolution" Is Transforming Global Corporations* (Gabriola Island, BC: New Society Publishers, 2007).

10. Michele Micheletti, *Political Virtue and Shopping: Individuals, Consumerism, and Collective Action* (New York: Palgrave Macmillan, 2003), xi.

11. Jeremy Brecher, *Strike!* (Boston: South End Press, 1999); Braydon G. King, "The Tactical Disruptiveness of Movements: Sources of Market and Mediated Disruption in Corporate Boycotts," *Social Problems* 48 (2011): 491–517; Braydon G. King and Mary Hunter-McDonnell, "Keeping Up Appearances: Reputation Threat and Prosocial Responses to Social Movement Boycotts," *Administrative Science Quarterly* 58, no. 3 (2013): 387–419.

12. Doug McAdam, *Political Process and the Development of Black Insurgency, 1930–1970* (Chicago: University of Chicago Press, 1999).

13. Robert H. Zieger and Gilbert Gall, *American Workers, American Unions* (Baltimore: Johns Hopkins University Press, 2002).

14. Micheletti, *Political Virtue and Shopping*.

15. Micheletti (ibid.) reports that some employers used the existence of the white label system as a justification for denying the need for unions—an unintended consequence of a system designed to ensure workers' well-being.

16. Underwriters Laboratories, http://www.ul.com/about/index.html (accessed July 4, 2010).

17. Bartley, "Certifying Forests and Factories."

18. For more discussion on the complex merger of social movements and private enterprise, see David J. Hess, "Organic Food and Agriculture in the U.S.; Object Conflicts in a Health-Environmental Social Movement," *Science as Culture* 13, no. 4 (December 2004): 493–513.

19. Braydon G. King and Nicholas A. Pearce, "The Contentiousness of Markets: Politics, Social Movements, and Institutional Change in Markets," *Annual Review of Sociology* 36 (2010): 249–267; Sarah A. Soule, *Contention and Corporate Social Responsibility* (Cambridge: Cambridge University Press, 2009); Tews, Busch, and Jorgens, "The Diffusion of New Environmental Policy Instruments"; Nella Van Dyke, Sarah A. Soule, and Verta A. Taylor, "The Targets of Social Movements: Beyond a Focus on the State," *Research in Social Movements Conflict and Change* 25 (2004): 27–51.

20. Hess, "Organic Food and Agriculture in the U.S."

21. See Arthur P. J. Mol, "Ecological Modernization: Industrial Transformations and Environmental Reform," in *International Handbook of Environmental Sociology*, ed. Michael Redclift and Graham Woodgate (Cheltenham, UK: Edward Elgar, 1997), 138–149; Arthur P. J. Mol and Gert Spaargaren, "Ecological Modernization Theory in Debate: A Review," in *Ecological Modernization around the World*, ed. Arthur P. J. Mol and David A. Sonnenfeld (London: Frank Cass, 2000), 17–49. For an example of an

organization promoting this approach, see Rocky Mountain Institute, http://www .rmi.org (accessed September 4, 2014).

22. Magnus Boström, "Environmental Organisations in New Forms of Political Participation: Ecological Modernisation and the Making of Voluntary Rules," *Environmental Values* 12, no. 2 (May 2003): 175–193; Micheletti, *Political Virtue and Shopping*.

23. Forest Stewardship Council, https://us.fsc.org/ (accessed December 22, 2013).

24. Bartley, "Certifying Forests and Factories"; Salzman, "Product and Raw Material Eco-Labeling."

25. Conroy, "Can Advocacy-Led Certification Systems Transform Global Corporate Practices?"

26. Ibid., 18.

27. Micheletti, *Political Virtue and Shopping*, 161.

28. Ibid.

29. Lawrence Busch, "The Private Governance of Food: Equitable Exchange or Bizarre Bazaar?" *Agriculture and Human Values* 28 (2011): 345–352; Braydon G. King, "Can Private Politics Effectively Replace Government Regulation?" (paper presented at Filling the Governance Gap: Aligning Enterprise and Advocacy, Kellogg School of Management/Aspen Institute Business and Society Leadership Summit, Evanston, IL, February 27–28, 2014).

30. Daniel Jaffe, *Brewing Justice: Fair Trade Coffee, Sustainability, and Survival* (Berkeley: University of California Press, 2007).

31. Fair Trade USA, http://fairtradeusa.org (accessed December 1, 2013).

32. Daniel Jaffee, "Weak Coffee: Certification and Co-optation in the Fair Trade Movement," *Social Problems* 59, no. 1 (2012): 94–116.

33. For example, Equal Exchange, a nonprofit workers' cooperative that was among the fair trade coffee pioneers, is critical of Fair Trade USA. See Equal Exchange, "Fair Trade," http://www.equalexchange.coop/fair-trade (accessed December 1, 2013).

34. Daniel Jaffee and Philip H. Howard, "Corporate Cooptation of Organic and Fair Trade Standards," *Agriculture and Human Values* 27 (2010): 387–399.

35. King, "Can Private Politics Effectively Replace Government Regulation?"

36. Michael Pollan quotes a spokesperson for Grimmway Farms, a conventional grower that bought up an organic operation, as explicitly stating the extent of the company's commitment to organic: "Whether we stay with organic for the long haul depends on profitability." Michael Pollan, *The Omnivore's Dilemma: A Natural History of Four Meals* (New York: Penguin, 2006), 174.

37. Organic Trade Association, "Industry Statistics and Projected Growth," June 2011, http://www.ota.com/organic/mt/business.html (accessed December 1, 2013).

38. Josée Johnston, "The Citizen-Consumer Hybrid: Ideological Tensions and the Case of Whole Foods Market," *Theory and Society* 37, no. 3 (2008): 229–270.

39. Salzman, "Product and Raw Material Eco-Labeling," 8.

40. Garth Youngberg and Suzanne P. DeMuth, "Organic Agriculture in the United States: A 30-Year Retrospective," *Renewable Agriculture and Food Systems* 28, no. 4 (May 2013): 1–35, doi:10.1017/S1742170513000173.

41. King, "Can Private Politics Effectively Replace Government Regulation?"

42. Youngberg and DeMuth, "Organic Agriculture in the United States."

43. For a more comprehensive critique of the way in which political consumerism can undermine collective efforts to secure social good, see Andrew Szasz, *Shopping Our Way to Safety: How We Changed from Protecting the Environment to Protecting Ourselves* (Minneapolis: University of Minnesota Press, 2009).

44. Marion Nestle, *Food Politics: How the Food Industry Influences Nutrition and Health* (Berkeley: University of California Press, 2007).

45. Patricia Allen and Martin Kovach, "The Capitalist Composition of Organic: The Potential of Markets in Fulfilling the Promise of Organic Agriculture," *Agriculture and Human Values* 17, no. 3 (September 2000): 221–232.

46. Ibid.

47. Kathleen Merrigan, "Negotiating Identity within the Sustainable Agriculture Advocacy Coalition" (PhD diss., Massachusetts Institute of Technology, September 2000).

48. Kennet S. C. Lynggaard, "The Farmer within an Institutional Environment: Comparing Danish and Belgian Organic Farming," *Sociologia Ruralis* 41, no. 1 (January 2001): 85–111.

49. Debra Friedman and Doug McAdam, "Collective Identities and Activism: Networks, Choices, and the Life of a Social Movement," in *Frontiers of Social Movement Theory*, ed. Aldon D. Morris and Carol Mueller (New Haven, CT: Yale University Press, 1992), 156–173; Hank Johnston, Enrique Laraña, and Joseph R. Gusfield, "Identities, Grievances, and New Social Movements," in *New Social Movements: Perspectives and Issues*, ed. Enrique Laraña, Hank Johnston, and Joseph R. Gusfield (Philadelphia: Temple University Press, 1994), 3–35; Alberto Melucci, "A Strange Kind of Newness: What's 'New' in New Social Movements?" in *New Social Movements: From Ideology to Identity*, ed. Enrique Laraña, Hank Johnston, and Joseph R. Gusfield (Philadelphia: Temple University Press, 1994), 101–130; Verta A. Taylor and Nancy Whittier, "Collective Identity in Social Movement Communities: Lesbian

Feminist Mobilization," in *Frontiers in Social Movement Theory*, ed. Aldon Morris and Carol McClurg Mueller. (New Haven, CT: Yale University Press, 1992), 104–130.

50. While market pressure significantly curtailed the use of synthetic bovine growth hormone in the United States, many other countries, including most of Europe, have prevented all use by banning it, thereby demonstrating the effectiveness of state policy relative to private, market-based efforts.

51. Nestle, *Food Politics*.

52. Pollan, *The Omnivore's Dilemma*; Thomas A. Lyson, *Civic Agriculture: Reconnecting Farm, Food, and Community* (Medford, MA: Tufts University Press, 2012); Wendell Berry, *Bringing It to the Table* (Berkeley, CA: Counterpoint, 2009).

53. Kate Clancy, "Reconnecting Farmers and Citizens in the Food System," in *Visions of American Agriculture*, ed. William Lockeretz (Ames: Iowa State University Press, 2000); K. A. Dahlberg, "Regenerative Food Systems: Broadening the Scope and Agenda of Sustainability," in *Food for the Future: Conditions and Contradictions of Sustainability*, ed. Patricia Allen (New York: John Wiley and Sons, 1993), 75–102.

# Bibliography

Adams, Damien C., and Matthew J. Salois. "Local versus Organic: A Turn in Consumer Preferences and Willingness-to-Pay." *Renewable Agriculture and Food Systems* 25, no. 4 (December 2010): 331–341.

Agricultural Justice Project. "History: The Development of Food Justice Certified." http://agriculturaljusticeproject.org/?page_id=18 (accessed December 21, 2013).

Agriculture of the Middle. "Case Studies Profile Mid-Scale Food Enterprises." http://www.agofthemiddle.org/archives/2009/11/value_chain_cas.html (accessed September 3, 2014).

Agriculture of the Middle. "Characterizing Ag of the Middle and Values-Based Supply Chains." http://www.agofthemiddle.org/archives/2012/01/characterizing .html (accessed September 3, 2014).

Agyeman, Julian, and Jesse McEntee. "Moving the Field of Food Justice Forward through the Lens of Urban Political Ecology." *Geography Compass* 8, no. 3 (2014): 211–220.

Alkon, Alison Hope. "Paradise or Pavement: The Social Constructions of the Environment in Two Urban Farmers' Markets, and Their Implications for Environmental Justice and Sustainability." *Local Environment* 13, no. 3 (April 2008): 271–289.

Alkon, Alison Hope, and Julian Agyeman, eds. *Cultivating Food Justice: Race, Class, and Sustainability*. Cambridge, MA: MIT Press, 2011.

Alkon, Alison Hope, and Kari Marie Norgaard. "Breaking the Food Chains: An Investigation of Food Justice Activism." *Sociological Inquiry* 79, no. 3 (August 2009): 289–305.

Allen, Patricia, and Martin Kovach. "The Capitalist Composition of Organic: The Potential of Markets in Fulfilling the Promise of Organic Agriculture." *Agriculture and Human Values* 17, no. 3 (September 2000): 221–232.

Altieri, Miguel. "Ecological Impacts of Industrial Agriculture and the Possibilities for Truly Sustainable Farming." In *Hungry for Profit: The Agribusiness Threat to Farmers,*

*Food, and the Environment*, edited by Fred Magdoff, John Bellamy Foster, and Frederick H. Buttel, 77–92. New York: Monthly Review Press, 2000.

Associated Press. "Target Rolling Out Organic, Natural Grocery Brand." June 7, 2013. http://www.cbsnews.com/8301-505145_162-57588227/target-rolling-out-organic-natural-grocery-brand (accessed August 29, 2014).

Audirac, Ivonne, ed. *Rural Sustainable Development in America*. New York: John Wiley and Sons, 1997.

Avery, Alex, and Dennis Avery. "Organic Food Campaign Goes Sharply Negative." Hudson Institute, November 16, 2002. http://www.hudson.org/index.cfm?fuseaction=publication_details&id=2075 (accessed August 27, 2014).

Avery, Dennis. "The Hidden Dangers in Organic Food." Hudson Institute, November 1, 1998. http://www.hudson.org/index.cfm?fuseaction=publication_details &id=1196 (August 27, 2014).

Avery, Dennis. "The Silent Killer in Organic Foods." Hudson Institute, April 8, 1999. http://www.hudson.org/index.cfm?fuseaction=publication_details&id=299 (accessed August 27, 2014).

Bacharach, Samuel B., and Edward J. Lawler. *Power and Politics in Organizations*. San Francisco: Jossey-Bass, 1980.

Badgley, Catherine, Jeremy Moghtader, Eileen Quintero, Emily Zakem. M. Jahi Chappell, Katia Avilés-Vázquez, Andrea Samulon, and Ivette Perfecto. "Organic Agriculture and the Global Food Supply." *Renewable Agriculture and Food Systems* 22, no. 2 (June 2007): 86–108.

Balfour, Eve. *The Living Soil*. London: Faber and Faber, 1943.

Balfour, Eve. "Towards a Sustainable Agriculture—The Living Soil." Talk at the International Federation of Organic Agriculture Movements, Sissach, Switzerland, 1977. http://www.soilandhealth.org/01aglibrary/010116Balfourspeech.html (accessed July 28, 2014).

Bartley, Tim. "Certifying Forests and Factories: States, Social Movements, and the Rise of Private Regulation in the Apparel and Forest Products Fields." *Politics and Society* 32, no. 3 (2003): 433–464.

Batie, Sandra S. *Soil Erosion: Crisis in America's Croplands?* Washington, DC: Conservation Foundation, 1983.

Beeson, Kenneth. "Spring Gardens: What about the 'Organic Way'?" *New York Times*, April 16, 1972, 33.

Belasco, Warren James. *Appetite for Change: How the Counterculture Took on the Food Industry*. Ithaca, NY: Cornell University Press, 2007.

Bell, Michael Meyerfield. *Farming for Us All: Practical Agriculture and the Cultivation of Sustainability*. State College: Pennsylvania State University Press, 2004.

Benbrook, Charles, Xin Zhao, Jaime Yáñez, Neal Davies, and Preston Andrews. *New Evidence Confirms the Nutritional Superiority of Plant-Based Organic Foods*. Washington, DC: Organic Center, March 2008.

Berry, Wendell. *Bringing It to the Table*. Berkeley, CA: Counterpoint, 2009.

Bittman, Mark. "That Flawed Stanford Study." *New York Times*, October 2, 2012, http://opinionator.blogs.nytimes.com/2012/10/02/that-flawed-stanford-study/ (accessed August 28, 2014).

Black, Jane. "What's in a Number? How the Press Got the Idea That Food Travels 1,500 Miles from Farm to Plate." *Slate*, September 17, 2008, http://www.slate.com/articles/life/food/2008/09/whats_in_a_number.html (accessed September 4, 2014).

Block, Fred. *Postindustrial Possibilities: A Critique of Economic Discourse*. Berkeley: University of California Press, 1990.

Blum, Andrea. "Organic Farming's Labor Problem." *Common Ground*, February 2006, http://www.columbia.org/pdf_files/cainstituteforruralstudies.pdf (accessed September 1, 2014).

Born, Branden, and Mark Purcell. "Avoiding the Local Trap Scale and Food Systems in Planning Research." *Journal of Planning Education and Research* 26, no. 2 (2006): 195–207.

Boström, Magnus. "Environmental Organisations in New Forms of Political Participation: Ecological Modernisation and the Making of Voluntary Rules." *Environmental Values* 12, no. 2 (May 2003): 175–193.

Brady, Diane. "The Organic Myth." *Business Week*, October 16, 2006, 50–56.

Brandt, Kirsten, Carlo Leifert, Roy Sanderson, and Chris Seal. "Agroecosystem Management and Nutritional Quality of Plant Foods: The Case of Organic Fruits and Vegetables." *Critical Reviews in Plant Sciences* 30, no. 1–2 (2011): 177–197.

Brecher, Jeremy. *Strike!* Boston: South End Press, 1999.

Breines, Wini. *Community and Organization in the New Left, 1962–1968: The Great Refusal*. New Brunswick, NJ: Rutgers University Press, 1981.

Brown Rosen, Emily. "The Devil Is in the Details: or Why Organic Standards Matter." *Natural Farmer*, Spring 2006, http://www.nofa.org/tnf/2006spring/The%20Devil%20Is%20In%20the%20Details.pdf (accessed August 30, 2014).

Brulle, Robert. "Environmental Discourse and Social Movement Organizations: A Historical and Rhetorical Perspective on the Development of US Environmental Organizations." *Sociological Inquiry* 66, no. 1 (January 1996): 58–83.

Buck, Daniel, Christina Getz, and Julie Guthman. "From Farm to Table: The Organic Vegetable Commodity Chain in Northern California." *Sociologia Ruralis* 44, no. 3 (1997): 3–20.

Burros, Marian. "Obamas to Plant Vegetable Garden at White House." *New York Times*, March 19, 2009, A1.

Busch, Lawrence. "The Private Governance of Food: Equitable Exchange or Bizarre Bazaar?" *Agriculture and Human Values* 28 (2011): 345–352.

Buttel, Frederick H. "Agricultural Change, Rural Society, and the State in the Late Twentieth Century: Some Theoretical Observations." In *Agricultural Restructuring and Rural Change in Europe*, edited by David Symes and Anton J. Jansen, 13–31. Wageningen: Agricultural University, 1994.

Buttel, Frederick H. "Environmentalization: Origins, Processes, and Implications for Rural Social Change." *Rural Sociology* 57, no. 1 (Spring 1992): 1–27.

Buttel, Frederick H. "The Treadmill of Production: An Appreciation, Assessment, and Agenda for Research." *Organization and Environment* 17, no. 3 (2004): 323–336.

Bystydzienski, Jill M., and Steven P. Schacht, eds. *Forging Radical Alliances across Difference: Coalition Politics for the New Millennium*. Lanham, MD: Rowman and Littlefield, 2001.

California Certified Organic Farmers. "Our History." http://www.ccof.org/ccof/history (accessed March 9, 2012).

Campbell, Hugh, and Ruth Liepins. "Naming Organics: Understanding Organic Standards in New Zealand as a Discursive Field." *Sociologia Ruralis* 41, no. 1 (2001): 21–39.

Canning, Patrick. *A Revised and Expanded Food Dollar Series: A Better Understanding of Our Food Costs*. Economic Research Report no. ERR-114. Washington, DC: US Department of Agriculture, Economic Research Service, February 2011. http://www.ers .usda.gov/publications/err-economic-research-report/err114.aspx#.UrToM7W1F8E (accessed August 31, 2014).

Canning, Patrick, Ainsley Charles, Sonya Huang, and Karen R. Polenske. *Energy Use in the U.S. Food System*. Economic Research Report no. 94. Washington, DC: US Department of Agriculture, Economic Research Service, March 2010.

Cantor, Alida, and Ron Strochlic. *Breaking Down Market Barriers for Small and Mid-Sized Organic Growers*. Davis: California Institute for Rural Studies, November 2009. http://www.ams.usda.gov/AMSv1.0/getfile?dDocName=STELPRDC5081306 (accessed September 3, 2014).

Center for Agroecology and Sustainable Food Systems. "Sustainable Agriculture at UC Santa Cruz." http://casfs.ucsc.edu/about/history/sustainable-agriculture-at-uc-santa-cruz (accessed August 27, 2014).

Certified Naturally Grown. "Brief History of Certified Naturally Grown." http://www.naturallygrown.org/about-cng/brief-history-of-certified-naturally-grown (accessed September 2, 2014).

Chop, Hilary. "Whose Label?" *Alternatives Journal* 29, no. 4 (2003): 19–20.

Cienfuegos, Paul. "The Organic Foods Movement: Led by Heinz Corporation or We the People?" CommonDreams.org, May 31, 2004, http://www.commondreams.org/views04/0531-11.htm (accessed August 6, 2014).

Clancy, Kate. "Reconnecting Farmers and Citizens in the Food System." In *Visions of American Agriculture*, edited by William Lockeretz, 47–58. Ames: Iowa State University Press, 2000.

Clunies-Ross, Tracey. "Organic Food: Swimming against the Tide?" In *Political, Social, and Economic Perspectives on the International Food System*, edited by Terry Marsden and Jo Little, 200–214. Aldershot, UK: Avebury Press, 1990.

Clunies-Ross, Tracey, and Graham Cox. "Challenging the Productivist Paradigm: Organic Farming and the Politics of Agricultural Change." In *Regulating Agriculture*, edited by Philip Lowe, Terry Marsden, and Sarah Whatmore, 53–74. London: David Fulton Publishers, 1994.

Cochrane, William W. *The Development of American Agriculture: A Historical Analysis*. Minneapolis: University of Minnesota Press, 1979.

Cohen, Roger. "The Organic Fable." *New York Times*, September 6, 2012, http://www.nytimes.com/2012/09/07/opinion/roger-cohen-the-organic-fable.html?_r=1&emc=tnt&tntemail1=y (accessed August 27, 2014).

Coleman, Eliot. "The Benefits of Growing Organic Food." *Mother Earth News*, December 2001–January 2002, http://www.motherearthnews.com/real-food/benefits-of-growing-organic-food-zmaz01djzgoe.aspx?PageId=1#axzz2hAIokGcY (accessed August 6, 2014).

Committee on the Role of Alternative Farming Methods in Modern Production Agriculture, National Research Council. *Alternative Agriculture*. Washington, DC: National Academies Press, 1989.

Community Involved in Sustaining Agriculture. "Our Programs." http://www.buylocalfood.org/about/our-programs/ (accessed September 3, 2014).

Cone, Cynthia, and Andrea Myhre. "Community-Supported Agriculture: A Sustainable Alternative to Industrial Agriculture?" *Human Organization* 59, no. 2 (Summer 2000): 187–197.

Conford, Philip. *The Origins of the Organic Movement*. Edinburgh: Floris Books, 2001.

Connell, David J., John Smothers, and Alun Joseph. "Farmers' Markets and the 'Good Food' Value Chair: A Preliminary Study." *Local Environment: The International Journal of Justice and Sustainability* 13, no. 3 (2008): 169–186.

Conroy, Michael E. *Branded! How the "Certification Revolution" Is Transforming Global Corporations.* Gabriola Island, BC: New Society Publishers, 2007.

Conroy, Michael E. "Can Advocacy-Led Certification Systems Transform Global Corporate Practices? Evidence and Some Theory." Working paper, Political Economy Research Institute, University of Massachusetts at Amherst, September 2001.

Constance, Douglas H., Jin Youn Choi, and Holly Lyke-Ho-Gland. "Conventionalization, Bifurcation, and Quality of Life: Certified and Non-Certified Organic Farmers in Texas." *Southern Rural Sociology* 23, no. 1 (2008): 208–234.

Consumer Reports. GreenerChoices. http://www.greenerchoices.org/home.cfm (accessed December 21, 2013).

Coombes, Brad, and Hugh Campbell. "Dependent Reproduction of Alternative Modes of Agriculture: Organic Farming in New Zealand." *Sociologia Ruralis* 38, no. 2 (1998): 127–145.

Cornucopia Institute. "The Organic Watergate—White Paper; Connecting the Dots: Corporate Influence at the USDA's National Organic Program." Cornucopia, WI: Cornucopia Institute, 2012. http://www.cornucopia.org/USDA/OrganicWatergate WhitePaper.pdf (accessed August 7, 2014).

Cronon, William. "The Trouble with Wilderness; or Getting Back to the Wrong Nature." *Environmental History* 1(1) (1996): 7–28.

Cuddeford, Vijay. "When Organics Go Mainstream." *Alternatives Journal* 29, no. 4 (2003): 14–19.

Dahlberg, K. A. "Regenerative Food Systems: Broadening the Scope and Agenda of Sustainability." In *Food for the Future: Conditions and Contradictions of Sustainability*, edited by Patricia Allen, 75–102. New York: John Wiley and Sons, 1993.

Danbom, David. "Past Visions of American Agriculture." In *Visions of American Agriculture*, edited by William Lockeretz, 3–16. Iowa City: Iowa State University Press, 2000.

Dangour, Alan D., Sakhi Dodhia, Arabella Hayter, Elizabeth Allen, Karen Lock, and Ricardo Uauy. "Nutritional Quality of Organic Foods: A Systematic Review." *American Journal of Clinical Nutrition* 90, no. 3 (2009): 680–685.

Dangour, Alan D., Karen Lock, Arabella Hayter, Andrea Aikenhead, Elizabeth Allen, and Ricardo Uauy. "Nutrition-Related Health Effects of Organic Foods: A Systematic Review." *American Journal of Clinical Nutrition* 92, no. 1 (2010): 203–210.

DeLind, Laura B. "Are Local Food and the Local Food Movement Taking Us Where We Want to Go? Or Are We Hitching Our Wagons to the Wrong Stars?" *Agriculture and Human Values* 28, no. 2 (June 2011): 273–283.

DeLind, Laura B. "Close Encounters with a CSA: The Reflections of a Bruised and Somewhat Wiser Anthropologist." *Agriculture and Human Values* 16 (1999): 3–9.

DeLind, Laura B. "Transforming Organic Agriculture into Industrial Organic Products: Reconsidering National Organic Standards." *Human Organization* 59, no. 2 (2000): 198–208.

DeLind, Laura B., and Jim Bingen. "Place and Civic Culture: Re-Thinking the Context for Local Agriculture." *Journal of Agricultural and Environmental Ethics* 21 (2008): 127–151.

Demeter Association, Inc. "Biodynamic Farm Standard." http://demeter-usa.org/ downloads/Demeter-Farm-Standard.pdf (accessed July 28, 2014).

DeMuth, Suzanne P. *Community Supported Agriculture (CSA): An Annotated Bibliography and Resource Guide.* Beltsville, MD: Alternative Farming Systems Information Center, September 1993. http://www.nal.usda.gov/afsic/pubs/csa/at93-02.shtml (accessed September 2, 2014).

DiMatteo, Katherine, and Grace Gershuny. "The Organic Trade Association." In *Organic Farming: An International History*, edited by William Lockeretz, 253–263. Oxfordshire, UK: CABI, 2007.

Dimitri, Carolyn, Anne Effland, and Neilson Conklin. *Transformation of U.S. Agriculture and Farm Policy.* Economic Information Bulletin no. 3. Washington, DC: US Department of Agriculture, Economic Research Service, June 2005.

Dimitri, Carolyn, and Lydia Oberholtzer. "Market-Led versus Government-Facilitated Growth: Development of the U.S. and EU Organic Agricultural Sectors." US Department of Agriculture, Economic Research Service, WRS-05–05, August 2005. http://www.ers.usda.gov/publications/wrs-international-agriculture-and-trade-outlook/wrs0505.aspx#.UopmG8QqhAQ (accessed August 7, 2014).

Domestic Fair Trade Association. "What We Do: Evaluations." http://www.thedfta .org/what-we-do/evaluations (accessed December 21, 2013).

Dunau, Mark, and Elizabeth Henderson. "Growing the Farmer's Pledge." *Natural Farmer* (Spring 2006): 44.

DuPuis, E. Melanie, and David Goodman. "Should We Go Home to Eat? Toward a Reflexive Politics of Localism." *Journal of Rural Studies* 21, no. 3 (2005): 359–371.

DuPuis, E. Melanie, David Goodman, and Jill Harrison. "Just Values or Just Value? Remaking the Local in Agro-Food Studies." In *Between the Local and the Global: Confronting Complexity in the Contemporary Agri-Food Sector*, edited by Terry Marsden and Jonathan Murdoch, 241–268. Bingley, UK: Emerald, 2006.

Duscha, Julius. "Up, Up, Up—Butz Makes Hay Down on the Farm." *New York Times*, April 16, 1972, SM34.

Ecolabel Index. http://www.ecolabelindex.com (accessed June 25, 2014).

Eder, Klaus. *The New Politics of Class: Social Movements and Cultural Dynamics in Advanced Societies*. Thousand Oaks, CA: SAGE, 1993.

Environmental Protection Agency. "Agriculture: Demographics." http://www.epa.gov/agriculture/ag101/demographics.html (accessed August 31, 2014).

Environmental Protection Agency. "What Is a CAFO?" http://www.epa.gov/region7/water/cafo (accessed August 27, 2014).

Epstein, Barbara. *Political Protest and Cultural Revolution: Nonviolent Direct Action in the 1970s and 1980s*. Berkeley: University of California Press, 1991.

Equal Exchange. "Fair Trade." http://www.equalexchange.coop/fair-trade (accessed December 1, 2013).

Fair Trade USA. http://fairtradeusa.org (accessed December 1, 2013).

The Farmer's Pledge. *Natural Farmer* (Fall 2003): 35.

Feagan, Robert, and Amanda Henderson. "Devon Acres CSA: Local Struggles in a Global Food System." *Agriculture and Human Values* 26 (2009): 203–217.

Federal Trade Commission. *Federal Register*. Washington, DC: Government Printing Office, 1978).

Federal Trade Commission. *Federal Register* 39842. Washington, DC: Government Printing Office, November 11, 1974.

Federal Trade Commission. *Proposed Trade Regulation Rule on Food Advertising, 16 CFR Part 437, Phase I: Staff Report and Recommendations*. Washington, DC: Government Printing Office, 1978.

Fernandez-Cornejo, Jorge. *Off-Farm Income, Technology Adoption, and Farm Economic Performance*. Economic Research Report no. ERR-36. Washington, DC: US Department of Agriculture, Economic Research Service, February 2007. http://www.ers.usda.gov/publications/err-economic-research-report/err36.aspx#.UrThs8r8IyE (accessed August 31, 2004).

Fetter, T. Robert, and Julie A. Caswell. "Variation in Organic Standards Prior to the National Organic Program." *American Journal of Alternative Agriculture* 17, no. 2 (2002): 55–75.

Fitzgerald, Deborah. *Every Farm a Factory: The Industrial Ideal in American Agriculture*. New Haven, CT: Yale University Press, 2003.

Food Alliance. "About Food Alliance." http://foodalliance.org/about (accessed September 2, 2014).

Forest Stewardship Council. https://us.fsc.org (accessed December 22, 2013).

Foster, John Bellamy. "The Treadmill of Accumulation: Schnaiberg's Environment and Marxian Political Economy." *Organization and Environment* 18, no. 1 (2005): 7–18.

Friedland, William. "The New Globalization: The Case of Fresh Produce." In *From Columbus to Conagra: The Globalization of Agriculture and Food*, edited by Alessandro Bonanno, Lawrence Busch, William Friedland, Lourdes Gouveia, and Enzo Mingione, 210–231. Lawrence: University of Kansas Press, 1994.

Friedman, Debra, and Doug McAdam. "Collective Identities and Activism: Networks, Choices, and the Life of a Social Movement." In *Frontiers of Social Movement Theory*, edited by Aldon D. Morris and Carol Mueller, 156–173. New Haven, CT: Yale University Press, 1992.

Frisch, Tracy. "Up Close: Liana Hoodes." *Valley Table* 58 (June–August 2012), http://www.valleytable.com/article.php?article=008+Up+Close%2FLiana+Hoodes (accessed September 3, 2014).

Fromartz, Samuel. *Organic, Inc.: Natural Foods and How They Grew*. Orlando, FL: Harcourt, 2007.

Gamson, William. *The Strategy of Social Protest*. 2nd ed. Belmont, CA: Wadsworth, 1990.

Gassmann, Aaron J., Jennifer L. Petzold-Maxwell, Ryan S. Keweshan, and Mike W. Dunbar. "Field-Evolved Resistance to Bt Maize by Western Corn Rootworm." *PLoS ONE* 6, no. 7 (July 29, 2011), doi:10.1371/journal.pone.0022629.

Geier, Bernward. "IFOAM and the History of the International Organic Movement." In *Organic Farming: An International History*, edited by William Lockeretz, 175–186. Oxfordshire, UK: CABI, 2007.

Gericke, Richard. "The Forum: Know Your Organic Food Producers." *Organic Gardening* (September 1951): 4–6.

Gilder, Joshua. "Science Reporting on Organic Food Is Out to Lunch." *U.S. News and World Report*, March 2, 2012, http://www.usnews.com/opinion/blogs/joshua-gilder/2012/03/02/science-reporting-on-organic-food-is-out-to-lunch (accessed August 27, 2014).

Gitlin, Todd. "The Left, Lost in the Politics of Identity." *Harper's Magazine*, September 1993: 16.

Goldschmidt, Walter. "Large-Scale Farming and the Rural Social Structure." *Rural Sociology* 43, no. 3 (1978): 362–366.

Goodman, David. "Organic and Conventional Agriculture: Materializing Discourse and Agro-Ecological Managerialism." *Agriculture and Human Values* 17, no. 3 (September 2000): 215–219.

Goldman, Lynn R., Daniel F. Smith, Raymond R. Neutra, L. Duncan Saunders, Esther M. Pond, James Stratton, Kim Waller, Richard J. Jackson, and Kenneth W. Kizer. "Pesticide Food Poisoning from Contaminated Watermelons in California, 1985." *Archives of Environmental Health: An International Journal* 45, no. 4 (1990): 229–236.

Goodwin, Jeff, and James M. Jasper. *The Social Movements Reader: Cases and Concepts.* 2nd ed. West Sussex, UK: Blackwell Publishing, 2009.

Gordon, Robert. "Poisons in the Fields: The United Farm Workers, Pesticides, and Environmental Politics." *Pacific Historical Review* 68, no. 1 (1999): 51–77.

Goss, Kevin F., Richard D Rodefeld and Frederick H. Buttel. "The Political Economy of Class Structure in US Agriculture: A Theoretical Outline." In *The Rural Sociology of the Advanced Societies: Critical Perspectives*, edited by Frederick H. Buttel and Howard Newby, 83–132. Montclair, NJ: Allanheld Osmun, 1980.

Gottlieb, Robert. *Environmentalism Unbound.* Cambridge, MA: MIT Press, 2001.

Gottlieb, Robert. *Forcing the Spring: The Transformation of the American Environmental Movement.* Washington, DC: Island Press, 2005.

Gottlieb, Robert, and Anupama Joshi. *Food Justice.* Cambridge, MA: MIT Press, 2010.

Gould, Kenneth, David Pellow, and Allan Schnaiberg. "Interrogating the Treadmill of Production: Everything You Wanted to Know about the Treadmill but Were Afraid to Ask." *Organization and Environment* 17, no. 3 (2004): 296–313.

Gray, Margaret. *Labor and the Locavore: The Making of a Comprehensive Food Ethic.* Berkeley: University of California Press, 2013.

Greene, Catherine, Carolyn Dimitri, Biing-Hwan Lin, William McBride, Lydia Oberholtzer, and Travis Smith. *Emerging Issues in the U.S. Organic Industry.* Economic Information Bulletin no. 55. Washington, DC: US Department of Agriculture, Economic Research Service, June 2009. http://www.ers.usda.gov/publications/eib-economic-information-bulletin/eib55.aspx#.UpVLvbW1F8E (accessed August 27, 2014).

Greene, Catherine, and Amy Kremen. *US Organic Farming in 2000–2001: Adoption of Certified Systems.* Agriculture Information Bulletin no. 780. Washington, DC: US Department of Agriculture, Economic Research Services, Resource Economics Division, 2003.

Greene, Wade. "Guru of the Organic Food Cult." *New York Times*, June 6, 1971, SM30.

Growing Power. "Grow." http://www.growingpower.org/growing.htm (accessed September 1, 2014).

Gurian-Sherman, Doug, and Margaret Mellon. "The Rise of Superweeds and What to Do about It." Union of Concerned Scientists Policy Brief, 2013. http://www.ucsusa

.org/assets/documents/food_and_agriculture/rise-of-superweeds.pdf (accessed September 4, 2014).

Gussow, Joan Dye. "Can an Organic Twinkie Be Certified?" http://joansgarden.org/ Twinkie.pdf. (accessed August 11, 2014).

Guthman, Julie. *Agrarian Dreams: The Paradox of Organic Farming in California*. Berkeley: University of California Press, 2004.

Guthman, Julie. "Fast Food/Organic Food: Reflexive Tastes and the Making of 'Yuppie Chow.'" *Social and Cultural Geography* 4, no. 1 (2003): 45–58.

Guthman, Julie. "Regulating Meaning, Appropriating Nature: The Codification of California Organic Agriculture." *Antipode* 30, no. 2 (April 1998): 135–154.

Guthman, Julie. "The Trouble with 'Organic Lite': A Rejoinder to the 'Conventionalization' Debate." *Sociologia Ruralis* 44, no. 3 (2004): 301–316.

Haddad, Abigail. "The Problem with Organic Food." *American*, June 8, 2008, http:// www.american.com/archive/2008/june-06-08/the-problem-with-organic-food (accessed August 27, 2014).

Haines, Herbert. "Black Radicalization and the Funding of Civil Rights, 1957–1970." *Social Problems* 32, no. 1 (October 1984): 31–43.

Hall, Alan, and Veronika Mogyorody. "Organic Farmers in Ontario: An Examination of the Conventionalization Argument." *Sociologia Ruralis* 41, no. 4 (October 2001): 399–422.

Hannaford, Stephen. *Market Domination! The Impact of Industry Consolidation on Competition, Innovation, and Consumer Choice*. Westport, CT: Praeger Publishers, 2007.

Hartman Group. "Beyond Organic and Natural 2010: Resolving Confusion in Marketing Foods and Beverages." Hartman Group Syndicated Survey, February 2010.

Hartman Group. "The Many Faces of Organic 2008." http://www.hartman-group .com/publications/reports/the-many-faces-of-organic-2008 (accessed August 11, 2014).

Hatamiya, Lon S., to Michael V. Dunn. "Informational Memorandum for the Department Secretary," May 1, 1997, US Department of Agriculture, Agricultural Marketing Services. http://www.motherjones.com/politics/1998/05/organic-engineering-memo-6 (accessed August 21, 2014).

Haydu, Jeffrey. "Cultural Modeling in Two Eras of U.S. Food Protest: Grahamites (1830s) and Organic Advocates (1960s–70s)." *Social Problems* 58, no. 3 (2011): 461–487.

Heap, Ian. "International Survey of Herbicide Resistant Weeds." WeedScience.org, http://www.weedscience.org/summary/home.aspx (accessed August 27, 2014).

Heckman, Joseph. "A History of Organic Farming: Transitions from Sir Albert Howard's War in the Soil to USDA National Organic Program." *Renewable Agriculture and Food Systems* 21, no. 3 (September 2006): 143–150, doi:10.1079/RAF2005126.

Hendrickson, Mary K., and William D. Heffernan. "Opening Spaces through Relocalization: Locating Potential Resistance in the Weaknesses of the Global Food System." *Sociologia Ruralis* 42, no. 4 (October 2002): 347–369.

Hess, David J. "Declarations of Independents: On Local Knowledge and Localist Knowledge." *Anthropological Quarterly* 83, no. 1 (2010): 153–176.

Hess, David J. *Localist Movements in a Global Economy: Sustainability, Justice, and Urban Development in the United States.* Cambridge, MA: MIT Press, 2009.

Hess, David J. "Organic Food and Agriculture in the U.S.: Object Conflicts in a Health-Environmental Social Movement." *Science as Culture* 13, no. 4 (December 2004): 493–513.

Hewitt, Jean. "Organic Food Fanciers Go to Great Lengths for the Real Thing." *New York Times*, September 7, 1970, 23.

Hinckley, Barbara. *Coalitions and Politics.* New York: Harcourt Brace Jovanovich, 1981.

Hine, Rachel, and Jules Pretty. *Organic Agriculture and Food Security in Africa.* New York: United Nations, 2008.

Hinrichs, C. Clare. "Embeddedness and Local Food Systems: Notes on Two Types of Direct Agriculture Market." *Journal of Rural Studies* 16 (2000): 295–303.

Hinrichs, C. Clare, and Patricia Allen.. "Selective Patronage and Social Justice: Local Food Consumer Campaigns in a Historical Context." *Journal of Agriculture and Environmental Ethics* 21, no. 4 (2008): 329–352.

Hofer, Karin. 2000. "Labelling of Organic Food Products." In *The Voluntary Approach to Environmental Policy: Joint Environmental Policy-Making in Europe*, edited by Arthur P. J. Mol, Volkmar Lauber, and Duncan Liefferink, 156–191. Oxford: Oxford University Press, 2000.

Hojnacki, Marie. "Interest Groups' Decisions to Join Alliances or Work Alone." *American Journal of Political Science* 41 (January 1997): 61–87.

Hoodes, Liana, Michael Sligh, Harriet Behar, Roger Blobaum, Lisa J. Bunin, Lynn Coody, Elizabeth Henderson, Faye Jones, Mark Lipson, and Jim Riddle. *National Organic Action Plan: From the Margins to the Mainstream—Advancing Organic Agriculture in the U.S.* Pittsboro, NC: Rural Advancement Foundation International—USA, January 2010. http://www.rafiusa.org/docs/noap.pdf (accessed August 29, 2014).

Howard, Albert. "The Cause of Plant Disease." *Organic Gardening and Farming* 2, no. 1 (December 1942): 4.

Howard, Albert. "The Good Earth." *Organic Farming and Gardening* 1, no. 2 (June 1942): 5.

Howard, Albert. *Organic Gardening* 3, no. 2 (July 1943): 16.

Howard, Albert. 2011. *The Soil and Health: A Study of Organic Agriculture*. Louisville: University of Kentucky Press (originally published 1945).

Howard, Albert. "The Wheel of Life," *Organic Gardening* 3, no. 5 (October 1943): 20.

Howard, Philip. "Consolidation in the North American Organic Food Processing Sector, 1997–2007." *International Journal of Agriculture and Food* 16, no. 1 (2009): 13–30.

Hula, Kevin W. *Lobbying Together: Interest Group Coalitions in Legislative Politics*. Washington, DC: Georgetown University Press, 1999.

Imhoff, Daniel. *CAFO: The Tragedy of Industrial Animal Factories*. San Rafael, CA: Earth Aware Editions, 2010.

Imhoff, Daniel, "Organic Incorporated." *Whole Earth* 92, no. 4 (1998).

International Federation of Organic Agriculture Movements. *The IFOAM Norms for Organic Production and Processing Version*. Bonn: Die Deutsche Bibliothek, 2012.

Jacobson, Michael F. "Feeding the People, Not Food Producers." *New York Times*, August 31, 1972, 33.

Jacques, Sean, and Lyn Collins. "Community Supported Agriculture: An Alternative to Agribusiness." *Geography Review* 16, no. 5 (2003): 30–33.

Jaffe, Daniel. *Brewing Justice: Fair Trade Coffee, Sustainability, and Survival*. Berkeley: University of California Press, 2007.

Jaffe, Daniel. "Weak Coffee: Certification and Co-optation in the Fair Trade Movement." *Social Problems* 59, no. 1 (2012): 94–116.

Jaffe, Daniel, and Philip H. Howard. "Corporate Cooptation of Organic and Fair Trade Standards." *Agriculture and Human Values* 27 (2010): 387–399.

Janssen, Brandi. "Local Food, Local Engagement: Community-Supported Agriculture in Eastern Iowa." *Culture and Agriculture* 32, no. 1 (2010): 4–16.

Jarvis, Michael, and Billy Cox. "USDA Issues Final Rule on Organic Access to Pasture." *USDA Agricultural Marketing Service*, February 12, 2010, http://www.ams.usda .gov/AMSv1.0/ams.fetchTemplateData.do?template=TemplateU&navID=&page= Newsroom&resultType=Details&dDocName=STELPRDC5082658&dID=126904&wf =false&description=USDA+Issues+Final+Rule+on+Organic+Access+to+Pasture+&top Nav=Newsroom&leftNav= (accessed August 30, 2014).

Johnston, Hank, Enrique Laraña, and Joseph R. Gusfield. "Identities, Grievances, and New Social Movements." In *New Social Movements: From Ideology to Identity*, edited by Enrique Laraña, Hank Johnston, and Joseph R. Gusfield, 3–35. Philadelphia: Temple University Press, 1994.

Johnston, Josée. "The Citizen-Consumer Hybrid: Ideological Tensions and the Case of Whole Foods Market." *Theory and Society* 37, no. 3 (2008): 229–270.

Kaplan, Melanie D. G. "Stonyfield Farm CEO: How an Organic Yogurt Business Can Scale." *Smart Planet*, May 17, 2010, http://www.smartplanet.com/blog/pure-genius/stonyfield-farm-ceo-how-an-organic-yogurt-business-can-scale/3638 (accessed August 31, 2014).

Karst, Tom. "Time to Push for USDA Certification of 'Sustainably Grown,'" *Packer*, September 12, 2012, http://www.thepacker.com/opinion/fresh-talk-blog/Time-to-push-for-USDA-certification-of--sustainably-grown-169457096.html (accessed September 2, 2014).

Keller, Bill. "Coalitions and Associations Transform Strategy, Methods of Lobbying in Washington." *Congressional Quarterly Weekly Report* 40 (1982): 119–123.

King, Braydon G. "Can Private Politics Effectively Replace Government Regulation?" Paper presented at Filling the Governance Gap: Aligning Enterprise and Advocacy, Kellogg School of Management/Aspen Institute Business and Society Leadership Summit, Evanston, IL, February 27–28, 2014.

King, Braydon G. "Reputational Dynamics of Private Regulation." *Socioeconomic Review* 12 (2014): 200–206.

King, Braydon G. "The Tactical Disruptiveness of Movements: Sources of Market and Mediated Disruption in Corporate Boycotts." *Social Problems* 48 (2011): 491–517.

King, Braydon G., and Mary Hunter-McDonnell. "Keeping Up Appearances: Reputation Threat and Prosocial Responses to Social Movement Boycotts." *Administrative Science Quarterly* 58, no. 3 (2013): 387–419.

King, Braydon G., and Nicholas A. Pearce. "The Contentiousness of Markets: Politics, Social Movements, and Institutional Change in Markets." *Annual Review of Sociology* 36 (2010): 249–267.

Kittredge, Jack. "Sligh: 'Stay the Course.'" *Natural Farmer*, Spring 2006, 38, http://www.nofa.org/tnf/2006spring/Sligh%20-%20Stay%20the%20Course.pdf (accessed August 17, 2014).

Klonsky, Karen. "Forces Impacting the Production of Organic Foods." *Agriculture and Human Values* 17 (2000): 233–243.

Kloppenburg, Jack, Jr., John Hendrickson, and G. W. Stevenson. "Coming into the Foodshed." *Agriculture and Human Values* 13, no. 3 (1996): 33–42.

Laraña, Enrique, Hank Johnston, and Joseph R. Gusfield, eds. *New Social Movements: From Ideology to Identity*. Philadelphia: Temple University Press, 1994.

LaTrobe, Helen, and Tim Acott. "Localizing the Global Food System." *International Journal of Sustainable Development and World Ecology* 7, no. 4 (2000): 309–320.

Lawler, Edward J., and George A. Youngs Jr. "Coalition Formation: An Integrative Model." *Sociometry* 38, no. 1 (March 1975): 1–17.

Lawson, Laura J. *City Bountiful: A Century of Community Gardening in America*. Berkeley: University of California Press, 2005.

Lichtenstein, Grace. "'Organic' Food Study Finds Pesticides." *New York Times*, December 2, 1972, 39.

Lipson, Mark. "Searching for the 'O-Word': Analyzing the USDA Current Research Information System for Pertinence to Organic Farming." Santa Cruz, CA: Organic Farming Research Foundation, 1997.

Lobao, Linda, and Katherine Meyer. "The Great Agricultural Transition: Crisis, Change, and Social Consequences of Twentieth Century US Farming." *Annual Review of Sociology* 27 (2001): 103–124.

Lockeretz, William, ed. *Organic Farming: An International History*. Oxfordshire, UK: CABI, 2007.

Lockeretz, William. *Visions of American Agriculture*. Iowa City: Iowa State University Press, 2000.

Lockie, Stewart, and Darren. Halpin. "The 'Conventionalization' Thesis Reconsidered: Structural and Ideological Transformation of Australian Organic Agriculture." *Sociologia Ruralis* 45, no. 4 (2005): 284–307.

Lockie, Stewart, Kristen Lyons, and Geoffrey Lawrence. "Constructing 'Green' Foods: Corporate Capital, Risk, and Organic Farming in Australia and New Zealand." *Agriculture and Human Values* 17, no. 4 (December 2000): 315–322.

Lockie, Stewart, Kristen Lyons, Geoffrey Lawrence, and Kerry Mummery. "Eating 'Green': Motivations behind Organic Food Consumption in Australia." *Sociologia Ruralis* 42, no. 1 (2002): 23–41.

Loomis, Burdett. "Coalitions of Interests: Building Bridges in the Balkanized State." In *Interest Group Politics*, edited by Allen J. Cigler and Burdett Loomis, 258–274. 2nd ed. Washington, DC: CQ Press, 1986.

Low, Sarah A., and Stephen Vogel. *Direct and Intermediated Marketing of Local Foods in the United States*. Economic Research Report no. 128. Washington, DC: US Department of Agriculture, Economic Research Service, November 2011.

Lu, Chensheng, Dianne Knutson, Jennifer Fisker-Andersen, and Richard A. Fenske. "Biological Monitoring Survey of Organophosphorus Pesticide Exposure among Preschool Children in the Seattle Metropolitan Area." *Environmental Health Perspectives* 109, no. 3 (March 2001): 299–303.

Lupkin, Sydney. "Pesticides in Tap Water Linked to Food Allergies." *ABC News*, December 3, 2012, http://abcnews.go.com/blogs/health/2012/12/03/pesticides-in-tap-water-linked-to-food-allergies/ (accessed August 27, 2014).

Lynggaard, Kennet S. C. "The Farmer within an Institutional Environment: Comparing Danish and Belgian Organic Farming." *Sociologia Ruralis* 41, no. 1 (January 2001): 85–111.

Lyons, Kristen. "Corporate Environmentalism and Organic Agriculture in Australia: The Case of Uncle Toby's." *Rural Sociology* 64, no. 2 (1999): 251–266.

Lyson, Thomas A. *Civic Agriculture: Reconnecting Farm, Food, and Community*. Medford, MA: Tufts University Press, 2012.

Lyson, Thomas A., Robert J. Torres, and Rick Welsh. "Scale of Agricultural Production, Civic Engagement, and Community Welfare." *Social Forces* 80, no. 1 (September 2001): 311–327.

Magdoff, Fred, John Bellamy Foster, and Frederick H. Buttel, eds. *Hungry for Profit: The Agribusiness Threat to Farmers, Food, and the Environment*. New York: Monthly Review Press, 2000.

Magnusson, Maria, Anne Arvola, Ulla-Kaisa Koivisto Hursti, Lars Aberg, and Per-Olow Sjoden. "Choice of Organic Foods Is Related to Perceived Consequences for Human Health and to Environmentally Friendly Behaviour." *Appetite* 40, no. 2 (2003): 109–118.

Majka, Linda, and Theo Majka. "Organizing US Farmworkers: A Continuous Struggle." In *Hungry for Profit: The Agribusiness Threat to Farmers, Food, and the Environment*, edited by Fred Magdoff, John Bellamy Foster, and Frederick H. Buttel, 161–174. New York: Monthly Review Press, 2000.

Mark, Jason. "Workers on Organic Farms Are Treated as Poorly as Their Conventional Counterparts." *Grist*, August 2, 2006, http://grist.org/article/mark/ (accessed September 1, 2014).

Marriott, Emily E., and Michelle M. Wander. "Total and Labile Soil Organic Matter in Organic and Conventional Farming Systems." *Soil Science Society of America* 70 (May 2006): 950–959.

Martin, Andrew. "How to Add Oomph to Organic." *New York Times*, August 19, 2007, http://www.nytimes.com/2007/08/19/business/yourmoney/19feed.html?_r=0 (accessed August 30, 2014).

Mazmanian, Daniel, and Michael Kraft. *Toward Sustainable Communities: Transitions and Transformations in Environmental Policy*. Cambridge, MA: MIT Press, 1999.

McAdam, Doug. "Conceptual Origins, Current Problems, Future Directions." In *Comparative Perspectives on Social Movements*, edited by Doug McAdam, John D. McCarthy, and Mayer Zald, 23–40. Cambridge: Cambridge University Press, 1996.

McAdam, Doug. *Political Process and the Development of Black Insurgency, 1930–1970.* Chicago: University of Chicago Press, 1999.

McCarthy, John D., David Britt, and Mark Wolfson. "The Institutional Channeling of Social Movements by the State in the United States." *Research in Social Movements, Conflicts, and Change* 13 (1991): 45–76.

McCarthy, John D., and Mayer N. Zald. "Resource Mobilization and Social Movements: A Partial Theory." *American Journal of Sociology* 82, no. 6 (May 1977): 1212–1241.

McCarthy, John D., and Mayer N. Zald. *The Trend of Social Movements in America: Professionalization and Resource Mobilization.* Morristown, NJ: General Learning Press, 1973.

McClintock, Nathan. "From Industrial Garden to Food Desert: Demarcated Devaluation in the Flatlands of Oakland, California." In *Cultivating Food Justice: Race, Class, and Sustainability*, edited by Alison Hope Alkon and Julian Agyeman, 89–120. Cambridge, MA: MIT Press, 2011.

McPherson Miller, Lynn Smith-Lovin, and Matthew E. Brashears. "Social Isolation in America: Changes in Core Discussion Networks over Two Decades." *American Sociological Review* 71 (2006): 353–375.

McWilliams, James. "Organic Crops Alone Can't Feed the World." *Slate*, March 10, 2011, http://www.slate.com/articles/health_and_science/green_room/2011/03/organic_crops_alone_cant_feed_the_world.html (accessed August 27, 2014).

Melucci, Alberto. "A Strange Kind of Newness: What's 'New' in New Social Movements?" In *New Social Movements: From Ideology to Identity*, edited by Enrique Laraña, Hank Johnston, and Joseph R. Gusfield, 101–130. Philadelphia: Temple University Press, 1994.

Merrigan, Kathleen. "Negotiating Identity within the Sustainable Agriculture Advocacy Coalition." PhD diss., Massachusetts Institute of Technology, September 2000.

Meyer, David S., and Catherine Corrigall-Brown. "Coalitions and Political Context: U.S. Movements against Wars in Iraq." *Mobilization* 10, no. 3 (2005): 327–344.

Meyer, David S., Valerie Jenness, and Helen Ingram. *Routing the Opposition: Social Movements, Public Policy, and Democracy.* Minneapolis: University of Minnesota Press, 2005.

Micheletti, Michele. *Political Virtue and Shopping: Individuals, Consumerism, and Collective Action.* New York: Palgrave Macmillan, 2003.

Michels, Robert. *Political Parties: A Sociological Study of the Oligarchical Tendencies of Modern Democracy.* New York: Evergreen Review, Inc., 2008.

Michelsen, Johannes. "Organic Farming in a Regulatory Perspective." *Sociologia Ruralis* 41, no. 1 (2001): 62–84.

Michelsen, Johannes. "Recent Development and Political Acceptance of Organic Farming in Europe." *Sociologia Ruralis* 41, no. 1 (January 2001): 3–20.

Minick, Jim. "Beyond Organic." *Counterpunch*, November 27–29, 2004, http://www.counterpunch.org/2004/11/27/beyond-organic/ (accessed September 3, 2014).

Mol, Arthur P. J. "Ecological Modernization: Industrial Transformations and Environmental Reform." In *International Handbook of Environmental Sociology*, edited by Michael Redclift and Graham Woodgate, 138–149. Cheltenham, UK: Edward Elgar, 1997.

Mol, Arthur P. J. *The Refinement of Production: Ecological Modernization Theory and the Dutch Chemical Industry*. Utrecht: Jan van Arkel/International Books, 1995.

Mol, Arthur P. J., and Gert Spaargaren. "Ecological Modernization Theory in Debate: A Review." In *Ecological Modernization around the World*, edited by Arthur P. J. Mol and David A. Sonnenfeld, 17–49. London: Frank Cass, 2000.

Mollison, Andrew. "Georgia Rep. Nathan Deal behind Attack on Organics." *Atlanta Journal and Constitution*, February 15, 2003.

Moore, Hilmar. "Rudolf Steiner: A Biographical Introduction for Farmers." *Biodynamics* 214 (November–December 1997): 29–32.

Murdoch, Jonathan, Terry Marsden, and Jo Banks. "Quality, Nature, and Embeddedness: Some Theoretical Considerations in the Context of the Food Sector." *Economic Geography* 76, no. 2 (2000): 107–125.

National Conference on Citizenship. "America's Civic Health Index: Broken Engagement." Washington, DC: National Conference on Citizenship, September 18, 2006.

National Conference on Citizenship. "Civic Health Assessment." Washington, DC: National Conference on Citizenship, 2010.

National Organic Coalition. "About Us." http://www.nationalorganiccoalition.org/about (accessed December 7, 2013).

National Organic Standards Board. Joint Committee Meetings, May 1–2, 4–6, 1992. http://www.ams.usda.gov/AMSv1.0/getfile?dDocName=STELPRDC5057489 (accessed December 7, 2013).

National Sustainable Agriculture Coalition. "Organic Agriculture Research and Extension Initiative." http://sustainableagriculture.net/publications/grassrootsguide/sustainable-organic-research/organic-research-extension-initiative/ (accessed August 27, 2014).

Nestle, Marion. *Food Politics: How the Food Industry Influences Nutrition and Health.* Berkeley: University of California Press, 2007.

Neuman, William. "New Pasture Rules Issued for Organic Dairy Producers." *New York Times,* February 12, 2010, http://www.nytimes.com/2010/02/13/business/13organic.html (accessed August 30, 2014).

Northbourne, Lord. *Look to the Land.* London: J. M. Dent and Sons, 1940.

Nuestras Raices. http://www.nuestras-raices.org/home.html (accessed December 21, 2013).

Obach, Brian. *Labor and the Environmental Movement: The Quest for Common Ground.* Cambridge, MA: MIT Press, 2004.

Obach, Brian. "Political Opportunity and Social Movement Coalitions: The Role of Policy Segmentation and Non-Profit Tax Law." In *Strategic Alliances: Coalition Building and Social Movements,* edited by Nella Van Dyke and Holly McCammon, 197–218. Minneapolis: University of Minnesota Press, 2010.

Obach, Brian. "Theoretical Interpretations of the Growth in Organic Agriculture: Agricultural Modernization or an Organic Treadmill?" *Society and Natural Resources* 20, no. 3 (2007): 229–244.

Obach, Brian, and Kathleen Tobin. *Agriculture Supporting Community in the Mid-Hudson Region.* Discussion Brief no. 5. New Paltz: Center for Research, Regional Education, and Outreach, State University of New York, Spring 2011.

Obach, Brian, and Kathleen Tobin. "Civic Agriculture and Community Engagement." *Agriculture and Human Values* 31, no. 2 (2014): 307–322.

Oldenburg, Ray. *The Great Good Place.* New York: Paragon House, 1991.

Oregon Tilth. "Farm Bill Victories for Organic Farmers." http://tilth.org/news/farm-bill-victories-for-organic-farmers (accessed August 27, 2014).

Organic Center. http://www.organic-center.org (accessed August 27, 2014).

"Organic Club Notes." *Organic Gardening and Farming* (March 1957): 18–19.

Organic Consumers Association. "About the OCA: Who We Are and What We're Doing." http://www.organicconsumers.org/aboutus.cfm (accessed August 21, 2014).

Organic Consumers Association. "USDA Attempts to Pack Organic Standards Board with Corporate Agribusiness Reps: Organic Consumers Fight Hijacked Seats on NOSB." December 7, 2006. http://www.organicconsumers.org/articles/article_3526.cfm (accessed August 30, 2014).

Organic Farmers Association Council. "Original Invitation to Join OFAC." December 6, 1990. http://rogerblobaum.com/656/ (accessed August 4, 2014).

Organic Foods Production Act of 1990, 7 U.S.C. 6501.

"Organic Gardening Clubs of America." *Organic Gardening and Farming* 2 (October 1955): 76–79.

"Organic Market Opens." *Organic Gardening* (December 1951): 46. Organic Trade Association. "Industry Statistics and Projected Growth." June 2011. http://www.ota .com/organic/mt/business.html (accessed December 1, 2013).

Organic Trade Association. "Lawsuit Chronology." http://www.ota.com/Lawsuit Chronology.html (accessed August 30, 2014).

Organic Trade Association. "Organic Foods Production Act Backgrounder." http:// www.ota.com/pp/legislation/backgrounder.html (accessed August 30, 2014).

Organic Trade Association. "2009 U.S. Families' Organic Attitudes and Beliefs Study." June 2009. http://www.ota.com/pics/documents/2009OTA-KiwiExecutive Summary.pdf (accessed September 1, 2014).

Organic Trade Association. "2011 Organic Industry Survey Overview." http://www .ota.com/pics/documents/2011OrganicIndustrySurvey.pdf (accessed July 23, 2014).

Organic Trade Association. "American Appetite for Organic Products Breaks through $35 Billion Mark." Press release, Washington, DC, May 15, 2014.

Organic Valley. "Transparency: Additional Farm Standards." http://www .organicvalley.coop/about-us/transparency/additional-farm-standards/ (accessed September 1, 2014).

Organic Valley. "Transparency: Organic Valley Dairy Herd Size." http://www .organicvalley.coop/about-us/transparency/herd-chart/ (accessed September 1, 2014).

"Organic World." *Organic Gardening and Farming* (June 1958): 2–3.

"Organic World." *Organic Gardening and Farming* (August 1958): 3.

"Organic World." *Organic Gardening and Farming* (April 1959): 2.

Paarlberg, Robert. "Attention Whole Foods Shoppers." *Foreign Policy*, May–June 2010, http://www.foreignpolicy.com/articles/2010/04/26/attention_whole_foods_ shoppers (accessed August 27, 2014).

Pacey, Margaret D. "Nature's Bounty: Merchandizers of 'Health Foods' Are Cashing in on It." *Barron's National Business and Financial Weekly*, May 10, 1971, P5.

Padel, Susanne. "Conversion to Organic Farming: A Typical Example of the Diffusion of an Innovation?" *Sociologia Ruralis* 41, no. 1 (2001): 40–61.

Parr, James F. "USDA Research on Organic Farming: Better Late Than Never." *American Journal of Alternative Agriculture* 18, no. 3 (September 2003): 171–172.

Patel, Raj. *Stuffed and Starved: The Hidden Battle for the World Food System*. Brooklyn: Melville House Publishing, 2007.

Pearlstein, Joanna. "Surprise! Conventional Agriculture Can Be Easier on the Planet." *Wired*, May 19, 2008, http://www.wired.com/science/planetearth/magazine/16-06/ff_heresies_03organics (accessed August 27, 2014).

Petrini, Carlo. *Slow Food: The Case for Taste*. New York: Columbia University Press, 2003.

Philpott, Tom. "5 Ways the Stanford Study Sells Organics Short." *Mother Jones*, September 5, 2012, http://www.motherjones.com/tom-philpott/2012/09/five-ways-stanford-study-underestimates-organic-food. (accessed August 28, 2014).

Pimentel, David. "Impacts of Organic Farming on the Efficiency of Energy Use in Agriculture." Washington, DC: Organic Center, August 2006. http://organic-center.org/reportfiles/EnergyExecSummary.pdf (accessed August 27, 2014).

Piore, Michael J., and Charles F. Sabel. *The Second Industrial Divide: Possibilities for Prosperity*. New York: Basic Books, 1984.

Pirog, Rich, Timothy Van Pelt, Kamyar Enshayan, and Ellen Cook. *Food, Fuel, and Freeways: An Iowa Perspective on How Far Food Travels, Fuel Usage, and Greenhouse Gas Emissions*. Ames: Leopold Center for Sustainable Agriculture, Iowa State University, June 2001.

Piven, Frances Fox, and Richard Cloward. *Poor People's Movements*. New York: Pantheon, 1977.

Pole, Antoinette, and Margaret Gray. "Farming Alone? What's Up with the 'C' in Community Supported Agriculture?" *Agriculture and Human Values* 30, no. 1 (March 2013): 85–100.

Pollan, Michael. *The Omnivore's Dilemma: A Natural History of Four Meals*. New York: Penguin, 2006.

Pollan, Michael. "Vote for the Dinner Party: Is This the Year That the Food Movement Finally Enters Politics?" *New York Times*, October 10, 2012, http://www.nytimes.com/2012/10/14/magazine/why-californias-proposition-37-should-matter-to-anyone-who-cares-about-food.html?pagewanted=all&_r=0 (accessed September 4, 2014).

Polsby, Nelson W. "Coalition and Faction in American Politics: An Institutional View." In *Emerging Coalitions in American Politics*, edited by Seymour Martin Lipset, 103–126. San Francisco: Institute for Contemporary Studies, 1978.

Porter, David, and Chester L. Mirsky. *Megamall on the Hudson: Planning, Walmart, and Grassroots Resistance*. Victoria, BC: Trafford Publishing, 2003.

Prentice, Jessica. "The Birth of Locavore." Oxford University Press blog, November 20, 2007. http://blog.oup.com/2007/11/prentice/. (accessed September 3, 2014).

Putnam, Robert D. *Bowling Alone*. New York: Simon and Schuster, 2000.

Putnam, Robert. D. *Making Democracy Work: Civic Traditions in Modern Italy*. Princeton, NJ: Princeton University Press, 1993.

Rahmann, Gerold. "Biodiversity and Organic Farming: What Do We Know?" *Agriculture and Forestry Research* 3 (2011): 189–208.

Rangan, Urvashi. "Putting the Label on the Table." Keynote address at the Harvard Food Law Society Forum on Food Labeling, Cambridge, MA, March 2013.

Raupp, Joachim, Carola Pekrun, Meike Oltmanns, and Ulrich Köpkes, eds. *Long Term Field Experiments in Organic Farming*. Berlin: International Society of Organic Agriculture Research, 2006.

Reed, Matthew. "Fight the Future! How the Contemporary Campaigns of the UK Organic Movement Have Arisen from Their Composting Past." *Sociologia Ruralis* 41, no. 1 (2001): 131–145.

Reganold, John P., Preston K. Andrews, Jennifer R. Reeve, Lynne Carpenter-Boggs, and Christopher W. Schadt, J. Richard Alldredge, Carolyn F. Ross, Neal M. Davies, and Jizhong Zhou. "Fruit and Soil Quality of Organic and Conventional Strawberry Agroecosystems." *PLoS ONE* 5, no. 9 (September 2010), doi:10.1371/journal.pone.0012346.Reganold, John P., Jerry D. Glover, Preston K. Andrews, and Herbert R. Hinman, "Sustainability of Three Apple Production Systems." *Nature* 410 (April 2001), doi:10.1038/35073574.

Rich, Deborah. "Organic and Sustainable Up for Review ... Again." *Organic Farming Research Foundation Information Bulletin* (Fall 2008): 4–13, http://www.organic-center.org/reportfiles/ib16%20organic%20and%20sustainable%20up%20for%20review.pdf (accessed July 22, 2014).

Riker, William H. *The Theory of Political Coalitions*. New Haven, CT: Yale University Press, 1962.

Rocky Mountain Institute. http://www.rmi.org (accessed September 4, 2014).

Rodale, Jerome Irving. "Introduction to Organic Farming." *Organic Farming and Gardening* 1, no. 1 (May 1942): 3.

Rodale, Jerome Irving. "With the Editor: Do Chemical Fertilizers Kill Earthworms." *Organic Gardening* (February 1948): 12–17.

Rose, Fred. *Coalitions across the Class Divide: Lessons from the Labor, Peace, and Environmental Movements*. Ithaca, NY: Cornell University Press, 2000.

Rudolf Steiner Archive. http://www.rsarchive.org/ (accessed July 28, 2014).

Saba, Anna, and Federico Messina. "Attitudes towards Organic Foods and Risk/Benefit Perception Associated with Pesticides." *Food Quality and Preference* 14, no. 8 (2003): 637–646.

Sabatier, Paul A., and Hank C. Jenkins-Smith. "Evaluating the Advocacy Coalition Framework." *Journal of Public Policy* 14, no. 2 (April 1994): 175–203.

Salisbury, Robert. "The Paradox of Interest Groups in Washington: More Groups, Less Clout." In *The New American Political System*, edited by Antony King, 203–230. Washington, DC: AEI Press, 1990.

Salisbury, Robert, John Heinz, Edward Laumann, and Robert Nelson. "Who Works with Whom? Interest Group Alliances and Opposition." *American Political Science Review* 81 (1987): 1217–1234.

Salzman, Jim. "Product and Raw Material Eco-Labeling: The Limits of a Transatlantic Approach." Working paper no. 117, Berkeley Roundtable on the International Economy, University of California at Berkeley, May 1998.

Saunders, Caroline, Andrew Barber, and Greg Taylor. *Food Miles—Comparative Energy/Emissions Performance of New Zealand's Agriculture Industry.* Research Report no. 285. Lincoln, NZ: Agribusiness and Economics Research Unit, Lincoln University, July 2006).

Schnaiberg, Allan. *The Environment: From Surplus to Scarcity.* New York: Oxford University Press, 1980.

Schnaiberg Allan, and Kenneth Gould. *Environment and Society.* New York: St. Martin's Press, 1994.

Schneiberg, Marc, and Tim Bartley. "Organizations, Regulation, and Economic Behavior: Regulatory Dynamics and Forms from the Nineteenth to Twenty-First Century." *Annual Review of Law and Social Science* 4 (2008): 31–61.

Schneider, Keith. "Science Academy Recommends Resumption of Natural Farming." *New York Times*, September 8, 1989, A1.

Schnell, Steven M. "Food with a Farmer's Face: Community-Supported Agriculture in the United States." *Geographical Review* 97, no. 4 (October 2007): 550–564.

Scowcroft, Bob. "From the Director." *Organic Farming Research Foundation Information Bulletin*, Fall 2008, 4.

Scowcroft, Bob. "The Organic Conversation Begins Anew (Again)." *GreenMoney Journal*, http://archives.greenmoneyjournal.com/article.mpl?newsletterid=39&articleid=505 (accessed August 30, 2014).

SCS Global Services. "About SCS: Company." http://www.scsglobalservices.com/company (accessed September 2, 2014).

Severson, Kim. "More Choice, and More Confusion, in Quest for Healthy Eating." *New York Times*, September 8, 2012, http://www.nytimes.com/2012/09/09/us/would-be-healthy-eaters-face-confusion-of-choices.html (accessed August 27, 2014).

Shapiro, Laura, and Linda Wright. "Suddenly, It's a Panic for Organic." *Newsweek*, March 27, 1989, 24. Page 24

Sheets, J. A. "Food—By the People, for the People: The Study of the Relationship between Local Food Networks and Civic Engagement." Paper presented at the joint annual meetings of the Agriculture, Food, and Human Values Society, Association for the Study of Food and Society, and Society for Anthropology of Food and Nutrition, Missoula, MT, June 2011.

Shelton, Suzanne. "Executive Interview: Q&A with Katherine DiMatteo, Executive Director, OTA." *Engredea News and Analysis*, December 22, 2005, http://newhope360.com/supply-news-amp-analysis/executive-interview-qa-katherine-dimatteo-executive-director-ota (accessed August 30, 2014).

Shiva, Vandana. *Stolen Harvest: The Hijacking of the Global Food Supply*. Cambridge, MA: South End Press, 2000.

Shreck, Aimee, Christy Getz, and Gail Feenstra, "Farmworkers in Organic Agriculture: Toward a Broader Notion of Sustainability." *Newsletter of the University of California Sustainable Agriculture Research and Education Program* 17, no. 1 (Winter–Spring 2005): 1–3.

Shwartz, Mark. "New Study Confirms the Ecological Virtues of Organic Farming." *Stanford Report*, March 10, 2006, http://news.stanford.edu/news/2006/march15/organics-030806.html (accessed August 27, 2014).

Skocpol, Theda. *Diminished Democracy: From Membership to Management in American Civic Life*. Norman: University of Oklahoma Press, 2003.Sligh, Michael, and Carolyn Christman. "Who Owns Organic? The Global Status, Prospects, and Challenges of a Changing Organic Market." Pittsboro, NC: Rural Advancement Foundation International—USA, 2003.

Sligh, Michael, and Thomas Cierpka. "Organic Values." In *Organic Farming: An International History*, edited by William Lockeretz, 30–39. Oxfordshire, UK: CABI, 2007.

Slow Food. "About Us." http://slowfood.com/international/1/about-us?-session=query_session:47A9264213ff82D20BOm4DD0A54A (accessed August 19, 2012).

Smith, Jeffrey M. *Genetic Roulette: The Documented Health Risks of Genetically Engineered Foods*. Fairfield, IA: Yes! Books, 2007.

Smith, Joel Preston. "A Sustainable Agriculture Label: Coming to Foods near You?" *PCC Natural Markets*, September 2008, http://www.pccnaturalmarkets.com/sc/0809/sc0809-sus-ag-label.html (accessed September 2, 2014).

Smith, Pete, Daniel Martino, Zucong Cai, Daniel Gwary, Henry Janzen, Pushpam Kumar, Bruce McCarl, Stephen Ogle, Frank O'Mara, Charles Rice, Bob Scholes, and Oleg Sirotenko. "Agriculture." In *Climate Change 2007: Mitigation; Contribution of Working Group III to the Fourth Assessment Report of the Intergovernmental Panel on Climate Change*, edited by Bert Metz, Ogunlade Davidson, Peter Bosch, Rutu Dave, and Leo Meyer, 498–540. Cambridge: Cambridge University Press, 2007.

Smith-Spangler, Crystal, Margaret L. Brandeau, Grace E. Hunter, J. Clay Bavinger, Maren Pearson, Paul J. Eschbach, Vandana Sundaram, Hau Liu, Patricia Schirmer, Christopher Stave, Ingram Olkin, and Dena M. Bravata. "Are Organic Foods Safer or Healthier Than Conventional Alternatives? A Systematic Review." *Annals of Internal Medicine* 157, no. 5 (September 2012): 348–366.

Sonnenfeld, David A. "Contradictions in Ecological Modernization: Pulp and Paper Manufacturing in Southeast Asia." In *Ecological Modernization around the World*, edited by. Arthur P. J. Mol and David A. Sonnenfeld, 235–256. London: Frank Cass, 2000.

Soule, Sarah A. *Contention and Corporate Social Responsibility*. Cambridge: Cambridge University Press, 2009.

Southworth, George. "Natural/Organic Industry Outlook." *Cooperative Grocer*, September–October 2001, http://www.cooperativegrocer.coop/articles/index.php?id=333 (accessed August 29, 2014).

Staggenborg, Suzanne. "Coalition Work in the Pro-Choice Movement: Organizational and Environmental Opportunities and Obstacles." *Social Problems* 33, no. 5 (June 1986): 374–390.

Staggenborg, Suzanne. "The Consequences of Professionalization and Formalization in the Pro-Choice Movement." *American Sociological Review* 53, no. 4 (August 1988): 585–605.

Stewart, Emily. "Organic in the Valley: Few Producers, Growing Demand." *Poughkeepsie Journal*, June 2, 2013.

Stinner, D. H. "The Science of Organic Farming." In *Organic Farming: An International History*, edited by William Lockeretz, 40–72. Oxfordshire, UK: CABI, 2007.

Stolle, Dietlind, Marc Hooghe, and Michele Micheletti. "Politics in the Supermarket: Political Consumerism as a Form of Political Participation." *International Political Science Review* 26, no. 3 (2005): 245–269.

Stone, Pat. "Organic Agriculture: Turning a Movement into an Industry." *Mother Earth News*, September–October 1989, http://www.motherearthnews.com/homesteading-and-livestock/organic-agriculture-movement-industry-zmaz89sozshe.aspx#axzz2jeKbB1WH (accessed August 1, 2014).

Strom, Stephanie. "Has 'Organic' Been Oversized?" *New York Times*, July 7, 2012, http://www.nytimes.com/2012/07/08/business/organic-food-purists-worry-about-big-companies-influence.html?pagewanted=all&_r=0 (accessed August 30, 2014).

Swanton Berry Farm. http://www.swantonberryfarm.com (accessed September 1, 2014).

Szasz, Andrew. *Shopping Our Way to Safety: How We Changed from Protecting the Environment to Protecting Ourselves*. Minneapolis: University of Minnesota Press, 2009.

Tarrow, Sidney. "States and Opportunities." In *Comparative Perspectives on Social Movements*, edited by Doug McAdam, John D. McCarthy, and Mayer Zald, 41–61. Cambridge: Cambridge University Press, 1996.

Taylor, Verta A., and Nancy Whittier. "Collective Identity in Social Movement Communities: Lesbian Feminist Mobilization." In *Frontiers in Social Movement Theory*, edited by Aldon Morris and Carol McClurg Mueller, 104–130. New Haven, CT: Yale University Press, 1992.

Tews, Kerstin, Per-Olof Busch, and Helge Jorgens. "The Diffusion of New Environmental Policy Instruments." *European Journal of Political Research* 42, no. 4 (June 2003): 569–600.

Tocqueville, Alexis de. *Democracy in America*. London: Penguin Classics, 2003.

Tolbert, Charles M., II. "Minding Our Own Business: Local Retail Establishments and the Future of Southern Civic Community." *Social Forces* 83, no. 4 (2005): 1309–1328.

Traub, James. "Into the Mouths of Babes." *New York Times*, July 24, 1988, SM18.

Treadwell, D. D., D. E. McKinney, and N. G. Creamer. "From Philosophy to Science: A Brief History of Organic Horticulture in the United States." *HortScience* 38, no. 5, (2003): 1009–1014.

Trewavas, Anthony. "Urban Myths of Organic Farming." *Nature* 410 (March 22, 2001): 409–410.

Tucker, David M. *Kitchen Gardening in America: A History*. Iowa City: Iowa State University Press, 1993.

UN Food and Agriculture Organization. "Priority Areas for Interdisciplinary Action: Medium Term Plan, 2002–2007." August 2000. http://www.fao.org/docrep/X7572E/X7572e02.htm (accessed August 28, 2014).

Underwriters Laboratories. http://www.ul.com/about/index.html (accessed July 4, 2010).

Union of Concerned Scientists. "Genetic Engineering." http://www.ucsusa.org/food_and_agriculture/our-failing-food-system/genetic-engineering/risks-of-genetic-engineering.html (accessed August 27, 2014).

United Natural Foods, Inc. "UNFI History." https://www.unfi.com/Company/Pages/UNFIHistory.aspx (accessed September 1, 2014).

University of California at Santa Cruz, University Library. "Timeline: Cultivating a Movement, An Oral History Series on Organic Farming and Sustainable Agriculture on California's Central Coast." http://library.ucsc.edu/reg-hist/cultiv/timeline (accessed August 1, 2014).

US Department of Agriculture. "Know Your Farmer, Know Your Food: Our Mission." http://www.usda.gov/wps/portal/usda/usdahome?navid=KYF_MISSION (accessed August 31, 2014).

US Department of Agriculture. "Organic Production and Organic Food: Information Access Tools." http://www.nal.usda.gov/afsic/pubs/ofp/ofp.shtml (accessed July 24, 2014).US Department of Agriculture, Agricultural Marketing Service. "Creating Access to Healthy, Affordable Food: Food Deserts." http://apps.ams.usda.gov/fooddeserts/foodDeserts.aspx (accessed September 1, 2014).

US Department of Agriculture, Agricultural Marketing Service. "National Count of Farmers Markets, Directory Listing Graph: 1994–2013." http://www.ams.usda.gov/AMSv1.0/ams.fetchTemplateData.do?template=TemplateS&navID=WholesaleandFarmersMarkets&leftNav=WholesaleandFarmersMarkets&page=WFMFarmersMarketGrowth&description=Farmers%20Market%20Growth&acct=frmrdirmkt (accessed August 31, 2014).

US Department of Agriculture, Agricultural Marketing Service, National Organic Program. "National Organic Standards Board." http://www.ams.usda.gov/AMSv1.0/NOSB (accessed August 7, 2014).

US Department of Agriculture, Economic Research Services. "Organic Production: Table 2—U.S. Certified Organic Farmland Acreage, Livestock Numbers, and Farm Operations, 1992–2011." http://www.ers.usda.gov/data-products/organic-production.aspx#25762 (accessed August 28, 2014).

US Department of Agriculture, Food and Nutrition Service. "Women, Infants, and Children: Farmers' Market Nutrition Program (FMNP)." http://www.fns.usda.gov/fmnp (accessed September 1, 2014).

US Department of Agriculture, Food Safety and Inspection Service. "If a Label Bears a Halal or Kosher Statement, Does FSIS Have to Monitor the Production of the Product?" http://askfsis.custhelp.com/app/answers/detail/a_id/375/related/1/session/L2F2LzEvdGltZS8xMzc3ODc3NjY5L3NpZC9Ea2lHXzN6bbA%3D%3D (accessed August 4, 2014).

US Department of Agriculture, National Agricultural Statistics Service. *Census of Agriculture—State Data, 2007*. Washington, DC: US Department of Agriculture, February 2009.

US Department of Agriculture, National Organic Program. Organic Foods Production Act of 1990, 7 U.S.C. 6501. http://www.ams.usda.gov/AMSv1.0/getfile?dDocName=STELPRDC5060370&acct=nopgeninfo (August 8, 2014).

"USDA Abandons Three Contentious Issues in Proposed Organic Standards." *American Journal of Alternative Agriculture* 13, no. 2 (1998): 1.

"USDA Backs Down on Curtailing Group Certification for Coops and Grower Groups." *Sustainable Food News*, May 2, 2007, http://www.organicconsumers.org/articles/article_5061.cfm (accessed September 1, 2014).

USDA National Organic Program, USDA Science and Technology Programs. "2010–2011 Pilot Study: Pesticide Residue Testing of Organic Produce." Washington, DC: Agriculture Marketing Service, November 2012. http://www.ams.usda.gov/AMSv1.0/getfile?dDocName=STELPRDC5101234 (accessed August 28, 2014).

USDA Study Team on Organic Farming. "Report and Recommendations on Organic Farming." Washington, DC: US Department of Agriculture, July 1980. http://www.nal.usda.gov/afsic/pubs/USDAOrgFarmRpt.pdf (accessed August 4, 2014).

Van Dyke, Nella. "Crossing Movement Boundaries: Factors That Facilitate Coalition Protest by American College Students, 1930–1990." *Social Problems* 50 (2003): 226–250.

Van Dyke, Nella, and Holly McCammon, eds. *Strategic Alliances: Coalition Building and Social Movements*. Minneapolis: University of Minnesota Press, 2010.

Van Dyke, Nella, Sarah A. Soule, and Verta A. Taylor. "The Targets of Social Movements: Beyond a Focus on the State." *Research in Social Movements Conflict and Change* 25 (2004): 27–51.

Vogt, Gunter. "The Origins of Organic Farming." In *Organic Farming: An International History*, edited by William Lockeretz, 9–29. Oxfordshire, UK: CABI, 2007.

Von Liebig, Justus. *Organic Chemistry in Its Applications to Agriculture and Physiology*. London: Bradbury and Evans, 1840.

Vos, Timothy. "Visions of the Middle Landscape: Organic Farming and the Politics of Nature." *Agriculture and Human Values* 17 (2000): 245–256.

W. Rogowski Farm. http://www.rogowskifarm.com/index.html (accessed September 1, 2014).

Walmart. "Global Responsibility: Sustainable Agriculture." http://corporate.walmart.com/global-responsibility/environment-sustainability/sustainable-agriculture (accessed September 1, 2014).

Warren, Virginia Lee. "Organic Foods: Spotting the Real Thing Can Be Tricky." *New York Times*, April 9, 1972, 72.

Webb, Denise. "Eating Well: Food Isn't Organic Just Because the Label Says So." *New York Times*, June 7, 1989, C10.

Weiss, E. B.. "Look Out for Coming Scandal in Surging Organic Foods." *Advertising Age*, December 6, 1971, 44.

Whelan, Elizabeth M., and Frederick John Stare. *Panic in the Pantry: Food Facts, Fads, and Fallacies*. New York: Atheneum, 1977.

Wilcox, Wayne F. "Fire Blight." New York State Integrated Pest Management Program, Cornell University. http://www.nysipm.cornell.edu/factsheets/treefruit/diseases/fb/fb.asp (accessed August 30, 2014).

Winter, Michael. "Embeddedness, the New Food Economy, and Defensive Localism." *Journal of Rural Studies* 19 (2003): 23–32.

Woese, Katrin, Dirk Lange, Christian Boess, and Klaus Werner Bogl. "A Comparison of Organically and Conventionally Grown Foods: Results of a Review of the Relevant Literature." *Journal of the Science of Food and Agriculture* 74, no. 3 (July 1997): 281–293.

Wright, John R., and Arthur S. Goldberg. "Risk and Uncertainty as Factors in the Durability of Political Coalitions." *American Political Science Review* 79, no. 3 (September 1985): 704–718.

York, Richard. "The Treadmill of (Diversifying) Production." *Organization and Environment* 17, no. 3 (2004): 355–362.

York, Richard, and Eugene A. Rosa. "Key Challenges to Ecological Modernization Theory." *Organization and Environment* 16, no. 3 (2003): 273–288.

Youngberg, Garth, and Suzanne P. DeMuth. "Organic Agriculture in the United States: A 30-Year Retrospective." *Renewable Agriculture and Food Systems* 28, no. 4 (December 2013): 1–35, doi:10.1017/S1742170513000173.

Zieger, Robert H., and Gilbert Gall. *American Workers, American Unions*. Baltimore: Johns Hopkins University Press, 2002.

# Index

Agrarian, 7, 10, 29, 36–37, 41, 91, 162, 164–165, 169, 215, 219–220

Agribusiness, 1, 4, 11, 14, 16–17, 21, 26, 39, 49, 52, 68–70, 79, 81, 89, 106, 117–118, 125 129, 139, 146–147, 151, 159, 164, 169, 174, 184, 193, 199, 203, 209, 211, 214–215, 217, 219, 226, 230–232, 237

organic, 17, 169 (*see also* Big Organic; Corporate organic)

Agricultural Justice Project, 195

Alar, 65–66

Aldicarb, 65

Alta Dena, 142

American Bar Association, 222

American Enterprise Institute, 129

American Medical Association, 222

Americans for Safe Food Project, 71

Animal rights, 5, 15, 20, 22, 120, 156, 193, 231. *See also* Humane organizations

Antibiotics, 96, 156–157, 238–239

Aurora Dairy, 144, 167

Avery, Dennis, 129

Back to the land, 37, 41, 163, 184, 211, 213

Balfour, Eve, 34, 37, 88

Beeson, Kenneth, 50

Berry, Wendell, 51, 91

Beyond Pesticides, 69, 76, 90–91, 154, 186

Big box stores, 2, 11, 14, 89, 141, 200. *See also* Target; Walmart

Big Organic, 11, 17, 81–82, 90, 125, 127, 133, 139, 140, 150, 154, 158–159, 161, 168, 205, 217, 219. *See also* Agribusiness, organic

Big Three (controversial provisions of organic rule), 117–121, 125, 149, 185, 233

Biodiversity, 14, 132–133, 145

Biodynamic agriculture, 28, 33–35, 38, 44, 53, 73, 216. *See also* Demeter; Rudolf Steiner

Biotechnology, 48, 119, 129, 134, 137, 149, 193. *See also* Genetically modified organisms

Blobaum, Roger, 71, 93, 96 190, 191

Borsodi, Ralph, 37

Bovine Growth Hormone (BGH), 233

Boycotts, 40, 120, 176, 186, 221–222

Bracero Program, 175–176. *See also* Migrant labor

Brattleboro Food Coop, 43

Brook Farm Project, 198

Brown Rosen, Emily, 155

Bush, George H. W., 87

Bush, George W., 148–149

Butz, Earl, 48–50, 136, 231

California Certified Organic Farmers, 54, 60, 64–65, 68–69, 71–72, 82, 90, 98, 186

California Organic Food Act of 1979, 63–64,

California Organic Food Act of 1990, 69, 76

Capitalism, 10, 36, 218, 220, 224

Cargill, 14

Carrageenan, 152–153

Carson, Rachel, 40

Carter, Jimmy, 66, 231

Cascadian Farm, 59, 72, 125, 142, 147

Center for Agroecology and Sustainable Food Systems, 130

Center for Consumer Freedom, 129

Center for Food Safety, 154, 187

Center for Science in the Public Interest, 71, 93

Certification, 3, 15–19, 22–23, 25–26, 34, 44, 47, 53, 56–57, 59–65, 86, 98, 117, 122, 127, 144–145, 156, 169–170, 173, 183–184, 199, 201, 204–207, 210, 214, 236, 240–241
    alternative certification debates, 187–197
    certification controversies within the organic community, 103–112
    certification as a social change strategy 217–230
    development of organic certification, 69–79

Certified Naturally Grown, 4, 22, 189, 191–193, 208

Certified Organic, 3–4, 60, 65, 101–102, 138, 186, 201, 203–204, 207, 227, 229, 232

Chavez, Cesar, 40, 176

Chez Panisse, 43

Chilean nitrate, 166

Christakis, George, 50

Christianity, 28, 35–36

Civil rights movement, 41, 163, 221

Climate change, 132, 158

Clinton, William Jefferson, 76, 119, 149

Coalitions (in social movements), 1, 5–6, 18, 20–21, 73, 79, 81–85, 113–117, 119–120, 174, 184, 186–187, 210

Coca-Cola, 14, 142

Comité de Apoyo a los Trabajadores Agrícolas/Farmworker Support Committee, 195

Commoner, Barry, 51

Communes, 8, 41–43, 54, 175, 215

Communities in Support of Sustaining Agriculture (CISA), 204–205

Community gardens, 172. *See also* Urban agriculture

Community Supported Agriculture, 2–3, 9, 21, 23, 159, 172, 174, 183, 198–199, 202, 205, 234, 236, 241

Compost, 30, 33, 37–38, 53, 100, 131, 144

ConAgra, 14

Concentrated animal feeding operations, 132, 230, 238

Congress of Industrial Organizations, 176

Consumer groups, 1, 19, 56, 70–71, 73, 76, 78, 82–83, 93–95, 97, 99–100, 103, 105, 110, 112, 118, 124–126, 140, 147, 157, 170, 185, 187, 189, 191, 216, 231, 236

Consumers Union, 89, 121, 186, 191, 196

Contact substances, 153

Conventional industry, 4, 8–12, 16, 48, 50, 53, 65, 67–68, 78–79, 81–82, 91, 94–95, 106, 111, 113, 115–116, 119, 125, 139–143, 146–147, 149, 160–161, 166–167, 177, 179, 183, 194, 217–219, 227–229, 232–233, 239. *See also* Agribusiness

Conventionalization, 11–12, 22, 141, 143–144, 219

Coody, Lynn, 53, 55–57, 63, 66, 72–74, 97, 100, 168
Cornucopia Institute, 121–123, 144, 147–148, 186
Corporate organic, 11, 150, 188. *See also* Agribusiness, organic; Big Organic
Costco, 145
Counterculture, 8, 25, 40–42, 44, 47, 54, 67, 77, 88, 96, 162–163, 215
Country Natural Beef, 204
Craft, 82
CROPP Cooperative, 169
Cummins, Ronnie, 41, 121–122, 184, 195

Dannon, 158
Deal, Nathan, 150
Dean Foods, 82, 142, 147
DeFazio, Peter, 76
Demeter, 33, 53, 73. *See also* Biodynamic agriculture
Department of Agriculture and Markets, 51
Diggers, 41
DiMatteo, Katherine, 61–62, 89–90, 97, 102, 110, 154, 165, 205
Dole, 166
Domestic Fair Trade Association, 195
Dust Bowl, 30

Earthbound Farm, 166
Ecological modernization theory, 8–9, 12, 17, 217–218, 223
Eden Foods, 59
Energy crises, 66
Environmental activists, 1, 8, 19, 48, 75, 90, 120, 125
Environmental benefits (of organic agriculture), 7, 9, 12, 14, 16–17, 57, 81, 87–88, 90, 96–97, 109, 112–113, 128–129, 131–134, 136–138, 145, 161, 166, 178, 196–199, 206–207, 210, 217, 219, 229, 240

Environmental justice, 173
Environmental organizations, 5, 19, 20, 36, 71, 76, 78, 82–83, 90–91, 93–95, 111–113, 118, 120, 124–126, 140, 147, 163, 173, 185, 231, 236
Environmental preservation, 7, 58, 88, 97
Environmental problems, 3, 7, 8, 10, 17, 25, 40, 43–44, 48, 58, 65, 95–96, 103, 132–133, 158, 161, 196, 199, 206, 217, 219, 224, 237, 240
Environmental sustainability, 4, 9–10, 20–21, 28, 49, 88, 162, 192, 204, 223, 241
Environmental Working Group, 111

Fair trade, 15, 22, 194–196, 204, 210, 225–228
Fair Trade USA, 225
Fairfield Farms, 150
Family farming, 10, 66, 91, 109, 159, 162–165, 167–170, 173, 175, 195, 200, 203–204
Fantle, William, 121–122, 144, 167
Farm Aid, 164
Farm Bill, 174, 236
  Farm Bill of 1981, 67
  Farm Bill of 1985, 67
  Farm Bill of 1990, 76, 111
  Farm Bill of 2002, 130
  Farm Bill of 2008, 130
  Farm Bill of 2012 (extension), 149
Farmer's Pledge, 192–193, 207
Farmers' markets, 1, 3, 9, 11, 21, 23, 68, 73, 139, 141, 159, 172–173, 183, 198, 202, 205, 209–210, 234, 236, 241
Farming System Trial, 134
Farm plan, 108, 156, 190–191
Farm workers, 20, 40, 88, 136, 162, 170, 175–177, 180, 195
  farm worker health, 40, 88, 134, 177, 231

Farm-to-table, 23
Fascism, 29, 36
Federal Trade Commission, 51–52, 66
Feldman, Jay, 76, 90–91, 111
Florida Organic Growers/Quality
    Certification Services, 195
Food and Agriculture Organization, 3,
    137
Food Alliance, 192
Food and Drug Administration, 231
Food coops, 11, 14, 23, 41–43, 54, 58,
    89, 93, 120, 139, 141, 146, 168, 183,
    215, 232
Food deserts, 171
Food justice, 17, 172–175, 181, 195,
    198, 210–211, 214, 219, 234
Food miles, 206–207, 210
Food scares, 48, 58, 65, 68
Forest Stewardship Council, 224
Formalization (of social movement
    organizations), 5, 18, 19, 45, 55–56,
    58, 181
Fowler, Wyche, 70
Fraud, 3, 38–39, 47, 51–53, 57–58,
    62–65, 68–71, 78, 122, 205–206, 216,
    223
Fundación RENACE, 195

General Mills, 14, 59, 142, 147–148,
    227
Genetically modified organisms
    (GMOs), 88, 118–119, 121, 133–134,
    136, 207, 209–210, 218–219, 235,
    238–239
Gershuny, Grace, 59–61, 94, 97,
    117–119
Great Depression, 37, 175
Greenhouse gasses, 132
Grower groups, 169
Growing Power, 172–173
Guenther, Dan, 2–5, 13, 22–23, 183,
    197–199, 201–202
Gussow, Joan Dye, 99, 101–102

Harvey, Arthur, 153–155
Haughley experiment, 34
Health, 7, 9, 12, 14, 16–17, 25, 37, 40,
    58, 65, 71, 81, 83, 88, 95–97, 103,
    127–129, 131–133, 145, 152, 161,
    166, 182, 197–198, 201, 210, 217,
    219, 229, 239–240
  health benefits of organic, 134–137
  health food, 48–51, 232
Heinz Company, 49
Henderson, Elizabeth, 55, 72–74, 104,
    149, 192
High Falls Food Co-op, 183
Hightower, Jim, 51
Hirshberg, Gary, 158–159
Honest Weight Food Coop, 43
Hoods, Liana, 1–2, 4–5, 13, 22, 91,
    105, 108, 114, 116, 155, 183–188,
    196–197, 204
Horizon Dairy, 59, 142
Howard, Albert, 27, 32–38, 56, 88
Hudson Institute Center for Global
    Food Issues, 129
Humane organizations, 71, 83, 120,
    231. *See also* Animal rights

Identity (in social movements), 43,
    55–56, 113–117, 228, 233–234
Indore process, 33–34, 38
Industrial agriculture, 3, 6, 11–12, 15,
    35–36, 49, 55, 68, 116, 132, 137,
    179, 235, 239, 240
Industrialization, 30, 32, 36, 44, 231
Input substitution, 11, 144
Institute for Sustainable Agriculture,
    130
Integrated pest management, 67
International Federation of Organic
    Agriculture Movements, 54, 56–57,
    61, 177
International Organic Inspectors
    Association, 146
Irradiation (of food), 118, 121

Kahn, Gene, 59, 72, 147
Kashi, 142
Kellogg, 142
Khosla, Ron, 3–5, 13, 22, 183, 187–189, 191, 197
Kirschenmann, Fred, 74, 94, 98–99, 104, 202–203

Labels, 4, 7, 15, 22–23, 52–53, 61–62, 74, 90, 101–102, 107, 109, 111, 118, 122, 130, 145, 148, 151–154, 171, 177, 180–186, 188, 191–194, 196–197, 202–203, 208–210, 222, 224–226, 236, 238. *See also* Certification
Labor contracting (in agriculture), 176, 178, 180
Leahy, Patrick, 67, 69, 76, 78, 105, 150
Leopold Center for Sustainable Agriculture, 130, 206
Leavenworth, Kansas, 72–75, 94, 103–104
Lewin, Jake, 60, 64, 105
Libertarianism, 29, 191
Lipson, Mark, 65, 68–69, 72, 98, 213
Little, Amy, 73, 84
Livestock, 11, 32, 71, 86, 96, 132, 144, 150, 154–155, 185, 231, 238
Local agriculture, 2–4, 22, 199, 211. *See also* Small farmers
Local food, 23, 90, 199–210, 214, 234, 236–237, 241
Locavore, 23, 201
Low-income consumers, 20, 162, 171–175, 179
Low-Input Sustainable Agriculture Program, 67

Made with Organic Ingredients (in organic standards), 101
Madigan, Edward, 87
Maine Organic Farmers and Gardeners Association, 54, 56
Market based certification strategy, 15, 18, 25–26, 82, 127, 157, 159, 170, 175, 218–219, 221, 223–225, 228–230, 233
Mayer, Jean, 50
McDonald's, 43
Mendelson, Joe, 187, 205
Merrigan, Kathleen, 69–71, 73–75, 78, 81, 89, 104–105, 111, 113–117, 119, 137, 149, 194, 209, 233
Migrant labor, 162, 175–176, 180
Monsanto, 133, 146, 194
MorningStar Farms, 142
Muir Glen, 125, 142, 147

National Academy of Sciences, 68
National Campaign for Sustainable Agriculture, 1, 73, 84, 114–117, 184
National Coalition Against the Misuse of Pesticides, 69. *See also* Beyond Pesticides
National Consumers League, 222
National Cooperative Grocers Association, 154, 186
Nationalism, 29, 36
National Labor Relations Act, 176
National list (in organic standards), 86, 97, 99, 151–152, 157
*National Organic Action Plan*, 143–145, 148
National Organic Coalition, 1, 84, 117, 157, 184, 186, 235
National Organic Program, 1, 3, 9, 18, 20–22, 24–25, 62, 66–67, 75–79, 83–84, 86, 93, 95, 101, 106, 108, 125, 128, 133, 136, 138–150, 153, 155–157, 159–160, 164, 167, 170, 179, 181, 183–197, 199, 201, 207–208, 214, 216–217, 226–229, 240–241
National Organic Standards Board, 18, 20, 22, 25, 75–76, 86–87, 91, 98–103, 110, 112, 117–118, 125, 145–151, 153–154, 184, 186, 188
controversial appointments, 146–148
designated board seats, 75

Natural foods, 49, 51, 58, 65–66, 74,
    142, 168, 171, 182, 203, 216
New Deal, 49
New Paltz (New York), 2–3, 182–183,
    190, 198, 204
Nixon, Richard, 48–49, 129, 231
Northbourne, Walter, 34
Northeast Organic Dairy Producers
    Alliance, 186
Northeast Organic Farming Association,
    54, 60–61, 72–73, 106, 117, 186,
    195, 235
    Northeast Organic Farming
    Association, New York, 55, 72–73,
    104, 108, 192–193
    Northeast Organic Farming
    Association, Vermont, 57, 59, 64, 73
Nuestras Raices, 172
Nutrition, 48, 50, 71, 101, 130,
    134–135, 174

Obama, Barack, 137, 149, 209
Obama, Michelle, 136
Odwalla, 142
Off-farm inputs, 11, 14, 32–33, 131, 164
Office of Management and Budget,
    117–119
Oregon Tilth, 18, 55, 57, 72 97, 186
Organic, definition, 14, 18, 28, 44,
    52–53, 63, 96, 98–99
Organic acreage, 90–91, 112, 138–139,
    143, 160
Organic Agriculture Research and
    Extension Initiative, 130
Organic Center, 130
Organic certification, 15–16, 18, 56–57,
    59, 62, 70, 75–76, 98, 103, 105–106,
    109, 112, 117, 127, 144, 169, 173,
    189–191, 193, 197, 199, 201, 204,
    206, 217, 223, 226, 240
Organic certifiers, 4, 7, 20, 23, 59,
    60–64, 69, 85, 104–110, 123, 145,
    148, 155, 165, 191

Organic consumers, 47, 52, 59, 95,
    120–121, 131, 148, 154, 167, 178,
    182, 201, 205, 232
Organic Consumers Association, 41,
    120–123, 154, 186, 195, 235
Organic Crop Improvement
    Association, 60–61
Organic farmers, 20, 28–29, 52, 54, 56–
    59, 62–63, 71–73, 75–77, 82, 93–97,
    99, 103, 106–107, 111–112, 120,
    141, 143, 156, 163–164, 166, 168,
    174, 177, 189–193, 207, 215–216,
    232, 240
Organic Farmers Association Council,
    75, 94, 97
Organic farmers' associations, 54–59,
    62, 68, 73, 77, 82–85, 92–93, 103–
    105, 107, 109, 121, 146, 186–187,
    189, 191, 216, 225, 236
*Organic Farming and Gardening*
    (magazine), 27, 37–38, 42, 44, 47,
    52–53, 108, 161
Organic Farming Research Foundation,
    90, 98, 130, 147, 213
Organic Food Production Association
    of North America, 61–62, 64, 70,
    72, 94, 177. *See also* Organic Trade
    Association
Organic Foods Production Act, 19, 22,
    82–87, 89, 92–93, 100–103, 111, 116,
    118, 124–125, 127–128, 139–140,
    144, 147, 149, 153, 159, 164–165,
    187, 216–217, 240
    development and passage, 69–77
Organic gardening clubs, 28, 38–39, 54
Organic industry, 1, 5–6, 8–12, 14–16,
    21, 24, 48, 51, 54, 61, 63, 66, 77, 82,
    89, 94, 101–104, 106–107, 120–121,
    125–126, 128, 139–141, 146, 149–
    151, 153–155, 159–161, 164, 175,
    181, 187, 194, 223
Organic market, 6–7, 9, 11–12, 44, 47,
    51, 53, 58–61, 63, 68, 73, 78–79,

81–82, 88–89, 94, 102, 109–110, 125–128, 137–145, 152, 154, 159, 164–169, 174, 179, 184, 216–219, 233
entry of conventional firms, 139–143
market growth under the NOP, 137–139
Organic policy, 1, 3, 9, 21, 57, 59, 79, 84, 97, 118, 146, 193, 218, 235
Organic processors, 20, 59–62, 86, 91–92, 94–95, 100–103, 107–110, 126, 141, 144, 147, 218
Organic research, 29, 32–34, 37, 50–51, 66–68, 100, 128, 130–132, 134–137, 141, 171, 184, 206, 232, 239
Organic sales, 11, 24, 112, 138–139, 141, 149, 160
Organic Trade Association, 61–62, 72, 89, 91–92, 94, 97, 110, 121–122, 127, 130, 150, 154–155. *See also* Organic Food Production Association of North America
Organic Valley, 169, 195, 203

Pacific Organics, 64
*Panic in the Pantry*, 50
Participatory guarantee, 3, 189
Pasture rule, 144–145, 155–156. *See also* Livestock
People's Park, 41
Pepsi, 206
Phillies Bridge Farm Project, 198
Pluralism, 114–115
Political consumerism, 15, 208, 221, 224, 226, 229, 236
Pollan, Michael, 181, 209–210
Populist movement, 37
Poverty, 45, 170, 174–175, 179
Prefigurative politics, 16, 54, 58, 73, 78, 84–85, 93, 115–116, 215
Processed organic foods, 59, 61, 90, 100, 103, 141, 151, 154

Professionalization (of social movement organizations), 18–19, 77
Public Voice for Food and Health Policy, 71
PurePak, 147

Racial justice, 45, 173
Radical flank effect (in social movements), 124
Rainbow Grocery, 43, 59
Rangan, Urvashi, 89, 104, 191, 194, 196
Reagan, Ronald, 67, 176, 231
Red Tomato, 204
*Report and Recommendations on Organic Farming*, 66
Retzloff, Mark, 59
Riddle, Jim, 63, 146, 190–191, 207
Rodale, Jerome Irving, 19, 27–28, 37–39, 42, 51–54, 56, 60, 88, 96, 108, 161, 170, 213, 216, 223
Rodale, Robert, 19, 42, 213
Rodale Institute, 37, 60, 67, 129, 134
Rondout Valley Growers Association, 204
Rule of return, 33, 35. *See also* Compost; Indore process
Rural Advancement Foundation International-USA, 91, 143, 144, 154, 195, 237

Safeway, 142, 145
Science (in the study of organic agriculture), 25, 28, 32–35, 40, 44, 50, 99, 115, 129, 131, 197
Scientific Certification Systems, 193
Scowcroft, Bob, 68–69, 72, 90, 147
Sewage sludge, 118, 121
Sierra Club, 154, 210
*Silent Spring*, 40
Sligh, Michael, 91–92, 102, 140, 149, 195
Slow food, 23, 201
Slow Food USA, 201

Small farmers, 2–4, 11, 20–22, 37, 45, 73–74, 81, 91–92, 94–96, 107, 110, 112, 121, 142–143, 159, 162–164, 167–171, 175, 179, 188, 190–191, 195, 200, 202, 209–210, 213, 231, 235, 237, 239–240. *See also* Family farming
Socialism, 29, 36
Social justice, 7, 17–18, 22, 25, 29, 36, 44, 88, 91, 101, 134, 160–162, 164, 167, 170, 175, 177–180, 195, 199, 229. *See also* Food justice
Social movements, 5–6, 12–14, 18, 20, 23–24, 40, 55, 77–78, 83, 85, 123–124, 185, 209, 218, 223, 233
Soil Association (Britain), 34
Soil erosion, 31, 35, 66, 68, 131–132
Soil and Health Foundation (Rodale Institute), 37
Special Supplemental Nutrition Program, 173
Spirituality, 25, 28, 33, 35–38, 44
Spreaders, 17, 87–93, 95, 126, 139, 145, 152–153, 165, 225
Starbucks, 226–227
Stare, Frederick, 50
Steiner, Rudolf, 33–35
Stonyfield Farm, 158–159, 168
Subscription farms, 198. *See also* Community Supported Agriculture
Supermarkets, 6, 11, 58, 65, 68, 141–142, 171, 182, 203
Sustainable agriculture, 22–23, 33, 36, 41, 130, 173, 183–184, 206, 214, 219, 230–231, 240
Sustainable agriculture movement, 21, 23–24, 73, 113–115, 180–181, 231, 233–241
Sustainable Agriculture Research and Education Program, 67–68
Swanton Berry Farm, 178
Synthetic chemicals, 28, 32, 40, 48, 50, 52–53, 60, 66–67 91, 96–98,
100–103, 119, 131, 135–136, 145, 151–154, 161, 185, 188, 232
fertilizers, 3, 11, 28, 36, 38–41, 47, 50, 88, 132, 192, 240
inputs, 30, 32, 39, 65, 68, 98, 131
pesticides, 3, 11, 36, 39–41, 47, 50, 53, 67, 88, 97, 111, 132–133

Target (retail store), 141–142, 192, 240
Technology, 9–10, 30–32, 40–41, 48, 50, 108, 131–133, 136–137, 140, 163–164, 166, 197, 215, 238–239
Tillers, 17, 87–88, 91–93, 95, 101, 126, 139, 145, 152–153, 161
Trader Joe's, 142
Transition period, 59–60
Treadmill of production theory, 10, 12, 17, 217–218, 220, 227–228

Underwriter Laboratories, 222
United Farm Workers, 40, 176, 178
United Nations, 3, 137
United Natural Foods International, 167
Urban agriculture, 172–173
US Department of Agriculture, 1, 4, 9–11, 21–22, 24, 49, 51, 66–68, 70, 75, 84, 99, 102, 104–107, 109–110, 117–121, 125, 136–137, 141, 146–149, 153–156, 167, 171–173, 184–186, 188–194, 197, 201–203, 207–209, 226, 228, 231, 233
US Department of Agriculture accredited, 3, 7, 23, 76, 105, 156, 169, 189
US Department of Agriculture certified, 3, 199, 201, 208, 229
US Environmental Protection Agency, 68

Vassar College, 2, 198
Veneman, Ann, 153
Vilsack, Tom, 137
Von Liebig, Justus, 31

Walmart, 89, 141–142, 144–145, 158,
    168, 182
Walnut Acres, 38
Water contamination, 66, 68, 132
Waters, Alice, 43
Weakley, Craig, 147
Western Alliance of Certifying
    Organizations, 72
WhiteWave Foods, 142
Whole Earth Catalog, 42
Whole Foods Market, 58, 66, 89, 141,
    159, 168, 171, 182, 202
Wild Oats (natural foods retailer), 66
Wonnacott, Enid, 57, 64–65, 73–74,
    122, 207
Workers' rights, 40, 45, 170, 176–178,
    180, 192, 195
World Trade Organization, 224
World War I, 32, 36,
World War II, 32, 48, 163, 175

Youngberg, Garth, 66–67, 228–229